CHEMICALS FROM BIOMASS

Integrating Bioprocesses into
Chemical Production Complexes
for Sustainable Development

GREEN CHEMISTRY AND CHEMICAL ENGINEERING

Series Editor: Sunggyu Lee

Ohio University, Athens, Ohio, USA

Proton Exchange Membrane Fuel Cells: Contamination and Mitigation Strategies
Hui Li, Shanna Knights, Zheng Shi, John W. Van Zee, and Jiujun Zhang

Proton Exchange Membrane Fuel Cells: Materials Properties and Performance
David P. Wilkinson, Jiujun Zhang, Rob Hui, Jeffrey Fergus, and Xianguo Li

Solid Oxide Fuel Cells: Materials Properties and Performance
Jeffrey Fergus, Rob Hui, Xianguo Li, David P. Wilkinson, and Jiujun Zhang

**Efficiency and Sustainability in the Energy and Chemical Industries:
Scientific Principles and Case Studies, Second Edition**
Krishnan Sankaranarayanan, Jakob de Swaan Arons, and Hedzer van der Kooi

Nuclear Hydrogen Production Handbook
Xing L. Yan and Ryutaro Hino

Magneto Luminous Chemical Vapor Deposition
Hirotsugu Yasuda

Carbon-Neutral Fuels and Energy Carriers
Nazim Z. Muradov and T. Nejat Veziroğlu

Oxide Semiconductors for Solar Energy Conversion: Titanium Dioxide
Janusz Nowotny

Lithium-Ion Batteries: Advanced Materials and Technologies
Xianxia Yuan, Hansan Liu, and Jiujun Zhang

Process Integration for Resource Conservation
Dominic C. Y. Foo

**Chemicals from Biomass: Integrating Bioprocesses into Chemical Production Complexes
for Sustainable Development**
Debalina Sengupta and Ralph W. Pike

GREEN CHEMISTRY AND CHEMICAL ENGINEERING

CHEMICALS FROM BIOMASS

Integrating Bioprocesses into Chemical Production Complexes for Sustainable Development

DEBALINA SENGUPTA

RALPH W. PIKE

CRC Press
Taylor & Francis Group
Boca Raton   London   New York

CRC Press is an imprint of the
Taylor & Francis Group, an **informa** business

CRC Press
Taylor & Francis Group
6000 Broken Sound Parkway NW, Suite 300
Boca Raton, FL 33487-2742

First issued in paperback 2017

© 2013 by Taylor & Francis Group, LLC
CRC Press is an imprint of Taylor & Francis Group, an Informa business

No claim to original U.S. Government works

Version Date: 20120621

ISBN 13: 978-1-4398-7814-9 (hbk)
ISBN 13: 978-1-138-07334-0 (pbk)

**Visit the Taylor & Francis Web site at
http://www.taylorandfrancis.com**

**and the CRC Press Web site at
http://www.crcpress.com**

Contents

Series Preface

Green Chemistry and Chemical Engineering

The subjects and disciplines of chemistry and chemical engineering have encountered a new landmark in the way of thinking about, developing, and designing chemical products and processes. This revolutionary philosophy, termed "green chemistry and chemical engineering," focuses on the designs of products and processes that are conducive to reducing or eliminating the use and generation of hazardous substances. In dealing with hazardous or potentially hazardous substances, there may be some overlaps and interrelationships between environmental chemistry and green chemistry. While environmental chemistry is the chemistry of the natural environment and the pollutant chemicals in nature, green chemistry proactively aims to reduce and prevent pollution at its very source. In essence, the philosophies of green chemistry and chemical engineering tend to focus more on industrial application and practice rather than academic principles and phenomenological science. However, as both chemistry and chemical engineering philosophy, green chemistry and chemical engineering derives from and builds upon organic chemistry, inorganic chemistry, polymer chemistry, fuel chemistry, biochemistry, analytical chemistry, physical chemistry, environmental chemistry, thermodynamics, chemical reaction engineering, transport phenomena, chemical process design, separation technology, automatic process control, and more. In short, green chemistry and chemical engineering is the rigorous use of chemistry and chemical engineering for pollution prevention and environmental protection.

The Pollution Prevention Act of 1990 in the United States established a national policy to prevent or reduce pollution at its source whenever feasible. And adhering to the spirit of this policy, the Environmental Protection Agency (EPA) launched its Green Chemistry Program to promote innovative chemical technologies that reduce or eliminate the use or generation of hazardous substances in the design, manufacture, and use of chemical products. The global efforts in green chemistry and chemical engineering have recently gained a substantial amount of support from the international community of science, engineering, academia, industry, and governments in all phases and aspects.

Some of the successful examples and key technological developments include the use of supercritical carbon dioxide as green solvent in separation technologies; application of supercritical water oxidation for destruction of harmful substances; process integration with carbon dioxide sequestration steps; solvent-free synthesis of chemicals and polymeric materials; exploitation of biologically degradable materials; use of aqueous hydrogen peroxide for efficient oxidation; development of hydrogen proton exchange membrane (PEM) fuel cells for a variety of power generation needs; advanced biofuel production; devulcanization of spent tire rubber; avoidance of the use of chemicals and processes causing generation of volatile organic compounds (VOCs); replacement of traditional petrochemical processes by microorganism-based bioengineering processes; replacement of chlorofluorocarbons (CFCs) with nonhazardous alternatives; advances in the design of energy efficient processes; use of clean, alternative, and renewable energy sources in manufacturing; and much more. This list, even though it is only a partial compilation, is undoubtedly growing exponentially.

This book series on green chemistry and chemical engineering by CRC Press/Taylor & Francis Group is designed to meet the new challenges of the twenty-first century in the chemistry and chemical engineering disciplines by publishing books and monographs based on cutting-edge research and development to the effect of reducing adverse impacts on the environment by chemical enterprise. And in achieving this, the series will detail the development of alternative sustainable technologies that will minimize the hazard and maximize the efficiency of any chemical choice. The series aims at delivering the readers in academia and industry with an authoritative information source in the field of green chemistry and chemical engineering. The publisher and its series editor are fully aware of the rapidly evolving nature of the subject and its long-lasting impact on the quality of human life in both the present and future. As such, the team is committed to making this series the most comprehensive and accurate literary source in the field of green chemistry and chemical engineering.

Sunggyu Lee

Preface

The vision for this book is the development of new plants that are based on renewable resources that supply the needed goods and services for existing petrochemical plants. The vision includes converting existing plants to ones that are based on renewable resources requiring nonrenewable resource supplements. Identifying and designing new chemical processes that use renewable feedstock as raw materials and show how these processes can be integrated into existing chemical production complexes are key to having a sustainable chemical industry. Also, identifying and designing new industrial processes that use carbon dioxide as a raw material are an important option in mitigating the effects of global warming.

The existing plants in the chemical production complex in the Lower Mississippi River Corridor produce a wide range of basic and specialty chemicals, monomers, and polymers. They were used as a base case to demonstrate the integration of new, biomass-based plants into an existing infrastructure of plants. Potential bioprocesses were evaluated based on selection criteria, and simulations of these bioprocesses were performed in Aspen HYSYS®. The bioprocesses were then converted to input–output block models. A superstructure of plants was formed, which was optimized to obtain the optimal configuration of existing and new plants (chemical complex optimization).

The optimal configuration of plants was based on economic, environmental, and sustainable costs and credits (triple bottom line). The optimal solution to this mixed integer, multicriteria, nonlinear programming problem was obtained using global solvers. Detailed results were provided that showed a triple bottom line increase, raw material cost decrease, utility cost increase, and pure carbon dioxide vented to the atmosphere reduced to zero in the optimal structure. Five case studies were performed demonstrating the versatility of the analysis, and the optimization software, Chemical Complex Analysis System, can be downloaded.

A methodology for the optimal integration of bioprocesses in an existing chemical production complex was developed and demonstrated. This methodology can be used to evaluate energy-efficient and environmentally acceptable plants and have new products from greenhouse gases. Based on these results, the methodology could be applied to other chemical complexes for new bioprocesses, reduced emissions, and energy savings. Detailed process designs for fermentation, transesterification, anaerobic digestion, gasification, and algae oil production can be downloaded for modification, as needed, along with optimization programs.

This book can serve as a text for a senior or first-year graduate course in bioprocess engineering, since it covers essentially all aspects of this topic.

These include bioprocess raw materials and products, design of bioprocess, economic and sustainability analysis, optimization of chemical complexes, and applications to existing processes and chemical production complexes.

Practicing engineers in the bioprocessing industries will find that this rapidly growing field requires a stand-alone text like this that covers all parts of biomass conversion to products. This book describes the technical and scientific expertise needed to bring the engineer and scientist "up to speed" in this field. The importance and rapid expansion of this field is covered in the *Wall Street Journal's* feature article, "Just One Word: Bioplastics," in the October 18, 2010, issue that describes the "huge potential" of plastics from plant materials.

The material is organized into seven chapters, a postscript, and appendices. In Chapter 1, a description of the production of chemicals from renewable resources is given, with the research vision being to develop the methodology for new plants based on renewable resources that supply the needed goods and services for existing plants. The criteria for optimal configuration of plants and optimization theory are introduced in this chapter. In Chapters 2 and 3, detailed literature reviews and analyses are covered for biomass as feedstock and for the production of chemicals from biomass. Conceptual designs of bioprocesses are constructed, as described in Chapter 4, and include detailed information about the bioprocesses. Five processes are developed in Aspen HYSYS® with cost estimations from Aspen ICARUS®. Information from other process simulation software, for example, SuperPro Designer®, is applied for the corn-to-ethanol fermentation process. In Chapter 5, bioprocess plant models are formulated for optimization using input and output streams, equilibrium rate equations, parameters, and thermodynamic information from HYSYS plant models. Two other processes included in this chapter are for the production of algae oil from carbon dioxide and for the production of syngas from corn stover by steam reforming. Then interconnections in the bioprocesses are developed for optimization.

In Chapter 6, the superstructure of chemical and biochemical plants was formulated by integrating the bioprocess and carbon dioxide–consuming plants with the base case of existing chemical plants in the Lower Mississippi River Corridor. Carbon dioxide from the integrated chemical complex was utilized for the production of algae and for chemicals from carbon dioxide. The optimal structure was obtained by maximizing the triple bottom line, which included product sales, economic costs (raw material and utility), environmental costs (67% of raw material costs), and sustainable costs and credits. The optimal solution gave the plants that were included in the optimal structure. Comparisons between the base case and optimal structure were given for triple bottom line costs, the pure and impure carbon dioxide emissions, the energy requirements for plants, and the capacity of the plants. Multicriteria optimization was used to determine Pareto optimal solutions for the optimal structure. Monte Carlo simulation was used to determine the parameter sensitivity of the optimal solution. Comparisons of results with approaches were included in this chapter.

In Chapter 7, optimization was used to evaluate five cases. Case I was a modification to evaluate integration of bioprocesses in the existing base case without carbon dioxide being used for chemicals or algae oil production. The other four cases examine other aspects obtained from the optimization. The postscript gives the conclusions from this methodology and extensions that can be used for future developments.

The existing plants used for the base case were developed with industrial colleagues led by Tom A. Hertwig of Mosaic Corporation, and their assistance was invaluable. The assistance of the members of the Total Cost Assessment Users' Group at the AIChE, and especially Lise Laurin for guidance in using the Total Cost Assessment Methodology, is gratefully acknowledged.

MATLAB® is a registered trademark of The Mathworks, Inc. For product information, please contact:

The MathWorks, Inc.
3 Apple Hill Drive
Natick, MA, 01760-2098 USA
Tel: 508-647-7000
Fax: 508-647-7001
E-mail: info@mathworks.com
Web: www.mathworks.com

Authors

Debalina Sengupta received her bachelor of engineering degree in chemical engineering from Jadavpur University, Calcutta, India, in 2003. She worked as a software engineer in Patni Computer Systems from 2003 to 2004. In 2005, she joined the Department of Chemical Engineering at Louisiana State University, Baton Rouge, Louisiana. She received her doctor of philosophy degree in chemical engineering under the guidance of Professor Ralph W. Pike for her research entitled "Integrating bioprocesses into industrial complexes for sustainable development" in 2010. Her expertise is in optimization of industrial complexes and sustainability analysis using total cost assessment methodology. She is now working as an ORISE postdoctoral fellow at the United States Environmental Protection Agency. Her current research is focused on sustainable supply chain design of biofuels and includes life cycle assessment (LCA) for ethanol as biofuel. Her research interests include model development for ethanol biorefineries for LCA and assessing environmental releases from biorefineries.

Ralph W. Pike is the director of the Minerals Processing Research Division and is the Paul M. Horton Professor of Chemical Engineering at Louisiana State University. He received his doctorate and bachelor's degrees in chemical engineering from Georgia Institute of Technology. He is the author of a textbook entitled *Optimization for Engineering Systems* and coauthor of four other books on design and modeling of chemical processes. Pike has directed 15 doctoral dissertations and 16 master's theses in chemical engineering. He is a registered professional engineer in Louisiana and Texas. His research has been sponsored by federal and state agencies and private organizations, with 107 awards totaling $5.6 million, and has resulted in over 200 publications and presentations. His research specialties are optimization theory and applications for the optimal design of engineering systems, online optimization of continuous processes, optimization of chemical production complexes, and related areas of resources management, sustainable development, continuous processes for carbon nanotubes, and chemicals from biomass.

Pike is a fellow of the American Institute of Chemical Engineers and is chair of the Environmental Division and past chair of the Fuels and Petrochemicals Division. He is an active member of the Institute for Sustainability and the Safety and Chemical Engineering Education (SACHE) Committee of the Center for Chemical Process Safety. He was the meeting program chairman for the 74th Annual Meeting and has cochaired 66 sessions on optimization, sustainability, transport phenomena, and reaction engineering. He has held all of the positions in the Baton Rouge Section of the AIChE. Pike is also on the editorial boards of *Environmental Progress and Renewable Energy* and *Clean*

Technology and Environmental Policy. He has served as coeditor in chief of *Waste Management*, an international journal devoted to information on prevention, control, detoxification, and disposition of hazardous, radioactive, and industrial wastes. He is a member of the American Chemical Society and Sigma Xi, the scientific society.

1

Introduction

1.1 Introduction

The vision of this book is to demonstrate the development of new plants that are based on renewable resources that supply the needed goods and services of current petrochemical plants. The vision includes converting existing plants to ones that are based on renewable resources requiring nonrenewable resource supplements. This vision is an essential component of sustainable development. It embodies the concepts that sustainability is a path of continuous improvement, wherein the products and services required by society are delivered with progressively less negative impact upon the Earth. It is consistent with the Brundtland Commission report that defines "Sustainable Development" as "development which meets the needs of the present without sacrificing the ability of the future to meet its needs" (United Nations, 1987).

Identifying and designing new chemical processes that use renewable feedstock as raw materials and showing how these processes can be integrated into existing chemical production complexes are the keys to having a sustainable chemical industry. Also, identifying and designing new industrial processes that use carbon dioxide as a raw material are an important option in mitigating the effects of global warming.

Global warming and biotechnology are on a collision course because new processes for chemicals from biomass are energy intensive and generate carbon dioxide. Global warming is caused by accelerative accumulation of carbon dioxide and other greenhouse gases in the atmosphere. Industrial processes that use carbon dioxide as a raw material are an important option in mitigating the effects of global warming. Approximately, 110 million MT/year of carbon dioxide is used as a raw material for the production of urea, methanol, acetic acid, polycarbonates, cyclic carbonates, and specialty chemicals such as salicylic acid and carbamates in the United States (Arakawa et al., 2001). Other uses include enhanced oil recovery, solvent (supercritical carbon dioxide), refrigeration systems, carbonated beverages, fire extinguishers, and inert gas-purging systems. Recent developments and renewed interest in growing algae as feedstock for bioprocesses provide alternate methods for the utilization of carbon dioxide.

TABLE 1.1

Chemical Complexes in the World

Continent	Name and Site	Notes
North America	Gulf Coast petrochemical complex in Houston area (USA)	Largest petrochemical complex in the world, supplying nearly two-thirds of the nation's petrochemical needs
	Chemical complex in the lower Mississippi River corridor (USA)	
South America	Petrochemical district of Camacari-Bahia (Brazil)	Largest petrochemical complex in the Southern Hemisphere
	Petrochemical complex in Bahia Blanca (Argentina)	
Europe	Antwerp port area (Belgium)	Largest petrochemical complex in Europe and worldwide second only to Houston, Texas
	BASF in Ludwigshafen (Germany)	Europe's largest chemical factory complex
Asia	The Singapore petrochemical complex in Jurong Island (Singapore)	World's third largest oil refinery center
	Petrochemical complex of Daqing Oilfield Company Limited (China)	
	SINOPEC Shanghai Petrochemical Co. Ltd. (China)	
	Joint venture of SINOPEC and BP in Shanghai under construction (2005) (China)	Largest petrochemical complex in Asia
	Jamnagar refinery and petrochemical complex (India)	
	Haldia Petrochemical Complex (India)	
	Sabic company, based in Jubail Industrial City (Saudi Arabia)	
	Petrochemical complex in Yanbu (Saudi Arabia)	World's largest polyethylene manufacturing site
	Equate (Kuwait)	World's largest and most modern for producing ethylene glycol and polyethylene
Oceania	Petrochemical complex at Altona (Australia)	
	Petrochemical complex at Botany (Australia)	
Africa	Petrochemical industries complex at Ras El Anouf (Libya)	One of the largest oil complexes in Africa

Source: Xu, A., Chemical production complex optimization, pollution reduction and sustainable development, PhD dissertation, Louisiana State University, Baton Rouge, LA, 2004.

After identifying and designing new biochemical processes, these processes were integrated into an existing chemical production complex. In Table 1.1, some of these chemical production complexes in the world are tabulated. The chemical production complex in the lower Mississippi River corridor was used as a base case to demonstrate the integration of these new plants into an existing infrastructure. Potential bioprocesses were evaluated based on selection criteria, and simulations of these bioprocesses were performed in Aspen HYSYS®. The bioprocesses were then converted to input–output block models. A superstructure of plants was formed, which was optimized to obtain the optimal configuration of existing and new plants (chemical complex optimization).

Chemical complex optimization is a powerful methodology for plant and design engineers to convert their company's goals and capital to viable projects that meet economic, environmental, and sustainable requirements. The optimal configuration of plants in a chemical production complex is obtained by solving a mixed-integer, nonlinear programming problem (MINLP). This methodology is applicable to other chemical production complexes in the world, including the ones in the Houston area (largest in the world), Antwerp port area (Belgium), BASF in Ludwigshafen (Germany), petrochemical district of Camacari-Bahia (Brazil), the Singapore petrochemical complex in Jurong Island (Singapore), and Equate (Kuwait), among others.

1.2 Research Vision

The research vision leads to the development of new plants that are based on renewable resources. The vision includes converting existing plants to ones that are based on renewable resources that may require nonrenewable resource supplements.

An example is ethanol produced from corn that was grown with chemical fertilizers produced from fossil fuels. Ethanol reduces greenhouse gas emissions by 22% compared to gasoline (Bourne, 2007). Another example is a wind farm of turbines producing electricity, where the turbines were built with materials that required energy from fossil fuels. Wind is considered the largest source of renewable energy, and 10,000 MW (megawatts) have been installed in the United States selling for 4–7 cents/kWh, the least expensive source of energy.

This vision is a path to sustainable development for the chemical industry, a path of continuous improvement, where products and services required by society are delivered with progressively less negative impact upon the Earth. It leads to development that meets the needs of the present without sacrificing the ability of the future to meet its needs (United Nations, 1987).

1.3 New Frontiers

The world is in a transition not ever experienced in history. An example is the U.S. Gulf Coast region, where losses from natural disasters, plants relocating to other parts of the world, environmental deterioration, and competition from imports have played a major role in shaping the future of the region. Any change for the improvement of the region needs a new vision and direction. This effort is driven by a desire to understand how sustainable industries can evolve from ones based on nonrenewable resources. Chemical plants in the Gulf Coast, which rely exclusively on natural gas as a feedstock, faced closure when natural gas prices reached over $13 per thousand cubic feet. To remain operational, many of these plants must carefully evaluate migration to new feedstocks. The Gulf Coast is uniquely positioned to take advantage of bio-derived feedstocks. There is a strong agricultural industry in the region, and the Mississippi River provides deep-water ports to ensure continuous biofeedstocks throughout the year.

Existing natural-gas-intensive processes, such as agricultural chemical production, can be reconfigured as bio-derived chemical plants. For example, the Farmland Industries ammonia plant in Pineville, Louisiana, migrated from ammonia production to biodiesel production from soybean oil. Farmland Industries is one of the 14 companies that has closed 17 ammonia plants with a total capacity of 5.6 million tons/year (Byers, 2006), only to reopen some of these plants when the demand for fertilizer rapidly increased for increased production of corn for biofuels.

The Pineville example is both encouraging and discouraging for the Gulf Coast. The Pineville biodiesel facility is in operation but with substantially fewer employees, about 20 employees now compared to over 100 as in an ammonia plant. It is anticipated that new employees will be hired as the facility moves from 100,000 to 200,000 gal of biodiesel fuel per year. This is somewhat encouraging, but there is a net loss in jobs.

What was most disturbing for the region was the ultimate use of the remaining sections of the ammonia plant in Pineville. The new biodiesel plant was constructed by modifying the existing water treatment facility in the ammonia plant with some improvements to the control room. However, the majority of the plant, its reactors, separators, distillation columns, etc., were sold to China. This Louisiana facility was disassembled piece by piece and moved to mainland China, where it will be used to produce ammonia (Knopf, 2007).

The opportunity existed for this plant to be reconfigured to make value-added chemicals here in the United States, but this alternative probably was not considered. This work evaluates potential alternatives, including the expertise to help evaluate ethanol and biodiesel as feedstocks for existing chemical plants. However, the profitability of these migrated plants is inextricably linked to energy efficiency. Processing bio-derived chemicals requires large steam and electrical demands, which must be met through cogeneration and online optimization. There is virtually no chance for profitable operation if these plants have to purchase generated power.

Food security is moving into the hands of major agricultural-chemicals-exporting countries such as Saudi Arabia, Russia, the Ukraine, and Venezuela as high natural gas prices result in the outsourcing of the U.S. agricultural chemical industry. About 40% of U.S. food production comes from commercial fertilizers. Natural gas, the raw material for the production of nitrogen fertilizers, is 93% of the cost of production (Wilson, 2006). Also, imported phosphate from Morocco is shutting down U.S. production (Hertwig, 2006). Mosaic, Incorporated has announced intent to produce ammonia from petroleum coke that is available from processing heavy crude oil from Venezuela (Thrasher, 2006).

As the research moved forward, the focus was on scientific and technical questions that form the basis of sustainable industrial development supplemented with nonrenewable resources. Research priorities focused on products and industries for which there is a strong indication of a sustainable development component and for which there is high or increasing impact on the world's population. Quantifying sustainable costs was a key element in the use of the triple bottom line (economic, environmental, and sustainable costs) to improve all aspects of the region. Sustainable costs are costs to society to repair damage from emissions within environmental regulations as compared to economic and environmental costs borne by the company.

1.4 Chemical Industry in the Lower Mississippi River Corridor

The chemical industry in the lower Mississippi River corridor is typical of the chemical production complexes listed in Table 1.1. A map of the plants in the lower Mississippi River corridor is shown in Figure 1.1a. There are about 150 chemical plants producing a wide range of petrochemicals that are used in housing, automobiles, fertilizers, and numerous other consumer products, consuming 1.0 quad (10^{15} BTUs/year) of energy (Peterson, 2000). The state's chemical industry is the largest single employer with nearly 26,000 direct employees, a number that does not include the thousands of contract and maintenance employees that work at the plants year round. These jobs generate $5.9 billion in earnings and $125 million in state and local taxes on personal income. Over a billion dollars is spent in Louisiana annually with Louisiana suppliers according to the Louisiana Chemical Association (LCA, 2007).

Figure 1.1b shows a chemical production complex that was developed with the assistance of industrial collaborators and published sources (Xu, 2004). It is based on the plants in the agricultural chemical chain and the methanol

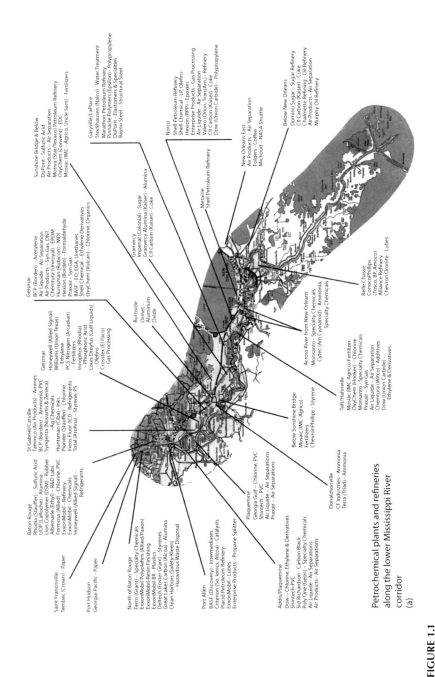

FIGURE 1.1

(a) Petrochemical plants in the lower Mississippi River corridor. (From Peterson, R.W., *Giants on the River*, Homesite Company, Baton Rouge, LA, 2000. With permission.)

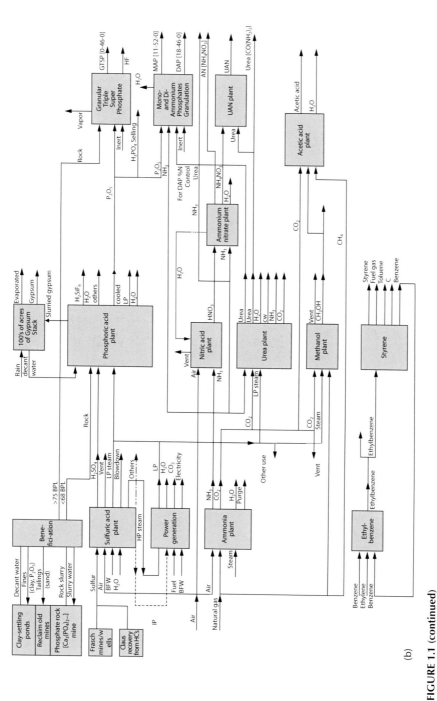

FIGURE 1.1 (continued)

(b) Base case of chemical plants.

and benzene chains in the lower Mississippi River corridor. This complex is representative of current operations and practices in the chemical industry and is called the base case of the existing plants. It includes the sources and consumers of carbon dioxide in the chemical production complex. This description of the chemical production complex was used in the research on bio-based chemicals, energy integration, and carbon dioxide utilization.

As shown in Figure 1.1b, this base case of chemical production complex has 13 production units plus associated utilities for power, steam, and cooling water and facilities for waste treatment. A production unit contains more than one plant. The phosphoric acid production unit contains four plants owned by three companies. The sulfuric acid production unit contains five plants owned by two companies (Hertwig, 2004). Here, ammonia plants produce 0.75 million tons/year of carbon dioxide, and methanol, urea, and acetic acid plants consume 0.14 million tons of carbon dioxide. This leaves a surplus of 0.61 million tons/year of high-purity carbon dioxide that is being vented to the atmosphere.

The raw materials used in the base case of the chemical production complex in Figure 1.1b include air, water, natural gas, sulfur, phosphate rock, ethylene, and benzene. The products are mono- and di-ammonium phosphate (MAP and DAP), granular triple super phosphate (GTSP), urea, ammonium nitrate, and urea ammonium nitrate (UAN) solution, phosphoric acid, ammonia, methanol, acetic acid, ethylbenzene, and styrene. Intermediates are sulfuric acid, phosphoric acid, ammonia, nitric acid, urea, carbon dioxide, and ethylbenzene. Ammonia is directly applied to crops and as a synthetic intermediate. MAP, DAP, UAN, and GTSP are directly applied to crops. Phosphoric acid can be used in other industrial applications. Methanol is used to produce formaldehyde, methyl esters, amines, and solvents along with many other organics, and acetic acid, ethylbenzene, and styrene are used as feedstock in other chemical processes. Emissions from the chemical production complex include sulfur dioxide, nitrogen oxides, ammonia, methanol, silicon tetrafluoride, hydrogen fluoride, and gypsum.

The plants in the base case are used as a starting point for the vision to convert industries based on nonrenewable resources to ones based on renewable resources. The bioprocesses were evaluated for the introduction of ethanol into the ethylene product chain and glycerin into the propylene chain. Ethanol is too valuable a commodity for the manufacture of plastics, detergents, fibers, films, and pharmaceuticals to be used as a motor fuel. Fatty acid methyl esters (FAME) from natural oils can be substitutes for polymers instead of producing biodiesel. Glycerin, a by-product from transesterification process for the production of FAME, is generated in large quantities and can be used in the propylene chain. By-products of agricultural production—bagasse, cane leaf materials, corn stover, rice husks, and poultry and hog wastes—are potential feedstocks and could fulfill some of the energy requirements of the plants.

1.5 Criteria for the Optimal Configuration of Plants

There are a number of methods that could be used as the criteria to determine the optimal configuration of new bio-based plants and existing plants. Some of these methods are summarized in the following, and they serve as the basis for selecting the triple bottom line that is based on economic, environmental, and sustainable costs.

Total cost assessment: Total cost assessment (TCA) is a methodology developed by industry professionals that was sponsored by the American Institute of Chemical Engineers (Constable et al., 1999; Laurin, 2007). TCA is a decision-making tool that provides cost information for internal managerial decisions. The TCA methodology identifies five types of costs including economic, environmental, and societal costs. These costs are described in detail in Appendix A. Dow Chemical, Monsanto, GlaxoSmithKline, and Eastman Chemical are industrial companies that have applied TCA methodology. TCA serves as the basis for the triple bottom line evaluation where the five types of costs are combined into economic, environmental, and sustainable costs and extended to sustainable credits. This methodology served as the basis for selecting the triple bottom line as the equation to be optimized since it incorporates economic and environmental cost to industry and sustainable cost and credits to society.

Life-cycle assessment: Life-cycle assessment (LCA) is a "cradle-to-grave" approach for assessing industrial systems (SAIC, 2006), which is described in detail in Appendix A. "Cradle-to-grave" begins with the gathering of raw materials from the earth to create the product and ends at the point when all materials are returned to the earth. LCA evaluates all stages of a product's life from the perspective that they are interdependent, meaning that one operation leads to the next. LCA enables the estimation of the cumulative environmental impacts resulting from all stages in the product life cycle, often including impacts not considered in more traditional analyses (e.g., raw material extraction, material transportation, and ultimate product disposal). By including the impacts throughout the product life cycle, LCA provides a comprehensive view of the environmental aspects of the product or process and a more accurate picture of the true environmental trade-offs in product and process selection. An LCA allows a decision maker to study an entire product system, hence, avoiding the suboptimization that could result if only a single process were the focus of the study.

Sustainability metrics: Sustainability metrics are intended to improve internal management decision making with respect to the sustainability of processes, products, and services. A leading developer of sustainability metrics was BRIDGES to Sustainability™, a not-for-profit organization that tested, adapted, and refined sustainability metrics (Tanzil et al., 2003). There are basic and complementary metrics under six impact categories: material, energy, water, solid wastes, toxic release, and pollutant effects. BRIDGES' sustainability metrics are constructed as ratios with environmental impacts in the numerator and a physically or financially meaningful representation of output in the denominator, the better process being the one with a smaller value for the ratio. The metrics are currently organized into six basic impact categories: material, energy, water intensities, solid waste to landfills, toxic releases, and pollutant effects. A detailed description of these metrics is given in Appendix A.

Sustainable process index: The concept of sustainable process index (SPI) is based on the sustainable flow of solar energy (Krotscheck and Narodoslawsky, 1996). The utilization of the solar energy is based on the area available. The area can be defined according to its use of land, water, and air. The production in these areas is denoted by production factors. Thus, with the dual function of area as a recipient of solar energy and as a production factor, the SPI can measure and relate the ecological impact of a process with respect to the quantity and the quality of the energy and the mass flow it induces. Processes needing more area for the same product or service are less competitive under sustainable economic conditions. SPI is the ratio of two areas in a given time period. One area is needed to embed the process to produce the service or product unit sustainability in the ecosphere and another is the area available for the sustainable existence of the product. Additional details on SPI are given in Appendix A including application to biodiesel.

Eco-efficiency analysis using SPI and LCA: Eco-efficiency analysis is a life-cycle tool that allows data to be presented in a concise format for use by decision makers. Ecological indicators are combined to provide an "ecological footprint," which is plotted against the life-cycle cost of process options, and the process that has the lowest of both measures is judged to have superior eco-efficiency. Additional details are given in Appendix A, including a case study comparing renewable resource versus petroleum-based polymers.

1.6 Optimization of Chemical Complex

The objective of optimization is to select the best possible decision for a given set of circumstances (Pike, 1986). Three basic components are required to optimize an industrial process. First, the process or a mathematical model of the process must be available, and the process variables that can be manipulated and controlled must be known. Second, an economic model of the process is required. This is an equation that represents the profit made from the sale of products and costs associated with their production, such as raw materials, operating costs, fixed costs, taxes, etc. Finally, an optimization procedure must be selected that locates the values of the independent variables of the process to produce the maximum profit or minimum cost as measured by the economic model. Also, the constraints in materials, process equipment, manpower, etc. must be satisfied as specified in the process model.

The statement for the optimization problem in the chemical production complex can be given as follows:

Optimize: Objective function
Subject to: Constraints from plant models

The first step of plant model formulation was achieved in Chapters 4 and 5. The process flow models were developed for the bioprocesses using Aspen HYSYS and then converted to input–output block models with mathematical relations. The constraint equations in the input–output block models describe relationships among variables and parameters in the processes, and they are material and energy balances, chemical reaction rates, thermodynamic equilibrium relations, and others.

The next requirement was the economic model for selecting the optimal configuration of plants from the new and existing plants. The optimization algorithm was formulated in Chapter 6. The TCA methodology discussed in Section 1.5 is the only method that incorporates costs for economic, environmental, and social criteria (sustainability). The concept of TCA was used for the economic model that optimized a triple bottom line equation given by Equation 1.1. The triple bottom line included a value-added economic model given by the profit in Equation 1.2. The Equation 1.1 also included environmental costs and sustainable costs. The objective function is to maximize the triple bottom line, based on the constraints from the plant model. Equation 1.3 shows the expanded form of Equation 1.1 that incorporates Equation 1.2:

$$\text{Triple bottom line} = \text{Profit} - \sum \text{Environmental costs}$$
$$+ \sum \text{Sustainable(Credits} - \text{Costs)} \tag{1.1}$$

$$\text{Profit} = \sum \text{Product sales} - \sum \text{Raw material costs} - \sum \text{Energy costs} \tag{1.2}$$

$$\text{Triple bottom line} = \sum \text{Product sales} - \sum \text{Raw material costs}$$
$$- \sum \text{Energy costs} - \sum \text{Environmental costs}$$
$$+ \sum \text{Sustainable(Credits} - \text{Costs)} \tag{1.3}$$

The third step was selecting an optimization procedure that maximized the triple bottom line. In the application of mathematical programming techniques to design and synthesis problems, it is always necessary to postulate a superstructure of alternatives (Grossmann et al., 1999). Thus, a superstructure of plants was constructed by integrating the bioprocess models into the base case of existing plants in the lower Mississippi River corridor. Binary variables were used to construct logical constraints for the selection of plants in the optimal structure. The model had linear and nonlinear constraint equations. Thus, a mixed-integer nonlinear programming problem was formulated, which required MINLP solvers for optimization. Global optimization solvers were used to optimize the triple bottom line subject to the constraints of plants in a superstructure. GAMS (general algebraic modeling system) interfaced with the chemical complex analysis system was the language used for optimization. Details on optimization theory and chemical complex analysis system are available in Appendix B.

The chemical complex analysis system was used to solve a multicriteria optimization problem formulated as given in the following. Multicriteria optimization theory is explained in Appendix B. The objective of optimization is to find optimal solutions that maximize industry's profits and minimize costs to society. This multicriteria optimization problem can be stated as in terms of industry's profit, P, and society's sustainable credits/costs, S, and these two objectives are given in Equation 1.4. To locate Pareto optimal solutions, multicriteria optimization problems are converted to a single criterion by applying weights to each objective and optimizing the sum of the weighted objectives (Equation 1.5).

$$\text{Max: } P = \sum \text{Product sales} - \sum \text{Economic costs} - \sum \text{Environmental costs}$$

$$S = \sum \text{Sustainable(Credits} - \text{Costs)} \tag{1.4}$$

Subject to: Multiplant material and energy balances, product demand, raw material availability, plant capacities

$$\text{Max: } w_1 P + w_2 S$$
$$w_1 + w_2 = 1 \tag{1.5}$$

Subject to: Multiplant material and energy balances, product demand, raw material availability, plant capacities

Details of this optimization are given in Chapter 6 for optimization model formulation and multicriteria optimization.

1.7 Contributions of This Methodology

This work developed a comprehensive methodology that encompasses bioprocesses development, sustainability analysis, and economic optimization techniques. It can be used to evaluate sustainable development quantitatively.

There have been very few reports on the development of bioprocesses for chemical production. The only notable report in this field was of screening 12 chemicals that may be produced from biomass sponsored by the Department of Energy (Werpy et al., 2004). These were conceptual methods of the processes for converting the biomass feedstock to chemicals. An important part of this research was to identify the chemical value of biomass, and the potential of using the renewable feedstock for chemicals. Fermentation, anaerobic digestion, transesterification, and gasification of renewable feedstock were identified as the bioprocesses that could potentially be integrated into existing industrial complexes.

The research developed detailed industrial-scale process designs for bioprocesses using the leading tools (Aspen HYSYS and Aspen ICARUS®) for industrial-scale design. There have been designs for fuels from biomass (Aden et al., 2002; Haas et al., 2006) but no one has provided a comprehensive approach to design processes for chemicals from renewable feedstock with an aim to integrate all the bioprocesses into a single platform.

The work evaluated introduction of ethanol into the ethylene product chain. Ethanol can be a valuable commodity for the manufacture of plastics, detergents, fibers, films, and pharmaceuticals. The introduction of glycerin into the propylene product chain was evaluated with cost-effective routes for converting glycerin to value-added products like propylene glycol. FAME were produced, which were starting materials for polymers. New methods to produce acetic acid from anaerobic digestion of biomass were developed, which were compared with existing processes for acetic acid production. Generation of synthesis gas for chemicals by hydrothermal gasification of biomass was included. The use of surplus carbon dioxide from chemical plants and refineries for algae oil production and new products were demonstrated.

There have been no reports to evaluate chemical complex optimization by integrating bioprocesses into an existing industrial plant complex, and the use of carbon dioxide from the complex for the production of algae and chemicals. This research was able to successfully demonstrate manufacturing chemicals using biomass as renewable feedstock and determine the optimal operation of integrated complex. The global optimization solvers in GAMS (SBB, DICOPT, and BARON) were effective in optimizing such a large system.

TCA methodology gives a quantitative approach for sustainability analysis. In this research, the TCA methodology was successfully employed for optimization of the triple bottom line. The optimal solution and case studies were provided, which demonstrate the ways in which the methodology can be used for varying parameters and show the effect on sustainability. Decisions regarding multicriteria optimization for maximizing economic profits with minimum societal costs were demonstrated in this methodology.

In summary, this research provides the decision maker with a methodology that can be followed for evaluating sustainable development. The choice of inclusion of a single process or several processes can be determined using triple bottom line criteria. The integration of bioprocesses was demonstrated on a base case of existing plants in the Mississippi River corridor, and this methodology can be applied to any chemical complex in the world.

1.8 Organization of Chapters

There are seven chapters and a postscript in this book, followed by references and relevant appendices. This section gives a brief overview of the organization of the chapters, with the key information from Chapters 2 through 6, highlighted in Figure 1.2.

Chapter 1 gives an introduction to chemical complex optimization, with the research vision for the production of chemicals from renewable resources. The criteria for optimal configuration of plants and the optimization theory used in the research are introduced in this chapter.

Chapters 2 and 3 are literature reviews of the feasibility of biomass as feedstock and the production of chemicals from biomass. Based on the literature, a conceptual design of bioprocesses is constructed, as shown in Figure 1.2. The units in the conceptual design are viewed from a top-down approach.

Chapter 4 starts with the conceptual design, and detailed information about the processes is gathered. A bottom-up approach is followed to develop five processes in Aspen HYSYS, with cost estimations in Aspen ICARUS. Three bioprocesses: fermentation, anaerobic digestion, and transesterification are modeled in this chapter. The chemicals from these bioprocesses include two designs, one for ethylene from ethanol (introduction of ethanol to the ethylene chain of chemicals) and the other for propylene glycol from glycerol (introduction of glycerol to the propylene chain of chemicals). Information from other process simulation software, for example, SuperPro Designer®, was applied for the corn-to-ethanol fermentation process. Figure 1.2 shows three of the processes, ethanol from fermentation of corn stover, ethylene from ethanol, and ethanol from fermentation of corn.

Chapter 5 formulates the bioprocess plant models for optimization. The bioprocesses described in Chapter 4 were converted to input–output block models as shown in Figure 1.2. Input and output streams, equilibrium rate equations, parameters, and thermodynamic information from HYSYS plant models were used to formulate the equality constraints and to validate the

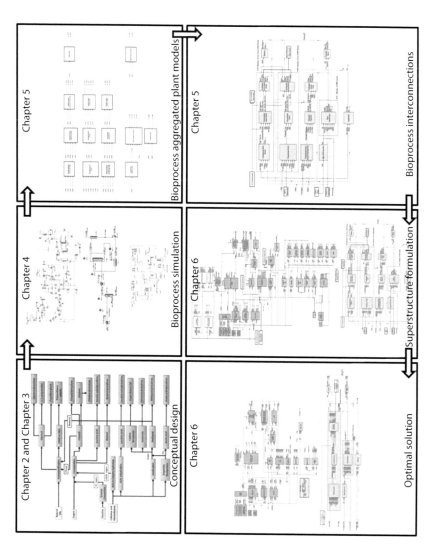

FIGURE 1.2
Organization of chapters.

models in this chapter. Two other processes were included in this chapter, one for the production of algae oil from carbon dioxide and the other for the production of syngas from corn stover by steam reforming. Then interconnections in the bioprocess models were developed for the optimization model, as shown in Figure 1.2.

Chapter 6 formulates the superstructure of chemical and biochemical plants. The bioprocesses from Chapter 5 were integrated into a base case of existing chemical plants in the lower Mississippi River corridor. The carbon dioxide from the integrated chemical complex was utilized for the production of algae and for chemicals from carbon dioxide. The inset for Chapter 6 on Figure 1.2 shows the plants in the superstructure. The units in green are the bioprocesses developed from Chapter 5. The units in blue are the plants in the existing base case. The units in red are new processes that utilize pure carbon dioxide for the production of chemicals.

Interconnections were developed for the integrated complex. Alternatives for production of chemicals were specified using binary variables and logical constraints for superstructure optimization. Inequality constraints for plant capacities and demand from each plant were also specified. The next step was constructing the objective function based on triple bottom line criteria. The triple bottom line included functions for product sales, economic costs (raw material and utility), environmental costs (67% of raw material costs), and sustainable costs and credits.

Chapter 6 then gives the results of the optimization of the superstructure. The optimal solution gave the plants that were included in the optimal structure. A comparison between the base case and optimal structure was given for triple bottom line costs, the pure and impure carbon dioxide emissions, the energy requirements for plants, and the capacity of the plants. Multicriteria optimization was used to determine Pareto optimal solutions for the optimal structure. Monte Carlo simulation was used for determining parameter sensitivity of the optimal solution. A comparison of results with previous investigations is also included in this chapter.

In Chapter 7, the superstructure described in Chapter 6 is used to demonstrate how it can be applied by the decision maker. Five cases were developed using the superstructure, as given in Table 1.2. Case I was a modification of the superstructure to study the integration of bioprocesses only in the existing base case. The carbon dioxide was not used for chemicals or algae oil production in this case. The impact of the addition of bioprocesses could be studied using this case.

Case II in Chapter 7 was a parametric study of sustainable costs and credits given for carbon dioxide, with the present scenario of zero carbon dioxide cost and credit as a reference. Carbon tax, cap-and-trade system, sequestration processes, etc., give probable costs for carbon dioxide. Some of these costs were used to construct cases for a $5, $25, $75, or $125 cost per MT of CO_2 for carbon dioxide emission, and $25 or $50 credit per MT of CO_2 for carbon dioxide consumption. The base case and the optimal structure were

TABLE 1.2

Case Studies Using the Superstructure in Chapter 7

Case Study	Description
Case Study I—Superstructure without carbon dioxide use	Aimed to study the optimal solution for integrating bioprocesses only, without reuse of carbon dioxide from the integrated complex
Case Study II—Effect of sustainable costs and credits on the triple bottom line	Aimed to study the optimal solution for various combinations of probable carbon dioxide costs for emission and credits for consumption
Case Study III—Effect of algae oil production costs on the triple bottom line	Aimed to study the optimal solution for various combinations of probable carbon dioxide costs for emission and credits for consumption
Case Study IV—Multicriteria optimization using 30% oil content algae production and sustainable costs/credits	Aimed to study the multicriteria solution for maximizing profit while minimizing sustainable cost when sustainable credits/costs and algae oil production costs are included
Case Study V—Effect of corn and corn stover costs and number of corn ethanol plants on the triple bottom line	Aimed to study the optimal solution for various combinations of corn and corn stover costs and number of corn ethanol plants

compared for these costs. The results of the optimal structure without carbon dioxide utilization from Case I were also given for comparison.

Case III in Chapter 7 was a parametric study of algae oil production costs. The superstructure considered zero algae oil production costs with new technology and algae strains used for oil production. This case incorporates costs for current technology using high, low, and average performance algae oil production plants and two strains of algae containing 30% and 50% oil content. Optimal structure results are presented with respect to triple bottom line costs and carbon dioxide utilization from the complex.

Case IV in Chapter 7 used the superstructure to construct a case for multicriteria optimization with parameters taken from Case II and Case III. A high carbon dioxide emission cost of $125 per MT of CO_2 for emission and $25 per MT of CO_2 for consumption were used in the model. The 30% oil content algae strain was used and the multicriteria optimization problem was solved for low-performance and high-performance algae oil plant performance. The Pareto optimal sets for maximizing company's profits and sustainable credits to the society are given as results of this case.

Case V in Chapter 7 used the superstructure to study variations in corn and corn stover costs. Corn costs have varied over the period from 2000 to 2010 with high costs of $160 per MT and low costs of $70 per MT. Corn stover costs ranged from $51 per MT to $72 per MT. Combinations of these costs were used to study the effect on the optimal solution. Also, combinations of two, three, or four corn ethanol plants and rest corn stover ethanol plants as constraints in the model were used to study the effect of inclusion of these in the optimal structure.

The postscript gives the conclusions of this methodology and future directions that may be undertaken from this research.

Appendix A gives a comprehensive review of methods for sustainable process evaluations. These include TCA, LCA, eco-efficiency analysis, and sustainability indices and metrics, among others. Carbon dioxide costs other than those mentioned in Chapter 7 are also included in this Appendix.

Appendix B gives a review of the optimization methods and solvers that are currently used. A comparison of computational results for the optimal structure using various global solvers is given in this Appendix.

Appendix C gives the price of raw materials and products used in the complex, with the source for the data collected. These include renewable raw materials like corn, corn stover, and soybean oil. The chemicals from the base case and new chemicals from biomass are included for raw material costs and product prices in this Appendix.

Appendix D gives a theoretical basis for estimating price elasticity of supply and demand, which was used for calculating cross-price elasticity of demand of ammonia in Chapter 7. The price elasticity of supply of corn, demand of corn, bioethanol, and ethylene were also given in this Appendix.

Appendix E gives an overview of the chemical complex analysis system, which was used for the superstructure formulation and optimization. A step-by-step guide to using the tool for the chemical complex optimization is given in this Appendix.

Appendix F gives detailed mass and energy balances for streams in the bioprocess designs. Appendix G gives the equipment mapping and costs of equipment from ICARUS for the bioprocess designs. Appendix H gives the molecular weight of the species used for the bioprocess design and model formulation.

1.9 Summary

A research vision is proposed that will lead to the development of new plants that are based on renewable resources supplying the needed goods and services of the current plants. The vision includes converting existing plants to ones that are based on renewable resources, requiring nonrenewable resource supplements. The objectives include developing a methodology used by decision makers that encompasses economic development, environmental improvements, and sustainable development. The methodology included identifying and designing new chemical processes that use biomass and carbon dioxide as raw materials and showed how these processes could be integrated into existing chemical production complexes. The research demonstrates how existing plants can transition to renewable feedstocks from nonrenewable feedstocks. The chemical production complex in the lower Mississippi River corridor was used to demonstrate the integration of these new plants into an existing infrastructure.

TCA is a methodology developed by industry professionals and sponsored by the American Institute of Chemical Engineers. It identifies five types of costs that include economic, environmental, and societal costs. TCA serves as the basis for the triple bottom line evaluation where the five types of costs are combined into economic, environmental, and sustainable costs and extended to sustainable credits.

2

Biomass as Feedstock

2.1 Introduction

The world is dependent heavily on coal, petroleum, and natural gas for energy, for fuel, and as feedstock for chemicals. These sources are commonly termed as fossil or nonrenewable resources. Geological processes formed fossil resources over a period of millions of years by the loss of volatile constituents from plant or animal matter. Human civilization has seen a major change in obtaining its material needs through abiotic environment only recently. Plant-based resources were the predominant source of energy, organic chemicals, and fibers in the Western world as recently as 200 years ago, and the biotic environment continues to play a role in many developing countries. The discovery of coal and its usage has been traced back to the fourth century BC. Comparatively, petroleum was a newer discovery in the nineteenth century and its main use was to obtain kerosene for burning oil lamps. Natural gas, a mixture containing primarily methane, is found associated with other fossil resources, for example, in coal beds. The historical, current, and projected use of fossil resources for energy consumption is given in Figure 2.1. Petroleum, coal, and natural gas constitute about 86% of resource consumption in the United States (EIA, 2010a). The remaining 8% comes from nuclear energy and 6% from renewable energy. Approximately 3% of total crude petroleum is currently used for the production of chemicals, the rest being used for energy and fuel.

Fossil resources are extracted from the Earth's crust, processed, and burnt or converted to chemicals. The proven reserves, in North America, for coal was 276,285 million ton (equivalent to 5382 EJ [exajoule = 10^{18} J]) in 1990, for oil was 81 billion barrels (equivalent to 476 EJ) in 1993, and for natural gas was 329×10^3 billon ft^3 (equivalent to 347 EJ) in 1993 (Klass, 1998). The United States has significant reserves of crude oil, but the country is also dependent on oil imports from other countries for meeting the energy requirements. The crude oil price has fluctuated over the past 40 years, the most recent price increase over $130 per barrel being in 2008. The EIA published a projection of the price of crude oil over the next 25 years, where a high and a low projection were given in addition to the usual projection of crude oil price as shown in Figure 2.2 (EIA, 2010a). The projection shows a steady increase in

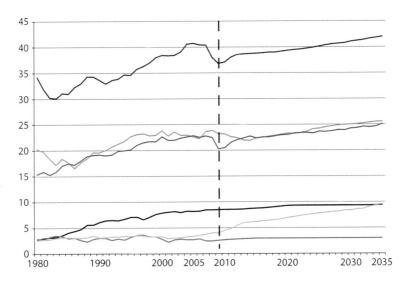

FIGURE 2.1
Energy consumption (quadrillion Btu) in the United States, 1980–2035. (From Energy Information Administration (EIA), Annual Energy Outlook 2010, Report No. DOE/EIA-0383(2010), 2010a.)

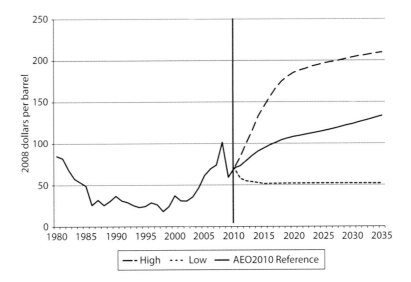

FIGURE 2.2
Oil prices (in 2008 dollars per barrel) historical data and projected data. (Adapted from Energy Information Administration (EIA), Annual Energy Outlook 2010, Report No. DOE/EIA-0383(2010), 2010a.)

the price of crude to above $140 per barrel in 2035. With a high price trend, crude can cost over $200 per barrel.

Fossil resources are burnt or utilized for energy, fuel, and chemicals. The process for combustion of fossil resources involves the oxidation of carbon and hydrogen atoms to produce carbon dioxide and water vapor, releasing heat from the reactions. Impurities in the resource, such as sulfur, produce sulfur oxides, and incomplete combustion of the resource produces methane. The Intergovernmental Panel on Climate Change (IPCC) identified that changes in atmospheric concentration of greenhouse gases (GHGs), aerosols, land cover, and solar radiation alter the energy balance of the climate system (IPCC, 2007). These changes are also termed as climate change. The GHGs include carbon dioxide, methane, nitrous oxide, and fluorinated gases. Atmospheric concentrations of carbon dioxide (379 ppm) and methane (1774 ppb) in 2005 were the highest amounts recorded on the earth (historical values computed from ice cores spanning many thousands of years) to date. The IPCC report states that global increases in CO_2 concentrations are attributed primarily to fossil resource use. In the United States, there was approximately 5814 million MT of carbon dioxide released into the atmosphere in 2008 and this amount is projected to increase to 6320 million MT in 2035 (EIA, 2010a), as shown in Figure 2.3.

The increasing trends in resource consumption, resource material cost, and consequent increase in carbon dioxide emissions from anthropogenic sources indicate that a reduction of fossil feedstock usage is necessary to

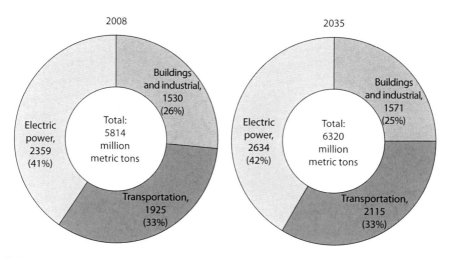

FIGURE 2.3
Carbon dioxide emissions in 2008 (current) and 2035 (projected) due to fossil feedstock use. (Adapted from Energy Information Administration (EIA), Annual Energy Outlook 2010, Report No. DOE/EIA-0383(2010), 2010a.)

address climate change. This has prompted world leaders, organizations, and companies to look for alternative ways to obtain energy, fuel, and chemicals.

Thus, carbon fixed naturally in fossil and nonrenewable resources over millions of years is released to the atmosphere by anthropogenic sources. A relatively faster way to convert the atmospheric carbon dioxide into useful resources is by photosynthetic fixation into biomass. The life cycle of the fossil resources show that coal, petroleum, and natural gas are all derivatives of decomposed biomass on the earth's surface trapped in geological formations. Thus, biomass, being a precursor to the conventional nonrenewable resources, can be used as fuel and can generate energy and produce chemicals with some modifications to existing processes.

TABLE 2.1

Heating Value of Biomass Components

Component	Heating Value (Gross) (GJ/MT Unless Otherwise Mentioned)
Bioenergy feedstocks	
Corn stover	17.6
Sweet sorghum	15.4
Sugarcane bagasse	18.1
Sugarcane leaves	17.4
Hardwood	20.5
Softwood	19.6
Hybrid poplar	19.0
Bamboo	18.5–19.4
Switchgrass	18.3
Miscanthus	17.1–19.4
Arundo donax	17.1
Giant brown kelp	10.0 MJ/dry kg
Cattle feedlot manure	13.4 MJ/dry kg
Water hyacinth	16.0 MJ/dry kg
Pure cellulose	17.5 MJ/dry kg
Primary biosolids	19.9 MJ/dry kg
Liquid biofuels	
Bioethanol	28
Biodiesel	40
Fossil Fuels	
Coal (low Rank; lignite/sub-bituminous)	15–19
Coal (high rank; bituminous/anthracite)	27–30
Oil (typical distillate)	42–45

Source: Klass, D.L., *Biomass for Renewable Energy, Fuels and Chemicals*, Academic Press, San Diego, CA, ISBN 0124109500, 1998; McGowan, T.F., *Biomass and Alternate Fuel Systems*, John Wiley & Sons Inc., Hoboken, NJ, ISBN 978-0-470-41028-8, 2009.

Biomass can be classified broadly as all the matter on the earth's surface of recent biological origin. Biomass includes plant materials such as trees, grasses, agricultural crops, and animal manure. Just as petroleum and coal require processing before use as feedstock for the production of fuel, chemicals, and energy, biomass also requires processing such that the resource potential can be utilized fully. As explained earlier, biomass is a precursor to fossil feedstock, and a comparison between the biomass energy content and fossil feedstock energy content is required. The heating value of fuel is the measure of heat released during the complete combustion of fuel at a given reference temperature and pressure. The higher or gross heating value is the amount of heat released per unit weight of fuel at the reference temperature and pressure, taking into account the latent heat of vaporization of water. The lower or net heating value is the heat released by fuel excluding the latent heat of vaporization of water. The higher heating value of some bioenergy feedstocks, liquid biofuels, and conventional fossil fuels are given in Table 2.1. It can be seen from the table that the energy content of the raw biomass species is lesser than the bioethanol, and the biodiesel compares almost equally to the traditional fossil fuels.

This chapter gives an outline for the use of biomass as feedstock. The following sections will discuss various methods for biomass formation, biomass composition, conversion technologies, and feedstock availability.

2.2 Biomass Formation

Biomass is the photosynthetic sink by which atmospheric carbon dioxide and solar energy is fixed into plants (Klass, 1998). These plants can be used to convert the stored energy in the form of fuel and chemicals. The primary equation of photosynthesis is given by Equation 2.1:

$$6CO_2 + 6H_2O + Light \rightarrow C_6H_{12}O_6 + 6O_2 \qquad (2.1)$$

The photosynthesis process utilizes inorganic material (carbon dioxide and water) to form organic compounds (hexose) and releases oxygen. The Gibbs free energy change for the process is +470 KJ/mol of CO_2 assimilated, and the corresponding enthalpy change is +470 KJ. The positive sign on the energy denotes that energy is absorbed in the process. Photosynthesis is a two-phase process comprising of the "light reactions" (in the presence of light) and the "dark reactions" (in the absence of light).

The light reactions are common to all plant types, where eight photons per molecule of carbon dioxide excite chlorophyll to generate ATP (adenosine triphosphate) and $NADPH_2$ (reduced nicotinamide adenosine dinucleotide phosphate) along with oxygen (Klass, 1998). The ATP and $NADPH_2$ react in the dark to reduce CO_2 and form the organic components in biomass via the dark reactions and regenerate ADP (adenosine diphosphate) and NADP (nicotinamide adenosine dinucleotide phosphate) for the light reactions.

The dark reactions can proceed in accordance with at least three different pathways, the Calvin–Benson cycle, the C4 cycle, and the CAM cycle, as discussed in the following sections.

2.2.1 Calvin–Benson Cycle

Plant biomass species, which use the Calvin–Benson cycle to form products, are called the C3 plants (Klass, 1998). This cycle produces the 3-carbon intermediate 3-phosphoglyceric acid and is common to fruits, legumes, grains, and vegetables. C3 plants usually exhibit low rates of photosynthesis at light saturation, low light saturation points, sensitivity to oxygen concentration, rapid photorespiration, and high CO_2 compensation points. The CO_2 compensation point is the CO_2 concentration in the surrounding environment below which more CO_2 is respired by the plant than is photosynthetically fixed. Typical C3 biomass species are alfalfa, barley, chlorella, cotton, Eucalyptus, Euphorbia lathyris, oats, peas, potato, rice, soybean, spinach, sugar beet, sunflower, tall fescue, tobacco, and wheat. These plants grow favorably in cooler climates.

2.2.2 C4 Cycle

In this cycle, CO_2 is initially converted to four-carbon dicarboxylic acids (malic or aspartic acids) (Klass, 1998). The C4 acid is transported to bundle sheath cells where decarboxylation occurs to regenerate pyruvic acid, which is returned to the mesophyll cells to initiate another cycle. The CO_2 liberated in the bundle sheath cells enter the C3 cycle described earlier, and it is in this C3 cycle where the CO_2 fixation occurs. The subtle difference between the C3 and C4 cycles is believed to be responsible for the wide variations in biomass properties. C4 biomass is produced in higher yields with higher rates of photosynthesis, high light saturation points, low levels of respiration, low carbon dioxide compensation points, and greater efficiency of water usage. Typical C4 biomass includes crops such as sugarcane, corn, sorghum, and tropical grasses such as Bermuda grass.

2.2.3 CAM Cycle

The CAM cycle is the Crassulacean acid metabolism cycle, which refers to the capacity of chloroplast-containing biomass tissues to fix CO_2 via phosphoenolpyruvate carboxylase in dark reactions leading to synthesis of free malic acid (Klass, 1998). The mechanism involves b-carboxylation of phosphoenolpyruvic acid by this enzyme and the subsequent reduction of oxaloacetic acid by maleate dehydrogenase. Biomass species in the CAM category are typically adapted to arid environments, have low photosynthesis rates, and higher water usage efficiencies. Plants in this category include cactus and succulents such as pineapple. The CAM has evolved so that the initial CO_2 fixation can take place in the dark with much less water loss than C3 or C4 pathways. CAM biomass also conserves carbon by recycling endogenously formed CO_2. CAM biomass species have not been exploited commercially for use as biomass feedstock.

Thus, different photosynthetic pathways produce different kinds of biomass. The following section discusses the different components in biomass.

2.3 Biomass Classification and Composition

Section 2.2 gave the mechanisms for the formation of biomass by photosynthesis. The classification and composition of biomass will be discussed in this section. Biomass can be classified into two major subdivisions, crop biomass and wood (forest) biomass. There are other sources of biomass, like waste from municipal areas and animal wastes, but these can be traced back to the two major sources. Crop biomass primarily includes corn, sugarcane, sorghum, soybean, wheat, barley, rice, etc. These contain carbohydrates, glucose, and starch as their primary constituents. Wood biomass is composed of cellulose, hemicellulose, and lignin. Examples of woody biomass include grasses, stalks, stover, etc. Starch and cellulose are both polymeric forms of glucose, a six-carbon sugar. Hemicellulose is a polymer of xylose. Lignin is composed of phenolic polymers, and oils are triglycerides. Other biomass components, which are generally present in minor amounts, include proteins, sterols, alkaloids, resins, terpenes, terpenoids, and waxes. These components are discussed in detail in the following.

2.3.1 Saccharides and Polysaccharides

Saccharides and polysaccharides are hydrocarbons with the basic chemical structure of CH_2O. The hydrocarbons occur in nature as five-carbon or six-carbon ring structures. The ring structures may contain only one or two connected rings, which are known as monosaccharides, disaccharides or simply as sugars, or they may be very long polymer chains of the sugar building blocks.

The simplest six-sided saccharide (hexose) is glucose. Long-chained polymers of glucose or other hexoses are categorized either as starch or cellulose. The characterization is discussed in the following sections. The simplest five-sided sugar (pentose) is xylose. Xylose forms long-chain polymers categorized as hemicellulose. Some of the common six-carbon and five-carbon monosaccharides are listed in Table 2.2.

TABLE 2.2

Common Six-Carbon and Five-Carbon Monosaccharides

Six-carbon sugars	Structure	Five-carbon sugars	Structure
D-Fructose		D-Xylose	
D-Glucose		D-Ribulose	
D-Glucose		D-Ribose	
D-Mannose		D-Arabinose	
D-Galactose			

2.3.2 Starch

Starch is a polymer of glucose as the monomeric unit (Paster et al., 2003). It is a mixture of α-amylose and amylopectin as shown in Figure 2.4. α-amylose is a straight chain of glucose molecules joined by α-1,4-glucosidic linkages as shown in Figure 2.4a. Amylopectin and amylase are similar except that short chains of glucose molecules branch off from the main chain (backbone) as shown in Figure 2.4b. Starches found in nature contain 10%–30% α-amylose and 70%–90% amylopectin. The α-1,4-glycosidic linkages are bent and prevent the formation of sheets and subsequent layering

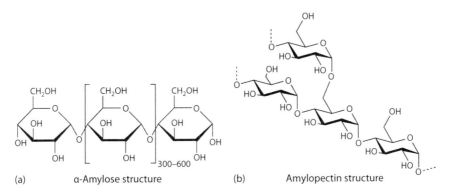

(a)   α-Amylose structure   (b)   Amylopectin structure

FIGURE 2.4

Structure of starch: (a) α-amylose and (b) amylopectin.

of polymer chains. As a result, starch is soluble in water and relatively easy to break down into utilizable sugar units.

2.3.3 Lignocellulosic Biomass

The non-grain portion of biomass (e.g., cobs and stalks), often referred to as agricultural stover or residues, and energy crops such as switchgrass are known as lignocellulosic biomass resources (also called cellulosic). These are comprised of cellulose, hemicellulose, and lignin (Paster et al., 2003). Generally, lignocellulosic material contains 30%–50% cellulose, 20%–30% hemicellulose, and 20%–30% lignin. Some exceptions to this are cotton (98% cellulose) and flax (80% cellulose). Lignocellulosic biomass is considered to be an abundant resource for the future bioindustry. Recovering the components in a cost-effective way requires pretreatment processes discussed in a later section.

Cellulose: Cellulosic biomass comprises 35%–50% of most plant material. Cellulose is a polymer of glucose with degree of polymerization of 1,000–10,000 (Paster et al., 2003). Cellulose is a linear unbranched polymer of glucose joined together by β–1,4-glycosidic linkages as shown in Figure 2.5. Cellulose can either be crystalline or amorphous. Hydrogen bonding between chains leads to chemical stability and insolubility and serves as a structural component in plant walls. The high degree of crystallinity of cellulose makes lignocellulosic materials much more resistant than starch to acid and enzymatic hydrolysis. As the core structural component of biomass, cellulose is also protected from environmental exposure by a sheath of lignin and hemicellulose. Extracting the sugars of lignocellulosics, therefore, involves a pretreatment stage to reduce the recalcitrance (resistance) of the biomass to cellulose hydrolysis.

Hemicellulose: Hemicellulose is a polymer containing primarily five-carbon sugars such as xylose and arabinose with some glucose and mannose dispersed throughout (Paster et al., 2003). The structure of xylose is shown in Figure 2.6. It forms a short-chain polymer that interacts with cellulose and lignin to form a matrix in the plant wall, thereby strengthening it. Hemicellulose is more easily hydrolyzed than cellulose. Much of the hemicellulose in lignocellulosic materials is solubilized and hydrolyzed to pentose

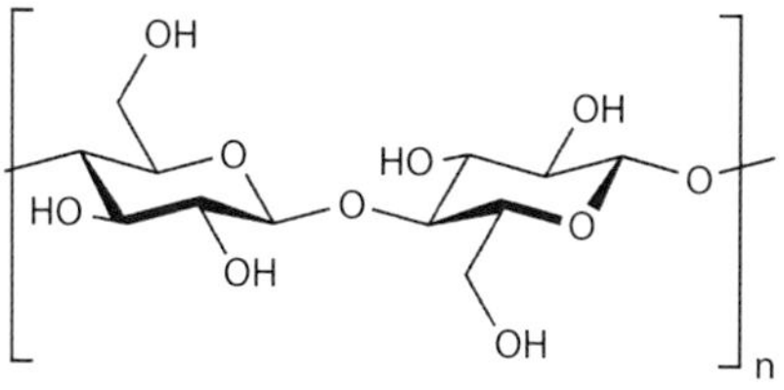

FIGURE 2.5
Structure of cellulose.

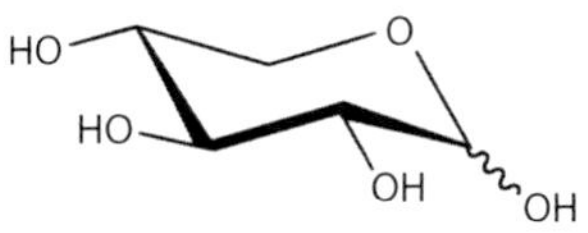

FIGURE 2.6
Structure of xylose, building block of hemicellulose.

and hexose sugars during the pretreatment stage. Some of the hemicellulose is too intertwined with the lignin to be recoverable.

Lignin: Lignin helps to bind the cellulose/hemicelluloses matrix while adding flexibility to the mixture. The molecular structure of lignin polymers is very random and disorganized and consists primarily of carbon ring structures (benzene rings with methoxyl, hydroxyl, and propyl groups) interconnected by polysaccharides (sugar polymers), as shown in Figure 2.7. The ring

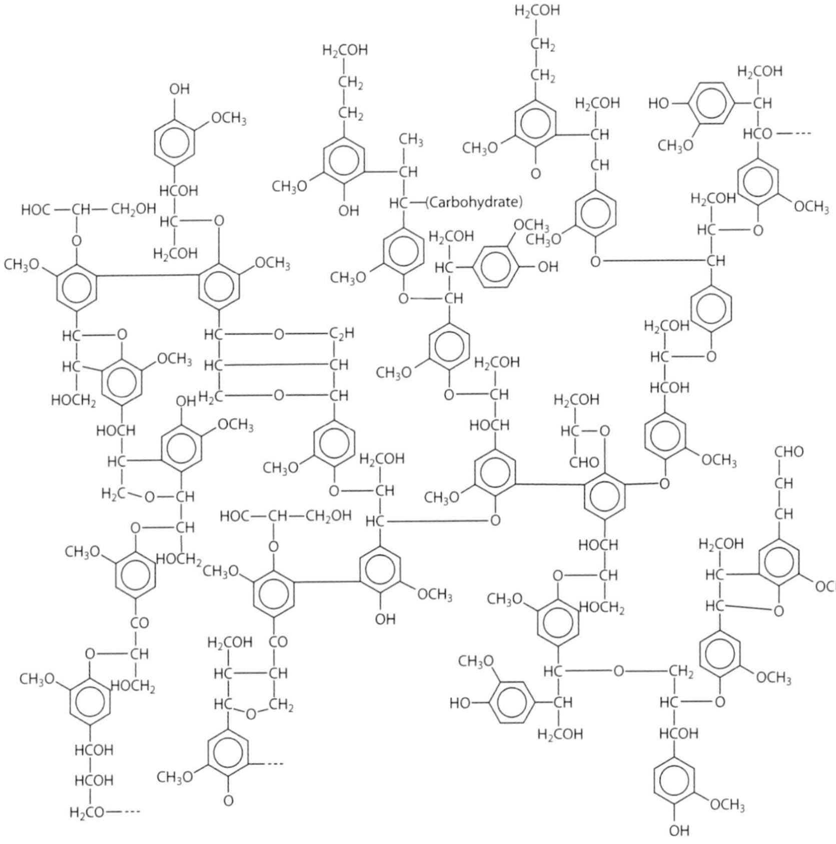

FIGURE 2.7
Structure of lignin. (From Glazer, A.W. and Nikaido, H., *Microbial Biotechnology: Fundamentals of Applied Microbiology*, W.H. Freeman & Company, San Francisco, CA, ISBN 0-71672608-4, 1995.)

structures of lignin have great potential as valuable chemical intermediates, mainly aromatic compounds. However, separation and recovery of the lignin is difficult. It is possible to break the lignin–cellulose–hemicellulose matrix and recover the lignin through treatment of the lignocellulosic material with strong sulfuric acid. Lignin is insoluble in sulfuric acid, while cellulose and hemicellulose are solubilized and hydrolyzed by the acid. However, the high acid concentration promotes the formation of degradation products that hinder the downstream utilization of the sugars. Pyrolysis can be used to convert the lignin polymers to valuable products, but separation techniques to recover the individual chemicals are lacking. Instead, the pyrolyzed lignin is fractionated into a bio-oil for fuel and high phenolic content oil that is used as a partial replacement for phenol in phenol–formaldehyde resins.

2.3.4 Lipids, Fats, and Oils

Oils can be obtained from oilseeds like soybean, canola, etc. Vegetable oils are composed primarily of triglycerides, also referred to as triacylglycerols. Triglycerides contain a glycerol molecule as the backbone with three fatty acids attached to glycerol's hydroxyl groups. The structure of a triglyceride is shown in Figure 2.8 with linoleic acid as the fatty acid chain. In this example, the three fatty acids are all linoleic acid, but triglycerides could be a mixture of two or more fatty acids. Fatty acids differ in chain length and degree of condensation. The fatty acid profile and the double bonds present determine the property of the oil. These can be manipulated to obtain certain performance characteristics. In general, the greater the number of double bonds, the lower the melting point of the oil.

2.3.5 Proteins

Proteins are polymers composed of natural amino acids, bonded together through peptide linkages (Klass, 1998). They are formed via condensation of the acids through the amino and carboxyl groups by removal of water to form polyamides. Proteins are present in various kinds of biomass as

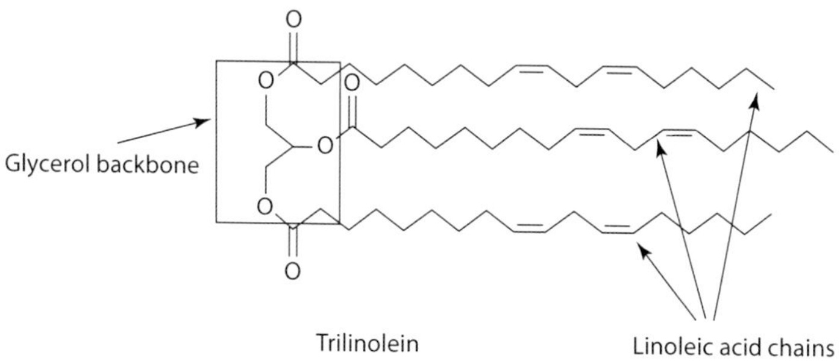

Glycerol backbone

Trilinolein   Linoleic acid chains

FIGURE 2.8
Formation of triglycerides (linoleic acid as representative fatty acid chain).

TABLE 2.3

Amino Acid Groups Present in Proteins

Family	Amino Acids
Glutamate	Glutamine, arginine, proline
Aspartate	Asparagine, methionine, threonine, isoleucine, lysine
Aromatic	Tryptophan, phenylalanine, tyrosine
Serine	Glycine, cysteine
Pyruvate	Alanine, valine, leucine

Source: Paster, M et al., Industrial Bioproducts: Today and Tomorrow. Department of Energy Report prepared by Energetics, Inc., http://www.energetics.com/resourcecenter/products/studies/Documents/bioproducts-opportunities.pdf, accessed May 8, 2010, 2003.

well as animals. The concentration of proteins may approach zero in different biomass systems but the importance of proteins arises while considering enzyme catalysis that promotes the various biochemical reactions. The apparent precursors of proteins are amino acids in which an amino group, or imino group in a few cases, is bonded to the carbon atom adjacent to the carboxyl group. Many amino acids have been isolated from natural sources, but only about 20 of them are used for protein biosynthesis. These amino acids are divided into five families: glutamate, aspartate, aromatic, serine, and pyruvate. The various amino acids under these groups are shown in Table 2.3.

Table 2.4 gives the composition of some biomass species based on the aforementioned components. The biomass types are marine, freshwater, herbaceous, woody, and waste biomass, and a representative composition is given in the table. Other components not included in the composition are ash and crude protein.

2.4 Biomass Conversion Technologies

The conversion of biomass involves the treatment of biomass so that the solar energy stored in the form of chemical energy in the biomass molecules can be utilized. Common biomass conversion routes begin with pretreatment in case of cellulosic and grain biomass and extraction of oil in case of oilseeds. Then the cellulosic or starch-containing biomass undergoes fermentation (anaerobic or aerobic), gasification, or pyrolysis. The oil in oilseeds is transesterified to get fatty acid esters. There are other process technologies including hydroformylation, metathesis, and epoxidation,

TABLE 2.4

Component Composition of Biomass Feedstocks

Name	Celluloses (dry wt%)	Hemicelluloses (dry wt%)	Lignins (dry wt%)
Corn stover	35	28	16–21
Sweet sorghum	27	25	11
Sugarcane bagasse	32–48	19–24	23–32
Hardwood	45	30	20
Softwood	42	21	26
Hybrid poplar	42–56	18–25	21–23
Bamboo	41–49	24–28	24–26
Switchgrass	44–51	42–50	13–20
Miscanthus	44	24	17
Arundo donax	31	30	21
RDF (refuse-derived fuel)	65.6	11.2	3.1
Water hyacinth	16.2	55.5	6.1
Bermuda grass	31.7	40.2	25.6
Pine	40.4	24.9	34.5

Source: Klass, D.L., *Biomass for Renewable Energy, Fuels and Chemicals*, Academic Press, San Diego, CA, ISBN 0124109500, 1998; McGowan, T.F., *Biomass and Alternate Fuel Systems*, John Wiley & Sons Inc., Hoboken, NJ, ISBN 978-0-470-41028-8, 2009.

related with direct conversion of oils to fuels and chemicals, the details of which are not included in this chapter.

2.4.1 Biomass Pretreatment

Biomass is primarily composed of cellulose, hemicelluloses, and lignin. The cellulose and hemicelluloses are polysaccharides of hexose and pentose. Any process that uses biomass needs to be pretreated so that the cellulose and hemicellulose in the biomass are broken down to their monomeric form. Pretreatment processes produce a solid pretreated biomass residue that is more amenable to enzymatic hydrolysis by cellulases and elated enzymes than native biomass. Biocatalysts like yeasts and bacteria can act only on the monomers and ferment them to alcohols, lactic acid, etc. The pretreatment process also removes the lignin in biomass which is not acted upon by enzymes or fermented further.

Pretreatment usually begins with a physical reduction in the size of plant material by milling, crushing, and chopping (Teter et al., 2006). For example, in the processing of sugarcane, the cane is first cut into segments and then fed into consecutive rollers to extract cane juice rich in sucrose and physically crush the cane, producing a fibrous bagasse having the consistency of sawdust. In the case of corn stover processing, the stover is chopped with knives or ball milled to increase the exposed surface area and improve wettability.

After the physical disruption process, the biomass may be chemically treated to remove lignin. Lignin forms a coating on the cellulose microfibrils in untreated biomass, thus making the cellulose unavailable for enzyme or acid hydrolysis. Lignin also absorbs some of the expensive cellulose-active enzymes.

The following pretreatment processes are employed for biomass conversion:

Hot wash pretreatment: The hot wash pretreatment process involves the passage of hot water through heated stationary biomass and is responsible for solubilization of the hemicellulose fraction (Teter et al., 2006). The hemicellulose is converted to pentose oligomers by this process, which needs to be further converted to respective monosaccharides before fermentation. The performance of this pretreatment process depends on temperature and flow rate, requiring about 8–16 min. About 46% of lignin is removed at high rates and temperatures. The hydrothermal process does not require acid-resistant material for the reactors, but water use and recovery costs are disadvantages to the process.

Acid hydrolysis: Hydrolysis is a chemical reaction or process where a chemical compound reacts with water. The process is used to break complex polymer structures into its component monomers. The process can be used for the hydrolysis of polysaccharides like cellulose and hemicelluloses (Katzen and Schell, 2006). When hydrolysis is catalyzed by the presence of acids like sulfuric, hydrochloric, nitric, or hydrofluoric acids, the process is called acid hydrolysis. The reactions for hydrolysis can be expressed as in reaction given by Equations 2.2 and 2.3.

$$\text{Cellulose (glucan)} \rightarrow \text{glucose} \rightarrow \text{5-hydroxymethylfurfural} \rightarrow \text{tars} \quad (2.2)$$

$$\text{Hemicellulose (xylan)} \rightarrow \text{xylose} \rightarrow \text{furfural} \rightarrow \text{tars} \quad (2.3)$$

The desired products of hydrolysis are the glucose and xylose. Under severe conditions of high temperature and acid concentrations, the product tends to hydroxymethylfurfural, furfural, and the tars.

Dilute sulfuric acid is inexpensive in comparison to the other acids. It has also been studied and the chemistry well known for acid conversion processes (Katzen and Schell, 2006). Biomass is mixed with a dilute sulfuric acid solution and treated with steam at temperatures ranging from 140°C to 260°C. Xylan is rapidly hydrolyzed in the process to xylose at low temperatures of 140°C–180°C. At higher temperatures, cellulose is depolymerized to glucose but the xylan is converted to furfural and tars.

Concentrated acids at low temperatures (100°C–120°C) are used to hydrolyze cellulose and hemicelluloses to sugars (Katzen and Schell, 2006). Higher yields of sugars are obtained in this case with lower conversion to tars. The viability of this process depends on low cost recovery of expensive acid catalysts.

Enzymatic hydrolysis: Acid hydrolysis explained earlier has a major disadvantage where the sugars are converted to degradation products like tars. This degradation can be prevented by using enzymes favoring 100% selective conversion of cellulose to glucose. When hydrolysis is catalyzed by such enzymes, the process is known as enzymatic hydrolysis (Katzen and Schell, 2006).

Enzymatic hydrolysis is carried out by microorganisms like bacteria, fungi, protozoa, insects, etc. (Teter et al., 2006). Advancement of gene sequencing in microorganisms has made it possible to identify the enzymes present in them, which are responsible for the biomass degradation. Bacteria like *Clostridium thermocellum*, *Cytophaga hutchinsonii*, *Rubrobacter xylanophilus*, etc. and fungi like *Trichoderma reesei* and *Phanerochaete chrysosporium* have revealed enzymes responsible for carbohydrate degradation.

Based on their target material, enzymes are grouped into the following classifications (Teter et al., 2006). Glucanases or cellulases are the enzymes that participate in the hydrolysis of cellulose to glucose. Hemicellulases are responsible for the degradation of hemicelluloses. Some cellulases have significant xylanase or xyloglucanase side activity which makes it possible for use in degrading both cellulose and hemicelluloses.

Ammonia fiber explosion: This process uses ammonia mixed with biomass in a 1:1 ratio under high pressure (1.4–3 atm) at temperatures of 60°C–110°C for 5–15 min, then explosive pressure release. The volatility of ammonia makes it easy to recycle the gas (Teter et al., 2006).

2.4.2 Fermentation

The pretreatment of biomass is followed by the fermentation process where pretreated biomass containing five-carbon and six-carbon sugars is catalyzed with biocatalysts to produce desired products. Fermentation refers to enzyme-catalyzed, energy-yielding chemical reactions that occur during the breakdown of complex organic substrates in the presence of microorganisms (Klass, 1998). The microorganisms used for fermentation can be yeast or bacteria. The microorganisms feed on the sucrose or glucose released after pretreatment and convert them to alcohol and carbon dioxide. The simplest reaction for the conversion of glucose by fermentation is given in Equation 2.4:

$$C_6H_{12}O_6 \rightarrow 2C_2H_5OH + 2CO_2 \qquad (2.4)$$

An enzyme catalyst is highly specific, catalyzes only one or a small number of reactions, and a small amount of enzyme is required. Enzymes are usually proteins of high molecular weight (15,000 < MW < several million Daltons) produced by living cells. The catalytic ability is due to the particular protein structure, and a specific chemical reaction is catalyzed at a small portion of the surface of an enzyme, called an active site (Klass, 1998). Enzymes have been used since early human history without knowing how they worked. Enzymes have been used commercially since the 1890s when fungal cell extracts were used to convert starch to sugar in brewing vats.

Microbial enzymes include cellulase, hemicellulase, catalase, streptokinase, amylase, protease, clipase, pectinase, glucose isomerase, lactase, etc. The type of enzyme selection determines the end product of fermentation. The growth of the microbes requires a carbon source (glucose, xylose, glycerol, starch, lactose, hydrocarbons, etc.) and a nitrogen source (protein, ammonia, corn steep liquor, diammonium phosphate, etc.). Many organic chemicals like ethanol, succinic acid, itaconic acid, lactic acid, etc. can be manufactured using live organisms that have the required enzymes for converting the biomass. Ethanol is produced by the bacteria *Zymomonous mobilis* or the yeast *Saccaromyces cervisiae*. Succinic acid is produced in high concentrations by *Actinobacillus succinogens* obtained from rumen ecosystem (Lucia et al., 2007). Other microorganisms capable of producing succinic acid include propionate-producing bacteria of the *Propionbacterium* genus, gastrointestinal bacteria such as *Escherichia coli,* and rumen bacteria such as *Ruminococcus flavefaciens*. Lactic acid is produced by a class of bacteria known as lactic acid bacteria (LAB) including the genera *Lactobacillus*, *Lactococcus*, *Leuconostoc*, *Enterococcus*, etc. (Axelsson, 2004).

Commercial processes for corn wet milling and dry milling operations and the fermentation process for lignocellulosic biomass through acid hydrolysis and enzymatic hydrolysis are discussed in detail in the Chapter 3.

2.4.3 Anaerobic Digestion

Anaerobic digestion of biomass is the treatment of biomass with a mixed culture of bacteria to produce methane (biogas) as a primary product. The four stages of anaerobic digestion are hydrolysis, acidogenesis, acetogenesis, and methanogenesis, as shown in Figure 2.9.

In the first stage, hydrolysis, complex organic molecules are broken down into simple sugars, amino acids, and fatty acids with the addition

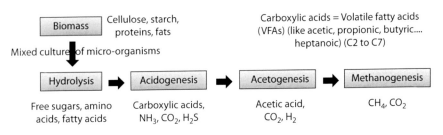

FIGURE 2.9
Anaerobic digestion process. (From Granda, The MixAlco process: Mixed alcohols and other chemicals from biomass, *Seizing Opportunity in an Expanding Energy Marketplace, Alternative Energy Conference*, LSU Center for Energy Studies, http://www.enrg.lsu.edu/Conferences/altenergy2007/granda.pdf, 2007.)

of hydroxyl groups. In the second stage, acidogenesis, volatile fatty acids (e.g., acetic, propionic, butyric, and valeric) are formed along with ammonia, carbon dioxide, and hydrogen sulfide. In the third stage, acetogenesis, simple molecules from acidogenesis are further digested to produce carbon dioxide, hydrogen, and organic acids, mainly acetic acid. Then in the fourth stage, methanogenesis, the organic acids are converted to methane, carbon dioxide, and water.

Anaerobic digestion can be conducted either wet or dry, where dry digestion has a solids content of 30% or more and wet digestion has a solids content of 15% or less. Either batch or continuous digester operations can be used. In continuous operations, there is a constant production of biogas while batch operations can be considered simpler and the production of biogas varies.

The standard process for anaerobic digestion of cellulose waste to biogas (65% methane–35% carbon dioxide) uses a mixed culture of mesophilic or thermophilic bacteria (Kebanli et al., 1981). Mixed cultures of mesophilic bacteria function best at 37°C–41°C and thermophilic cultures function best at 50°C–52°C for the production of biogas. Biogas also contains a small amount of hydrogen and a trace of hydrogen sulfide, and it is usually used to produce electricity. There are two by-products of anaerobic digestion: acidogenic digestate and methanogenic digestate. Acidogenic digestate is a stable organic material comprised largely of lignin and chitin resembling domestic compost, and it can be used as compost or to make low grade building products such as fiberboard. Methanogenic digestate is a nutrient-rich liquid, and it can be used as a fertilizer but may include low levels of toxic heavy metals or synthetic organic materials such as pesticides or PCBs, depending on the source of the biofeedstock.

Kebanli, et al. (1981) give a detailed process design along with pilot unit data for converting animal waste to fuel gas that is used for power generation. A first-order rate constant, 0.011 ± 0.003 per day, was measured for the conversion of volatile solids to biogas from dairy farm waste. In a biofeedstock, the total solids are the sum of the suspended and dissolved solids, and the total solids are composed of volatile and fixed solids. In general, the residence time for an anaerobic digester varies with the amount of feed material, type of material, and the temperature. Resident time of 15–30 days is typical for mesophilic digestion, and residence time for thermophilic digestion is about one-half of that for mesophilic digestion. The digestion of the organic material involves mixed culture of naturally occurring bacteria, each performs a different function. Maintaining anaerobic conditions and a constant temperature are essential for the viability of the bacterial culture.

Holtzapple et al. (1999) describe a modification of the anaerobic digestion process, the MixAlco process, where a wide array of biodegradable material is converted to mixed alcohols. Thanakoses et al. (2003b) describe the process of converting corn stover and pig manure to the third stage of carboxylic acid formation. In the MixAlco process, the fourth stage in anaerobic digestion of the conversion of the organic acids to methane, carbon dioxide, and water

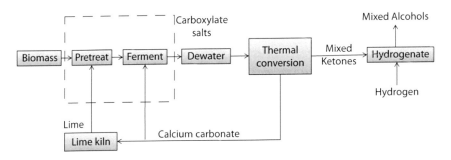

FIGURE 2.10
Flow diagram for the MixAlco process using anaerobic digestion. (From Granda, The MixAlco process: Mixed alcohols and other chemicals from biomass, *Seizing Opportunity in an Expanding Energy Marketplace, Alternative Energy Conference,* LSU Center for Energy Studies, http://www.enrg.lsu.edu/Conferences/altenergy2007/granda.pdf, 2007.)

is inhibited using iodoform (CHI_3) and bromoform ($CHBr_3$). Biofeedstocks to this process can include urban wastes, such as municipal solid waste and sewage sludge, agricultural residues, such as corn stover and bagasse. Products include carboxylic acids (e.g., acetic, propionic, and butyric acid), ketones (e.g., acetone, methyl ethyl ketone, and diethyl ketone), and biofuels (e.g., ethanol, propanol, and butanol). The process uses a mixed culture of naturally occurring microorganisms found in natural habitats such as the rumen of cattle to anaerobically digest biomass into a mixture of carboxylic acids produced during the acidogenic and acetogenic stages of anaerobic digestion. The fermentation conditions of the MixAlco process make it a viable process, since the fermentation involves mixed culture of bacteria obtained from animal rumen, which is available at lower cost compared to genetically modified organisms and sterile conditions required by other fermentation processes.

The Mixalco process is outlined in Figure 2.10 where biomass is pretreated with lime to remove lignin. Calcium carbonate is also added to the pretreatment process. The resultant mixture containing hemicellulose and cellulose is fermented using a mixed culture of bacteria obtained from cattle rumen. This process produces a mixture of carboxylate salts, which is then fermented. Carboxylic acids are naturally formed in the following places: animal rumen, anaerobic sewage digestors, swamps, termite guts, etc. The same microorganisms are used for the anaerobic digestion process and the acid products at different culture temperatures are given in Table 2.5.

The MixAlco process proceeds to form carboxylate salts with the calcium carbonate. Dewatering process removes water. Then the carboxylate salts are thermally decomposed to mixed ketones like acetone, diethyl ketone, and dipropyl ketones. The mixed ketones can then be converted to ethanol by hydrogenation using Raney nickel catalyst at a temperature of 130°C and a pressure of 12 atm in a stirred tank reactor for 35 min.

TABLE 2.5

Carboxylic Acid Products at Different
Culture Temperatures

Acid	40°C	55°C
C2—Acetic	41 wt%	80 wt%
C3—Propionic	15 wt%	4 wt%
C4—Butyric	21 wt%	15 wt%
C5—Valeric	8 wt%	<1 wt%
C6—Caproic	12 wt%	<1 wt%
C7—Heptanoic	3 wt%	<1 wt%
	100 wt%	100 wt%

Source: Granda, C., The MixAlco process: Mixed
alcohols and other chemicals from biomass, *Seizing Opportunity in an Expanding
Energy Marketplace, Alternative Energy
Conference*, LSU Center for Energy Studies,
http://www.enrg.lsu.edu/Conferences/
altenergy2007/granda.pdf, 2007.

2.4.4 Transesterification

Transesterification is the reaction of an alcohol with natural oil containing triglycerides to produce monoalkyl esters and glycerol (Meher et al., 2006). The
glycerol layer settles down at the bottom of the reaction vessel. Diglycerides
and monoglycerides are the intermediates in this process. The Figure 2.11
shows the general reaction for transesterification with an example for trilinolein as the representative triglyceride and methanol as the representative
alcohol. The alcohols that can be used for transesterification depend on the
type of esters desired. Methanol (CH_3OH) gives methyl esters and ethanol
(C_2H_5OH) produces ethyl esters.

A wide variety of vegetable oils and natural oils can be used for transesterification. In Table 2.6 a list of oils is given that can be used with their
respective constituent fatty acid content. Linoleic acid and oleic acid are the
main constituents for soybean oil.

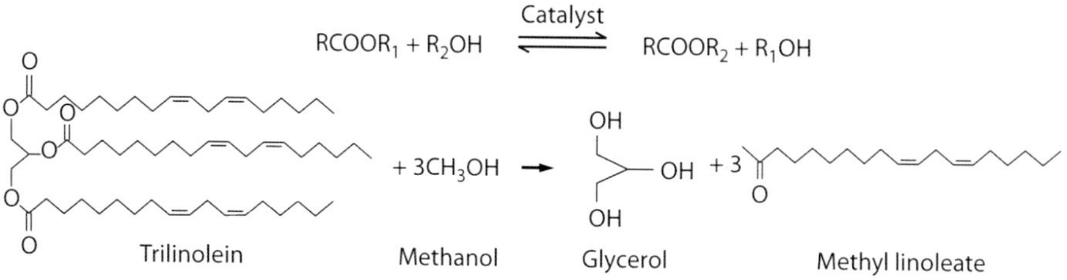

FIGURE 2.11

General transesterification reaction with example for $RCOOR_1$ as trilinolein and R_2OH as
methanol.

TABLE 2.6

Commonly Used Catalysts in Transesterification and Their Advantages and Disadvantages

Type	Commonly Used Compounds/ Enzymes	Advantages	Disadvantages
Alkali catalysts	NaOH, KOH, NaOCH$_3$, and KOCH$_3$ (other alkoxides are also used)	1. Faster than acid-catalyzed transesterification	1. Ineffective for high free fatty acid content and for high water content (problems of saponification) 2. Energy intensive 3. Recovery of glycerol difficult 4. Alkaline wastewater requires treatment
Acid catalysts	HCl, H$_2$SO$_4$, H$_3$PO$_4$, and sulfonic acid	1. Good for processes with high water content and free fatty acids	1. Slow process compared to alkali (alkoxides) 2. Require after treatment of triglycerides with alkoxides formed for purification purposes
Enzyme/lipase/ heterogeneous Catalysts	*M. miehi*, *C. antarctica*, *P. cepacia*, *C. rugosa*, and *P. fluorescens*	1. Possibility of regeneration and reuse of the immobilized residue 2. Free fatty acids can be completely converted to alkyl esters 3. Higher thermal stability of the enzyme due to the native state 4. Immobilization of lipase allows dispersed catalyst, reducing catalyst agglomeration 5. Separation of product and glycerol is easier using this catalyst	1. Some initial activity can be lost due to volume of the oil molecules 2. Number of support enzyme is not uniform 3. Biocatalyst is more expensive than the natural enzyme

Source: Ma, F. and Hanna, M.A., *Bioresour. Technol.*, 70(1), 1, 1999; Fukuda, H. et al., *J. Biosci. Bioeng.*, 92(5), 405, 2001; Meher, L.C. et al., *Renew. Sust. Energy Rev.*, 10(3), 248, 2006.

The catalyst used for transesterification may be an acid, a base, or a lipase. The commonly used catalysts are given in Table 2.7 along with their advantages and disadvantages (Ma and Hanna, 1999; Fukuda et al., 2001; Meher et al., 2006).

The mechanism of alkali-catalyzed transesterification is described in Figure 2.12. The first step involves the attack of the alkoxide ion to the carbonyl

TABLE 2.7

Fatty Acid Compositions of Common Oils (Percentages)

Fatty Acid	Soybean	Cottonseed	Palm	Lard	Tallow	Coconut
Lauric	0.1	0.1	0.1	0.1	0.1	46.5
Myristic	0.1	0.7	1	1.4	2.8	19.2
Palmitic	10.2	20.1	42.8	23.6	23.3	9.8
Stearic	3.7	2.6	4.5	14.2	19.4	3
Oleic	22.8	19.2	40.5	44.2	42.4	6.9
Linoleic	53.7	55.2	10.1	10.7	2.9	2.2
Linolenic	8.6	0.6	0.2	0.4	0.9	0

Source: Meher, L.C. et al., *Renew. Sust. Energy Rev.*, 10(3), 248, 2006.

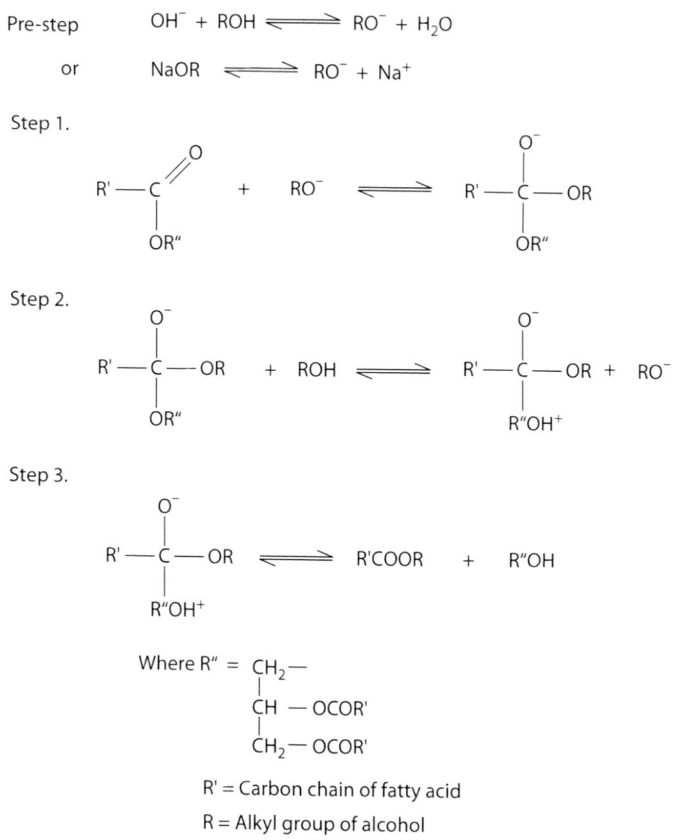

FIGURE 2.12
Mechanism of alkali-catalyzed transesterification. (Adapted from Meher, L.C. et al., *Renew. Sust. Energy Rev.*, 10(3), 248, 2006.)

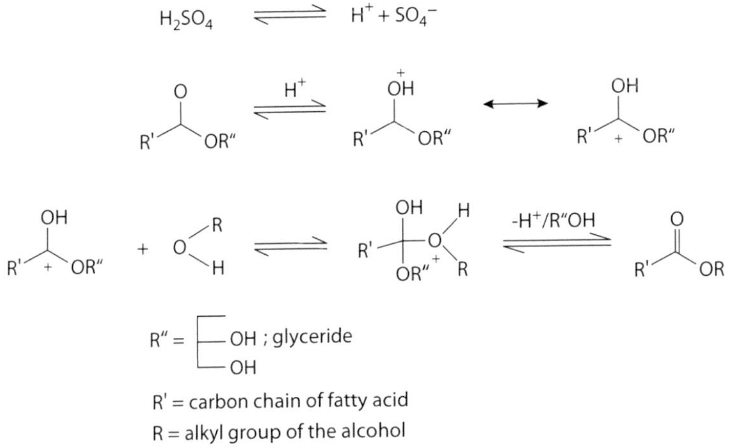

$$H_2SO_4 \rightleftharpoons H^+ + SO_4^-$$

R" = OH ; glyceride
    OH

R' = carbon chain of fatty acid

R = alkyl group of the alcohol

FIGURE 2.13

Mechanism of acid-catalyzed transesterification. (Adapted from Meher, L.C. et al., *Renew. Sust. Energy Rev.*, 10(3), 248, 2006.)

carbon of the triglyceride molecule, which results in the formation of a tetrahedral intermediate. The reaction of this intermediate with an alcohol produces the alkoxide ion in the second step. In the last step the rearrangement of the tetrahedral intermediate gives rise to an ester and a diglyceride.

The mechanism of acid-catalyzed transesterification of vegetable oil (for a monoglyceride) is shown in Figure 2.13. It can be extended to di- and triglycerides. The protonation of carbonyl group of the ester leads to the carbocation, which after a nucleophilic attack of the alcohol produces a tetrahedral intermediate. This intermediate eliminates glycerol to form a new ester and to regenerate the catalyst.

Both the triglycerides in vegetable oil and the methyl esters from the transesterification of vegetable oils can be used as monomers to form resins, foams, thermoplastics, and oleic methyl ester (Wool and Sun, 2005). A thermosetting polymer is formed by the polymerization of triglycerides with styrene using a free radical initiator and curing for 4h at 100°C that has very good tensile strength, rigidity, and toughness properties. Lignin can enhance toughness, and it can be molded to a material with an excellent ballistic impact resistance. Triglycerides can be functionalized to acrylated, epoxidized soybean oil that can be used for structural foam that has biocompatibility properties. Methyl esters can be functionalized to epoxidized oleic methyl ester and acrylated oleic methyl ester that can be polymerized with co-monomers methyl methacrylate and butyl acrylate to form oleic methyl ester. A monolithic hurricane-resistant roof has been designed using these materials.

Haas et al. (2006) describe an industrial-scale transesterification process for the production of methyl esters from the transesterification of soybean oil. Figure 2.14 gives a schematic overview of the process model. A two-reactor model was designed with crude degummed soybean oil as feedstock with

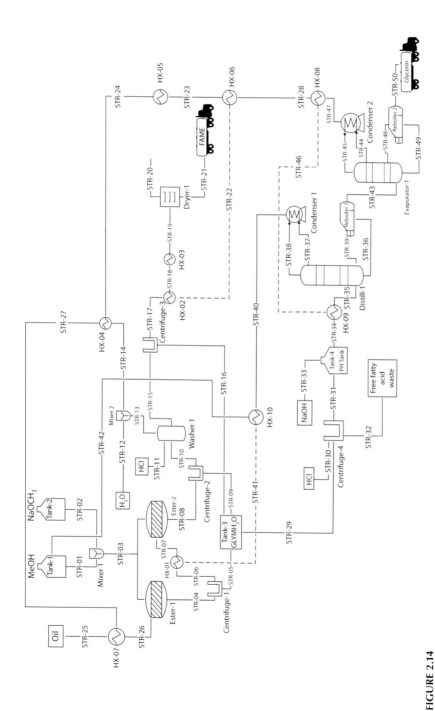

FIGURE 2.14

Process model for the production of fatty acid methyl esters (FAME) and glycerol. (Adapted from Haas, M.J. et al., *Biores. Technol.*, 97(4), 671, 2006.)

phospholipid content of less than 50 ppm and negligible fatty acids, sodium methoxide catalyst, and methanol as the alcohol. The design contained three sections: a transesterification section, a purification section, and a glycerol recovery section. The transesterification section consisted of two sequential reactors. The purification section had a centrifugation column that separated esters from the aqueous phase. The glycerol recovery and purification section also consisted of a centrifugal reactor and subsequent distillation and evaporation columns for 80% (w/w) glycerol as a by-product. The cost analysis of the overall process was done with a depreciable life of 10 years and an escalation rate of 1%. The annual production capacity for the methyl ester plant was set at 10×10^6 gal. With a feedstock cost of $0.236/lb of soybean oil, a production cost of $2.00/gal of methyl ester was achieved.

2.4.5 Gasification/Pyrolysis

Thermal conversion processes such as gasification and pyrolysis can be used to convert biomass to synthesis gas, a mixture of carbon monoxide and hydrogen. Pyrolysis is the direct thermal decomposition of the organic components in biomass in the absence of oxygen to yield an array of useful products like liquid and solid derivatives and fuel gases (Klass, 1998).

In biomass gasification, steam and oxygen are used to produce synthesis gas where the amount of steam and oxygen is controlled to produce carbon monoxide and hydrogen with a minimum amount of carbon dioxide and other products. Synthesis gas is a 1:1 mixture of carbon monoxide and hydrogen. In the 1800s, coal gasification was used to provide syngas used for lighting and heating. At the beginning of the twentieth century, syngas was used to produce fuel and chemicals. Many of the syngas conversion processes were developed in Germany during the first and second world wars at a time when natural resources were becoming scarce for the country and alternative routes for hydrogen production, ammonia synthesis, and transportation fuels were a necessity. With the development of the petroleum industry in the 1940s and beyond, the economics of many of these syngas routes became unfavorable and these were replaced by petroleum-based processes. The Fischer–Tropsch synthesis reactions for the catalytic conversion of a mixture of carbon monoxide and hydrogen into liquid alcohol fuels was one such process developed in Germany.

The United States has the highest proven reserves of coal amongst all its natural resources. Coal co-fired with biomass and complete biomass gasification processes are alternatives that are being considered for the production of syngas for fuels and chemicals. The U.S. DOE multiyear program plan for 2010 outlines the fuels, energy, and chemicals that can be produced from the thermochemical routes for biomass processing (DOE, 2010b). Biomass gasification technologies are similar to coal gasification, and both produce similar product gases. However, biomass contains more volatile matter, gasification occurs at lower temperatures and pressures than coal, and pyrolytic chars are more reactive than coal products. The increase in pressure lowers equilibrium concentrations for hydrogen and carbon monoxide and increases the carbon dioxide and methane concentrations. Biomass contains oxygen in cellulose and hemicellulose, which makes them more reactive than oxygen-deficient coal. Volatile matter in biomass is about 70%–90% in wood as compared to 30%–45% in coal.

Commercial biomass gasification facilities started worldwide in the 1970s and 1980s. Typically, gasification reactors are comprised of a vertical reactor that has drying, pyrolysis, and combustion zones. Synthesis gas leaves the top of the reactor and molten slag leaves the bottom of the reactor. The reactions that take place in the reactor are given in Equations 2.5 through 2.7 using cellulose as representative of biomass (Klass, 1998).

$$\text{Pyrolysis: } C_6H_{10}O_5 \rightarrow 5CO + 5H_2 + C \tag{2.5}$$

$$\text{Partial oxidation: } C_6H_{10}O_5 + O_2 \rightarrow 5CO + 5H_2 + CO_2 \tag{2.6}$$

$$\text{Steam reforming: } C_6H_{10}O_5 + H_2O \rightarrow 6CO + 6H_2 \tag{2.7}$$

Synthesis gas is used in the chemical production complex of the lower Mississippi River corridor to produce ammonia and methanol. Currently, ammonia and methanol are produced using synthesis gas from natural gas, naphtha, or refinery light gas. Nearly 12.2 billion lb of methanol are produced annually in the United States, and most of the methanol is converted to higher value chemicals such as formaldehyde (37%), methyl tertiary butyl ether (28%), and acetic acid (8%) (Paster et al., 2003). Ethanol can be produced from the synthesis gas, and Fischer–Tropsch chemistry is another approach to convert synthesis gas to chemicals and fuels. The chemicals that can be produced include paraffins, mono-olefins, aromatics, aldehydes, ketones, and fatty acids.

Pyrolysis is the direct thermal decomposition of the organic components in biomass in the absence of oxygen to yield an array of useful products like liquid and solid derivatives and fuel gases (Klass, 1998). Conventional pyrolysis is the slow, irreversible, thermal degradation of the organic components in biomass in the absence of oxygen and includes processes like carbonization, destructive distillation, dry distillation, and retorting. The products of pyrolysis under high pressure and temperature include mainly liquids with some gases and solids (water, carbon oxides, hydrogen, charcoal, organic compounds, tars, and polymers). The pyroligneous oil is the liquid product formed and mainly composed of water, settled tar, soluble tar, volatile acids, alcohols, aldehydes, esters, and ketones. Depending on pyrolysis conditions and feedstock, the liquid product contains valuable chemicals and intermediates. The separation of these intermediates in a cost-effective manner is required.

ConocoPhillips has funded a $22.5 million and 8 year research program at Iowa State University to develop new technologies for processing lignocellulosic biomass to biofuels (C&E News, 2007b). The company wants to investigate routes using fast pyrolysis to decompose biomass to liquid fuels.

Faustina Hydrogen Products LLC announced a $1.6 billion gasification plant in Donaldsonville, Louisiana. The plant will use petroleum coke and high sulfur coal as feedstocks instead of natural gas to produce anhydrous ammonia for agriculture, methanol, sulfur, and industrial-grade carbon dioxide. Capacities of the plant include 4,000 ton/day of ammonia, 1,600 ton/day of methanol, 450 ton/day sulfur, and 16,000 tons/day of carbon dioxide. Mosaic Fertilizer and Agrium Inc. have agreements to purchase the anhydrous ammonia from the plant. The carbon dioxide will be sold to Denbury Resources Inc. for use in enhanced oil recovery of oil left after conventional rig drilling processes in old oil fields in Southern Louisiana and the Gulf Coast. The rest of the carbon dioxide would be sequestered and sold as an industrial feedstock. The facility claims to have the technology to capture all the carbon dioxide during manufacturing process.

Eastman Chemical Company, a Fortune 500 company, will provide the Faustina gasification plant with necessary maintenance and services and plans to have a 25% equity position along with a purchase contract to buy the methanol produced in the plant. Eastman Chemicals will use methanol to make raw materials like propylene and ethylene oxide. Faustina is also backed by D.E. Shaw Group and Goldman Sachs.

Eastman Chemicals also plans to have 50% stake in a $1.6 billion plant to be built in Beaumont, Texas, in 2011 (Tullo, 2007). The plant will use gasification to produce syngas. Eastman will use the syngas to produce 225 million gal of methanol and 225,000 MT of ammonia per year at Terra Industries in Beaumont. Air Products & Chemicals will supply 2.6 million MT/year of oxygen to the gasifiers and market the hydrogen produced in the complex.

2.5 Biomass Feedstock Availability

The challenge with biomass feedstock usage is the availability of biomass on an uninterrupted basis. Biomass, as a feedstock, has a wide variation due to a number of causes. These include climate and environmental factors like insolation, precipitation, temperature, ambient carbon dioxide concentration, nutrients, etc.

The availability of land and water areas for biomass production is important for the sustainable growth of biomass. The land capability in the United States is classified according to eight classes (USDA, 2012) and is given in Table 2.8. There have been numerous studies on the availability of biomass as feedstock in the United States, the most recent survey and estimation being undertaken by Perlack et al. (2005). Their findings are summarized in this section for land biomass resources.

The land base of the United States is approximately 2263 million acres, including the 369 million acres of land in Alaska and Hawaii (Perlack et al., 2005).

TABLE 2.8

Land Capability Classification

Class	Description
Class I	Contains soils having few limitations for cultivation
Class II	Contains soils having some limitations for cultivation
Class III	Contains soils having severe limitations for cultivation
Class IV	Contains soils having very severe limitations for cultivation
Class V	Contains soils unsuited to cultivation, although pastures can be improved and benefits from proper management can be expected
Class VI	Contains soils unsuited to cultivation, although some may be used provided unusually intensive management is applied
Class VII	Contains soils unsuited to cultivation and having one or more limitations which cannot be corrected
Class VIII	Contains soils and landforms restricted to use as recreation, wildlife, water supply, or aesthetic purposes

Source: USDA, 2012, National Soil Survey Handbook, title 430-VI, http://soils.usda.gov/technical/handbook, U.S. Department of Agriculture, Natural Resources Conservation Service, accessed February 29, 2012.

The land area is classified according to forestland, grassland pasture and range, cropland, special uses, and other miscellaneous uses like urban areas, swamps, and deserts. The distribution of the land areas according to these categories is given in Figure 2.15. The land base in the lower 48 states having some potential for growth of biomass is about 60%.

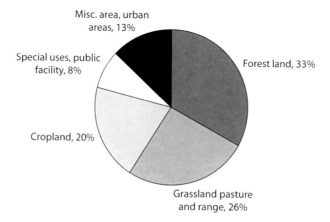

FIGURE 2.15

U.S. land base resource. (From Perlack, R.D. et al., Biomass as feedstock for a bioenergy and bioproducts industry: The technical feasibility of a billion-ton annual supply, USDA document prepared by Oak Ridge National Laboratory, Oak Ridge, TN, ORNL/TM-2005/66, 2005.)

The two major categories of biomass resources availability are based on forestland and cropland (or agricultural land). The detailed classification of the biomass resources are given in Figure 2.16. The primary resources are often referred to as virgin biomass and the secondary and tertiary are referred to as waste biomass. Currently, slightly more than 75% of biomass consumption in the United States (about 142 million dry tons) comes from forestlands. The remainder (about 48 million dry tons), which includes bio-based products, biofuels, and some residue biomass, comes from cropland.

2.5.1 Forest Resources

2.5.1.1 Forestland Base

The total forestland resource base in the United States is approximately 749 million acres (one-third of the total land resource). The forestland is grouped into unreserved "timberland," "reserved land," and "others." The 749 million acres is divided into 504 million acres of timberland capable of growing 20 ft^3/acre of wood annually, 166 million acres of forestland classified as "other" (incapable of growing 20 ft^3/acre of wood annually and, hence, used for of livestock grazing and extraction of some nonindustrial wood products), and 78 million acres of reserved forestland used for parks and wilderness. "Timberland" and the "other" land are considered as the resource base that can be utilized for forest biomass resources.

2.5.1.2 Types of Forest Resource

The primary forest resources include logging residues and excess biomass (not harvested for fuel treatments and fuelwood) from timberlands. Logging residues are the unused portions of growing-stock and nongrowing-stock trees cut or killed by logging and left in the woods. Fuelwood extracted from forestlands for residential and commercial use and electric utility use accounts for about 35 million dry tons of current consumption. In total, the amount of harvested wood products from timberlands in the United States is less than the annual forest growth and considerably less than the total forest inventory.

The processing of sawlogs and pulpwood harvested for forest products generate significant amounts of mill residues and pulping liquors. These are secondary forest resources and constitute the majority of biomass in use today. The secondary residues are used by the forest products industry to manage residue streams, produce energy, and recover chemicals. About 50% of current biomass energy consumption comes from the secondary residues.

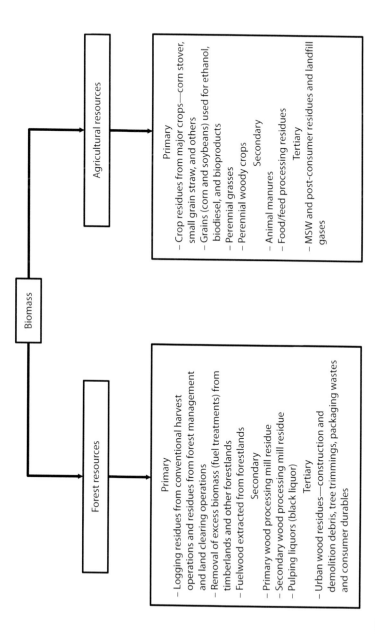

FIGURE 2.16
Biomass resource base (primary, secondary, and tertiary biomass). (Adapted from Perlack, R.D. et al., Biomass as feedstock for a bioenergy and bio-products industry: The technical feasibility of a billion-ton annual supply, USDA document prepared by Oak Ridge National Laboratory, Oak Ridge, TN, ORNL/TM-2005/66, 2005.)

The various categories in which primary and secondary forest resources can be grouped are given as follows:

- Logging residue: The recovered residues generated by traditional logging activities and residues generated from forest cultural operations or clearing of timberlands
- Fuel treatments (forestland): The recovered residues generated from fuel treatment operations on timberland and other forestland
- Fuelwood: The direct conversion of roundwood to energy (fuelwood) in the residential, commercial, and electric utility sectors
- Forest products industry residues and urban wood residues: Utilization of unused residues generated by the forest products industry
- Forest growth: Forest growth and increase in the demand for forest products

The estimate of currently recoverable forest biomass is given in Figure 2.17. The approximate total quantity is 368 million dry tons annually. This includes about 142 million dry tons of biomass currently being used primarily by the forest products industry and an estimated 89 million dry tons that could come from a continuation of demand-and-supply trends in the forest products industry.

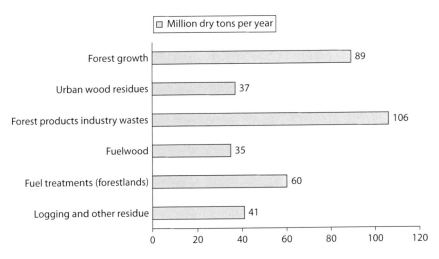

FIGURE 2.17
Estimate of sustainably recoverable forest biomass. (From Perlack, R.D. et al., Biomass as feedstock for a bioenergy and bioproducts industry: The technical feasibility of a billion-ton annual supply, USDA document prepared by Oak Ridge National Laboratory, Oak Ridge, TN, ORNL/TM-2005/66, 2005.)

2.5.1.3 Limiting Factors for Forest Resource Utilization

The 368 million tons of potential forest biomass feedstock base is constrained by some restrictions for exploitation. For forest resources inventory, development in forest utilization relationships and land ownership is expected to play a major role in utilizing the resource. There are three major limiting factors for forest residues from fuel treatment thinning resource, namely, accessibility (having roads to transport the material and operate logging/collection systems, avoiding adverse impacts on soil and water), economic feasibility (value of the biomass compared against the cost of removing the material), and recoverability (function of tree form, technology, and timing of the removal of the biomass from the forests).

Forest products industry processing residues include primary wood processing mills, secondary wood processing mills, and pulp and paper mills. Residues from these sources include bark, sawmill slabs and edgings, sawdust, and peeler log cores, residues from facilities that use primary products, and black liquor. A significant portion of this residue is burnt or combusted to produce energy for the respective industries. Excess amount of residue remains unutilized after the burning and combustion and can be used in biomass processes. Urban wood residues include municipal solid wastes, and construction and demolition debris. A part of it is recovered and a significant part is unexploited. Finally, future forest growth and increased demands in forest products are likely to affect the availability of forest resources for biomass feedstock base. In summary, all of these forest resources are sustainably available on an annual basis, but not currently used to its full potential due to the above constraints. For estimating the residue tonnage from logging and site-clearing operations and fuel treatment thinning, a number of assumptions were made by Perlack et al. (2005):

- All forestland areas not currently accessible by roads were excluded.
- All environmentally sensitive areas were excluded.
- Equipment recovery limitations were considered.
- Recoverable biomass was allocated into two utilization groups—conventional forest products and biomass for bioenergy and bio-based products.

2.5.1.4 Summary for Forest Resources

Thus, biomass derived from forestlands currently contributes about 142 million dry tons to the total annual consumption in the United Sates of 190 million dry tons. With increased use of potential and currently unexploited biomass, this amount of forestland-derived biomass can increase to approximately 368 million dry tons annually. The distribution of the forest resource potential is summarized in Figure 2.18.

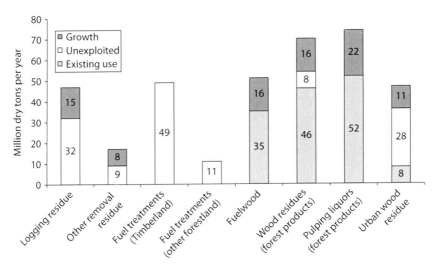

FIGURE 2.18
Summary of potentially available forest biomass resources. (From Perlack, R.D. et al., Biomass as feedstock for a bioenergy and bioproducts industry: The technical feasibility of a billion-ton annual supply, USDA document prepared by Oak Ridge National Laboratory, Oak Ridge, TN, ORNL/TM-2005/66, 2005.)

This estimate includes the current annual consumption of 35 million dry tons of fuelwood extracted from forestland for residential, commercial, and electric utility purposes; 96 million dry tons of residues generated and used by the forest products industry; and 11 million dry tons of urban wood residues. There are relatively large amounts of forest residue produced by logging and land-clearing operations that are currently not collected (41 million dry tons per year) and significant quantities of forest residues that can be collected from fuel treatments to reduce fire hazards (60 million dry tons per year). Additionally, there are unutilized residues from wood processing mills and unutilized urban wood. These sources total about 36 million dry tons annually. About 48% of these resources are derived directly from forestlands (primary resources). About 39% are secondary sources of biomass from the forest products industry. The remaining amount of forest biomass would come from tertiary or collectively from a variety of urban sources.

2.5.2 Agricultural Resources

2.5.2.1 Agricultural Land Base

The agricultural land resource base for the United States is approximately 455 million acres, approximately 20% of the total land base in the country. Out of this, 349 million acres is actively used for crop growth, 39 million

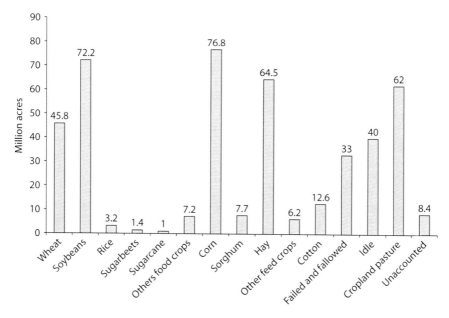

FIGURE 2.19
Summary of agricultural land use by major crops in United States. (From Perlack, R.D. et al., Biomass as feedstock for a bioenergy and bioproducts industry: The technical feasibility of a billion-ton annual supply, USDA document prepared by Oak Ridge National Laboratory, Oak Ridge, TN, ORNL/TM-2005/66, 2005.)

acres constitutes idle cropland, and 67 million acres is used for pasture. Cropland utilization is affected by soil and weather conditions, expected crop prices, and government incentives. Cropland is also lost due to conversion of the land for other uses like urban development, etc. The major food crops planted acreage constitutes wheat, soybean, and rice. The feed crops include corn, sorghum, and hay. The fallow and failed land is a part of cropland. Apart from cropland, there is idle land which includes acreage diverted from crops under the Acreage Reduction Program (ARP), the Conservation Reserve Program (CRP), and other federal ARPs. The cropland used only for pasture is also separately accounted for. The distribution of agricultural land base and planted crop acreages in the United States are shown in Figure 2.19.

2.5.2.2 Types of Agricultural Resource

The agricultural resource base is primarily comprised of grains and oilseeds in the United States. Currently, grains are primarily used for cattle feed. Grains, primarily corn, can be used for producing ethanol and oilseeds, primarily soybean, can be used to produce biodiesel. Approximately 93% of the total U.S. ethanol is produced from corn. Apart from these, agricultural residues, like corn stover, can also be used for producing ethanol. In the United States, approximately 428 million dry tons of annual crop residues, 377 million dry tons of perennial crops, 87 million dry tons of grains, and 106 million dry tons of animal manures, process residues, and other miscellaneous feedstocks can be produced on a sustainable basis (Perlack et al., 2005). This resource potential was evaluated based on changes in the yields of crops grown on active cropland, crop residue-to-grain or -seed ratios, annual crop residue collection technology and equipment, crop tillage practices, land use change to accommodate perennial crops (i.e., grasses and woody crops), biofuels (i.e., ethanol and biodiesel), and secondary processing and other residues. Three scenarios were evaluated for availability of crop biomass, and they are given in the following.

2.5.2.2.1 *Current Availability of Biomass Feedstocks from Agricultural Land*

The current availability scenario studies biomass resources current crop yields, tillage practices (20%–40% no-till for major crops), residue collection technology (~40% recovery potential), grain to ethanol and vegetable oil for biodiesel production, and use of secondary and tertiary residues on a sustainable basis. The amount of biomass currently available for bioenergy and bioproducts is about 194 million dry tons annually as shown in Figure 2.20. The largest source of this current potential is 75 million dry tons of corn residues or corn stover, followed by 35 million dry tons of animal manure and other residues.

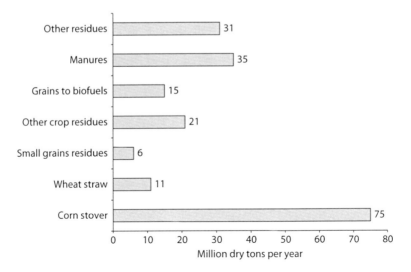

FIGURE 2.20
Current availability of biomass from agricultural lands. (From Perlack, R.D. et al., Biomass as feedstock for a bioenergy and bioproducts industry: The technical feasibility of a billion-ton annual supply, USDA document prepared by Oak Ridge National Laboratory, Oak Ridge, TN, ORNL/TM-2005/66, 2005.)

2.5.2.2.2 *Biomass Availability through Technology Changes in Conventional Crops with No Land Use Change*

This scenario analyzed the biomass availability of conventional crops achieved through technology changes. The land utilization for conventional crops projected for 2014 was used for this analysis. Technology changes to increase crop yields, improve collection equipment, and sustainable agricultural practices were considered in this scenario. Corn yields were assumed to increase by 25%–50% from 2001 values, while yields of wheat and other small grains, sorghum, soybean, rice, and cotton are assumed to increase at rates lower than for corn. The increased production of corn contributed to increase in corn stover as residue. Soybean contributed no crop residue under a moderate yield increase of about 13% but made a small contribution with a high yield increase of about 23%. The collection of these residues from crops was increased through better collection equipment capable of recovering as much as 60% of residue under the moderate yield increases and 75% under the high yield increases, but the actual removal amounts depend on the sustainability requirements. No-till cultivation method was assumed to be practiced on approximately 200 million acres under moderate yield increases and all of active cropland under high yields. The amount of corn and soybean available for ethanol, biodiesel production, or other bioproducts was calculated by subtracting amounts needed to meet food requirements plus feed and export requirements from total quantities. All remaining grain was assumed to be available for bioproducts. Further, about 75 million dry tons of manure and other secondary and tertiary residues and wastes, and 50% of the biomass produced on CRP lands (17–28 million dry tons) were assumed to be available for bioenergy production. Thus, this scenario for use of crop residue results in the annual production of 423 million dry tons per year under moderate yields and 597 million dry tons under high yields. In this scenario, about two-thirds to three-fourths of total biomass are from crop residues, as can be seen in Figure 2.21.

2.5.2.2.3 *Biomass Availability through Technology Changes in Conventional Crops and New Perennial Crops with Significant Land Use Change*

This scenario assumes the addition of perennial crops, land use changes, and changes in soybean varieties, as well as the technology changes assumed under the previous scenario. Technology changes are likely to increase the average residue-to-grain ratio of soybean varieties from 1.5 to a ratio of 2.0. The land use changes considered in this scenario included the conversion of land for growth of perennial crops on 40 million acres for moderate yield increase or 60 million acres for high yield increase. Woody crops produced for fiber were expanded from 0.1 million acres to 5 million acres, where they can produce an average annual yield of 8 dry tons per acre. About 25% of the wood fiber crops are assumed to be used for bioenergy and the remainder for other, higher-value conventional forest products.

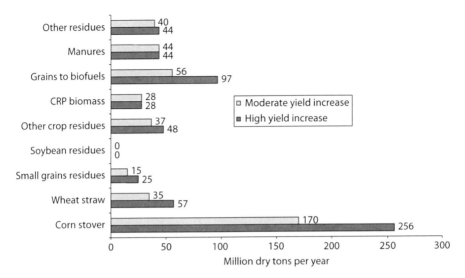

FIGURE 2.21
Availability of biomass for increased crop yields and technology changes. (From Perlack, R.D. et al., Biomass as feedstock for a bioenergy and bioproducts industry: The technical feasibility of a billion-ton annual supply, USDA document prepared by Oak Ridge National Laboratory, Oak Ridge, TN, ORNL/TM-2005/66, 2005.)

Perennial crops (trees or grasses) grown primarily for bioenergy expand to either 35 million acres at 5 dry tons per acre per year or to 55 million acres with average yields of 8 dry tons per acre per year. About 93% of the perennial crops are assumed available for bioenergy and the remainder for other products. A small fraction of the available biomass (10%) was assumed as lost during the harvesting operations. This scenario resulted in the production of 581 (moderate yield) to 998 (high yield) million dry tons as shown in Figure 2.22. The crop residues increased even though conventional cropland was less because of the addition of more soybean residue together with increased yields. The single largest source of biomass is the crop residue, accounting for nearly 50% of the total produced. Perennial crops account for about 30%–40%, depending on the extent of crop yield increase (i.e., moderate or high).

2.5.2.3 Limiting Factors for Agricultural Resource Utilization

The annual crop residues, perennial crops, and processing residues can produce 998 million tons of potential agricultural biomass feedstock. The limiting factors for the utilization of crop residues and growth of perennial crops for the purpose of feedstock generation will require significant changes in current crop yields, tillage practices, harvesting and collection technologies, and transportation. Agricultural residues serve as a land cover and prevent soil erosion after harvesting of crops. The removal of large quantities

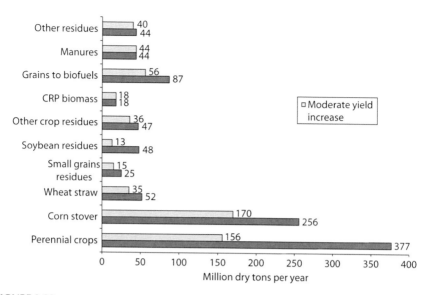

FIGURE 2.22
Availability of biomass for increased crop yields and technology changes, and inclusion of perennial crops. (From Perlack, R.D. et al., Biomass as feedstock for a bioenergy and bioproducts industry: The technical feasibility of a billion-ton annual supply, USDA document prepared by Oak Ridge National Laboratory, Oak Ridge, TN, ORNL/TM-2005/66, 2005.)

of the residue can affect the soil quality by the removal of soil carbon and nutrients and may need to be replenished with fertilizers. The fertilizers, in turn, require energy for production, and hence the optimum removal of the residues needs to be evaluated. Perennial crops require less nutrient supplements and are better choices for preventing soil erosion compared to annual crops, and they are considered for planting.

Important assumptions made for this evaluation of agricultural biomass availability by Perlack et al. (2005) included the following:

- Yields of corn, wheat, and other small grains were increased by 50%.
- The residue-to-grain ratio for soybean was increased to 2:1.
- Harvest technology was capable of recovering 75% of annual crop residues (when removal is sustainable).
- All cropland was managed with no-till methods.
- 55 million acres of cropland, idle cropland, and cropland pasture were dedicated to the production of perennial bioenergy crops.
- All manure in excess of that which can be applied on-farm for soil improvement under anticipated EPA restrictions was used for biofuels.
- All other available residues were utilized.

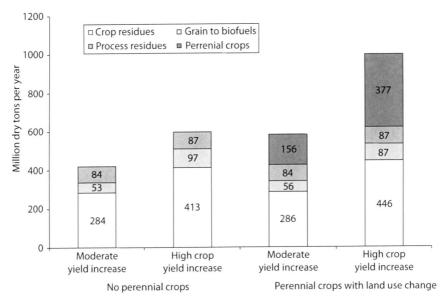

FIGURE 2.23
Summary of potentially available agricultural biomass resources. (From Perlack, R.D. et al., Biomass as feedstock for a bioenergy and bioproducts industry: The technical feasibility of a billion-ton annual supply, USDA document prepared by Oak Ridge National Laboratory, Oak Ridge, TN, ORNL/TM-2005/66, 2005.)

2.5.2.4 Summary for Agricultural Resources

Thus, biomass derived from agricultural lands currently available for removal on a sustainable basis is about 194 million dry tons. This amount can be increased to nearly one billion tons annual production through a combination of technology changes, adoption of no-till cultivation, and land use change to grow perennial crops. The amount of biomass available without the addition of perennial crops but high crop yield increase would be 600 million dry tons. Approximately, the same amount of biomass would be produced on agricultural lands with moderate crop yield increase and addition of perennial crops. The distribution of the agricultural resource potential is summarized in Figure 2.23.

2.5.3 Aquatic Resources

The previous sections discussed conventional biomass feedstock options grown on land. Apart from the crop and forest biomass resources, other organisms that undergo photosynthesis are cyanobacteria and algae. There are several ongoing attempts to find the ideal growth conditions for cultivating algae on a sustainable basis. Key areas of research interests in algae include high per-acre productivity compared to typical terrestrial oil-seed crops, nonfood-based feedstock resources, use of otherwise nonproductive, nonarable land for algae cultivation, utilization of a wide variety of water sources (fresh, brackish, saline, and wastewater), and production of both biofuels and valuable coproducts. The Energy Efficiency and Renewable Energy Laboratory at the Department of Energy (DOE) commissioned a working group to assess the current state of algae technology and to determine the next steps toward commercialization (DOE, 2010a). The workshop addressed the following topics and technical barriers in algal biology, feedstock cultivation, harvest and dewatering, extraction and fractionation of microalgae, algal biofuel conversion technologies, coproducts production, distribution and utilization of algal based-fuels, resources and siting, corresponding standards, regulation and policy, systems and techno-economic analysis of algal biofuel deployment, and public–private partnerships. This section aims to capture some of those efforts. A model algal lipid production system with algae growth, harvesting, extraction, separation, and uses is shown in Figure 2.24. Methods to convert whole algae into biofuels exist through anaerobic digestion to biogas, supercritical fluid extraction and pyrolysis to liquid or vapor fuels, and gasification process for production of syngas-based fuels and chemicals. Algae oil can be supplement refinery diesel in hydrotreating units, or be used as feedstock for the biodiesel process. The research on algae as a biomass feedstock is a very dynamic field currently, and the potential of algae seems promising as new results are presented continuously.

Methods to cultivate algae have been developed over the years. Recent developments in algae growth technology include vertical reactors (Hitchings, 2007) and bag reactors (Bourne, 2007) made of polythene mounted on metal frames, eliminating the need for land use for cultivation. The NREL Aquatic Species Program (Sheehan et al., 1998) mentions "raceway" ponds design for growth of algae. This method required shallow ponds built on land area and connected to a carbon dioxide source such as a power plant. Productivity in these pond designs were few grams/m²/d. Other designs include tubular cultivation facilities and the semi-continuous batch cultures gave improved

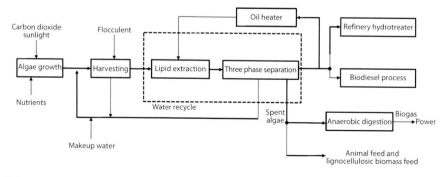

FIGURE 2.24
Model algae lipid production system. (Adapted from Pienkos, P.T. and Daezins, A., *Biofuel. Bioprod. Biorefin.*, 3(4), 431, 2009.)

TABLE 2.9

Comparison of Productivity between Algae and Soybean

		Algae		
		Low Productivity (10 g/m²/day)	Medium Productivity (25 g/m²/day)	High Productivity (50 g/m²/day)
Productivity	Soybean	15% TAG	25% TAG	50% TAG
Gallons/acre	48	633	2637	10,549
Total acres	63.6 million	63.6 million	25 million	6.26 million
Gallons/year	3 billion	40 billion	66 billion	66 billion
% petrodiesel	4.5%	61%	100%	100%

Source: Pienkos, P.T. and Daezins, A., *Biofuel. Bioprod. Biorefin.*, 3(4), 431, 2009.

production rates of algae. For example, the 3D Matrix System of Green Fuel Technologies Corporation has an average areal productivity of $98\,g/m^2/day$ (ash-free, dry-weight basis), with highs of over $170\,g/m^2/day$ achieved during a run time of 19 days (Pulz, 2007).

Algae have the potential for being an important source of oil and carbohydrates for the production of fuels, chemicals, and energy. Carbon dioxide and sunlight can be used to cultivate algae and produce algae with 60% triglycerides and 40% carbohydrates and protein (Pienkos and Daezins, 2009). A comparison in productivity between algae and soybean is given in Table 2.9. The table shows that even at low productivity of algae, yields are more than ten times in gallons per acre when all the U.S. soybean acreage is utilized for algae. Higher yields are obtained at medium and high productivity levels of algae (higher triacylglycerols) with reduced acreage requirements. The algae oil resulting from low productivity can replace approximately 61% of the total U.S. diesel requirements, as compared to only 4.5% for soybean-oil-based diesel. The other advantage is that, at these yields, algae can capture up to 2 billion ton of carbon dioxide during photosynthesis.

2.5.3.1 Recent Trends in Algae Research

The growth of algae on a large scale for the production of oil and chemicals seems to be the most important barrier at this stage. The following technologies developed seem promising ways to cultivate algae, apart from traditional open-pond systems. These are discussed on a per-case basis, with the companies that have developed these technologies. Some of the current research trends in algae bioreactor systems are presented in the following sections.

Raceway pond systems: "Raceway" design for algae growth included shallow ponds in which the algae, water, and nutrients circulate around a "racetrack" as shown in Figure 2.25 (inset) (Sheehan et al., 1998). Motorized paddles help

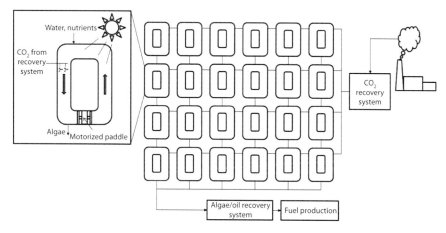

FIGURE 2.25
Algae raceway design (inset) and algae farm system for algae growth. (Adapted from Sheehan, J. et al., *A Look Back at the U.S. Department of Enegy's Aquatic Species Program—Biodiesel from Algae*, National Renewable Energy Laboratory, Golden, CO, NREL/TP-580-24190, 1998.)

to provide the flow and keep algae suspended in water and circulated back up to the surface on a regular frequency. The ponds are shallow to ensure maximum exposure of sunlight (sunlight cannot penetrate beyond certain depths). The ponds are operated as continuous reactors with water and nutrients fed to the pond and carbon dioxide bubbled through the system. The algae-containing water is removed at the other end of the pond. The algae is then harvested and processed for oil extraction.

The concept of the raceway design for algae growth can be extended to an algae farm as shown in Figure 2.25. This consists of numerous ponds similar to the raceway in which algae is grown and harvested. The size of these ponds is measured in terms of surface area (as opposed to volume) as the surface area is critical to capturing sunlight. The productivity is measured in terms of biomass produced per day per unit of available surface area. These designs required large acres of land and, thus, obtained the scale of farms.

Algenol Biofuels–direct to Ethanol™ process: Algenol Biofuels have developed metabolically engineered algae species to produce ethanol in closed bioreactor systems. The proprietary Capture Technology™ bioreactors hold single-cell cyanobacteria in closed and sealed plastic bag units preventing contamination, maximize ethanol recovery, and allow freshwater recovery. The advantage of the process lies in the fact that it is a one-step process where the cyanobacteria utilize the carbon dioxide to convert it to ethanol, and secrete the ethanol from the cell (Voith, 2009). There is a requirement for strict maintenance of growth parameters such as CO_2, nutrients, water, pH, temperature, salinity, and other environmental conditions for the engineered species of microorganism. The process of making ethanol from algae utilizes 1.5 million tons of carbon dioxide per 100 million gal of ethanol produced. Algenol, The Dow

Chemical Company, and the DOE have teamed to produce ethanol using this technology at Dow's Freeport, Texas site. Dow would contribute with 25 acres of their site, carbon dioxide source, and technical expertise for the $25 million project. Dow plans to utilize their expertise in film technology to device an ideal bioreactor for the system, with optimum sunlight penetration.

Exxon Mobil algae research: Exxon Mobil is funding $600 million for algae research partnered with Synthetic Genomics, Inc. to identify and develop algae strains to produce bio-oils at low costs (Kho, 2009). The research will also determine the best production systems for growing algal strains, for example, open ponds or closed photo-bioreactor systems. The company also plans for scale-up to large amounts of CO_2 utilization and developing integrated commercial systems.

Shell algae research: Shell and HR Biopetroleum formed a joint venture company in 2007, called Cellana, to develop an algae project for a demonstration facility on the Kona coast of Hawaii Island. The site was leased from the Natural Energy Laboratory of Hawaii Authority (NELHA) and is near existing commercial algae enterprises, primarily serving the pharmaceutical and nutrition industries. The facility will grow only non-modified, marine microalgae species in open-air ponds using proprietary technology. Algae strains used for the process are indigenous to Hawaii or approved by the Hawaii Department of Agriculture.

Green fuels technology: GreenFuel Technologies developed a process that grows algae in plastic bags using CO_2 from smokestacks of power plants via naturally occurring species of algae. The CO_2 source can also come from fermentation or geothermal gases. Algae can be converted to transportation fuels and feed ingredients or recycled back to a combustion source as biomass for power generation. Industrial facilities do not need any internal modifications to host a GreenFuel algae farm. In addition, the system does not require fertile land or potable water. Water used can be recycled, and wastewater can be used as compared to oilseed crops' high water demand. With high growth rates, algae can be harvested daily.

Valcent products: A vertical reactor system is being developed by Valcent Products, Inc. of El Paso, Texas, using the 340 annual days of sunshine and carbon dioxide available from power plant exhaust. Enhanced Biofuel Technology and A2BE Carbon Capture LLC are some of the firms that use the concept of the raceway pond design and algae farm for production of algae for biofuels. Research is under way to determine the species of algae for oil production and the best method of extracting the oil. Extraction methods being evaluated include expeller/press, hexane solvent extraction, and supercritical fluid extraction and are the more costly steps in the process. Approximately 70%–75% of algae oil can be extracted using expeller press, while 95% oil can be extracted by hexane solvent oil extraction and 100% oil extracted using supercritical fluid extraction.

2.5.3.2 Algae Species

Algae are plant-like microorganisms that preceded plants in developing photosynthesis, the ability to turn sunlight into energy. Algae range from small, single-celled organisms to multicellular organisms, some with fairly complex differentiated form. Algae are usually found in damp places or bodies of water and, thus, are common in terrestrial as well as aquatic environments. Like plants, algae require primarily three components to grow: sunlight, carbon-dioxide, and water. Microalgae are the most efficient in photosynthesis, with 60%–70% of each cell's volume capable of photosynthesis (Arnaud, 2008). The algae also do not have roots, stems, or leaves, which divert resources to produce hydrocarbons. Algae cells contain light-absorbing chloroplasts and produce oxygen through photosynthesis. Biologists have categorized microalgae in a variety of classes, mainly distinguished by their pigmentation, life cycle, and basic cellular structure. The four most important classes (in terms of abundance) are (Sheehan et al., 1998):

- The diatoms (Bacillariophyceae): These algae dominate the phytoplankton of the oceans but are also found in freshwater and brackish water. Approximately 100,000 species are known to exist. Diatoms contain polymerized silica (Si) in their cell walls. All cells store carbon in a variety of forms. Diatoms store carbon in the form of natural oils or as a polymer of carbohydrates known as chyrsolaminarin.
- The green algae (Chlorophyceae): These types of algae are abundant in freshwater, for example, in a swimming pool. They can occur as single cells or as colonies. Green algae are the evolutionary progenitors of modern plants. The main storage compound for green algae is starch, though oils can be produced under certain conditions.
- The blue-green algae (Cyanophyceae): This type of algae is closer to bacteria in structure and organization. These algae play an important role in fixing nitrogen from the atmosphere. There are approximately 2000 known species found in a variety of habitats.
- The golden algae (Chrysophyceae): This group of algae is similar to the diatoms. They have more complex pigment systems, and can appear yellow, brown, or orange in color. Approximately 1000 species are known to exist, primarily in freshwater systems. They are similar to diatoms in pigmentation and biochemical composition. The golden algae produce natural oils and carbohydrates as storage compounds.

A DOE program studied over 3000 strains of organisms, which were then narrowed down to about 300 species of microorganisms. The program concentrated not only on algae that produced significant amounts of oil but also on algae that grow under severe conditions—extremes of temperature, pH, and salinity (Sheehan et al., 1998).

TABLE 2.10

Percentage Composition of Protein, Carbohydrate, Lipids, and Nucleic
Acid Composition of Various Strains of Algae

Strain	Protein	Carbohydrates	Lipids	Nucleic Acid
Scenedesmus obliquus	50–56	10–17	12–14	3–6
Scenedesmus quadricauda	47	—	1.9	—
Scenedesmus dimorphus	8–18	21–52	16–40	—
Chlamydomonas rheinhardii	48	17	21	—
Chlorella vulgaris	51–58	12–17	14–22	4–5
Chlorella pyrenoidosa	57	26	2	—
Spirogyra sp.	6–20	33–64	11–21	—
Dunaliella bioculata	49	4	8	—
Dunaliella salina	57	32	6	—
Euglena gracilis	39–61	14–18	14–20	—
Prymnesium parvum	28–45	25–33	22–38	1–2
Tetraselmis maculata	52	15	3	—
Porphyridium cruentum	28–39	40–57	9–14	—
Spirulina platensis	46–63	8–14	4–9	2–5
Spirulina maxima	60–71	13–16	6–7	3–4.5
Synechoccus sp.	63	15	11	5
Anabaena cylindrica	43–56	25–30	4–7	—

Source: Sheehan, J., *A Look Back at the U.S. Department of Enegy's Aquatic Species Program—Biodiesel from Algae*, National Renewable Energy Laboratory, Golden, CO, NREL/TP-580-24190, 1998.

Algal biomass contains three main components: carbohydrates, proteins, and natural oils. Algae contain 2%–40% of lipids/oils by weight. The composition of various algal species is given in Table 2.10. These components in algae can be used for fuel or chemical production in three ways, mainly the production of methane via biological or thermal gasification, ethanol via fermentation, or conversion to esters by transesterification (Sheehan et al., 1998). *Botryococcus braunii* species of algae has been engineered to produce the terpenoid C30 botryococcene, a hydrocarbon similar to squalene in structure (Arnaud, 2008). The species has been engineered to secrete the oil, and the algae can be reused in the bioreactor. A further modification to the algae is smaller light collecting antennae, allowing more light to penetrate the algae in a polythene container reactor system. A gene, *tla1*, responsible for the number of chlorophyll antennae, can be modified to reduce the chlorophyll molecules from 600 to 130. Botryococcene is a triterpene and, unlike triglycerides, cannot undergo transesterification. It can be used as feedstock for hydrocracking in an oil refinery to produce octane, kerosene, and diesel.

Dry algae factor is the percentage of algae cells in relation with the media where they are cultured; for example, if the dry algae factor is 50%, one would need 2 kg of wet algae (algae in the media) to get 1 kg of algae cells.

Lipid factor is the percentage of vegetable oil in relation to the algae cells needed to get it; that is, if the algae lipid factor is 40%, one would need 2.5 kg of algae cells to get 1 kg of oil.

Carbon dioxide sources for algae growth can be from pipelines for CO_2, flue gases from power plants, or any other sources rich in carbon dioxide. The flue gases from power plants were previously not considered as suitable algae cultivation land, and cultivation was not found near power plants. However, with newer designs of algae reactors linked with power plants, the flue gases can be suitable sources for algae cultivation. Water usage for algae growth is also a concern for design. In an open-pond system, the loss of water is greater than in closed tubular cultivation or bag cultivation methods. The water can be local industrial water and recycled after harvesting algae.

2.6 Summary

This chapter aimed to give an overview of the use of biomass as the next-generation feedstock for energy, fuels, and chemicals. The formation of biomass gave the methods in which atmospheric carbon dioxide is fixed naturally to different types of biomass. The classification of biomass into starch, cellulose, hemicellulose, lignin, lipids and oils, and proteins helped to understand the chemical composition of biomass. Biomass species are available in nature as a combination of the components, and it is important to separate the components for use as energy, fuels, and chemicals. Various conversion technologies are employed for the separation of the components of biomass to make it more amenable, and these include pretreatment, fermentation, anaerobic digestion, transesterification, gasification, and pyrolysis.

The availability of biomass on a sustainable basis is required for the uninterrupted production of energy, fuels, and chemicals. The current forest biomass feedstock used per year is 142 million MT. This can be potentially increased to 368 million MT, which include currently unexploited and future growth of forest biomass. The agricultural biomass currently available per year on a sustainable basis is 194 million dry tons. This amount can be potentially increased to 423–527 million MT/year with technology changes in conventional crops and 581–998 million MT with technology and land use changes in conventional and perennial crops.

Apart from crop and forest biomass, research in algae and cyanobacteria are ongoing for the production of carbohydrate-based and oil-based feedstock. These processes are currently constrained primarily by successful scale-up to meet the biomass needs. However, recent advances in photobioreactors and algae ponds show considerable potential for large-scale growth of algae as biomass feedstock.

The biomass resource base is capable of producing feedstock for a sustainable supply of fuels, energy, and chemicals. However, technological challenges, market drivers, fossil feedstock cost fluctuations, and government policies and mandates play a significant role in utilizing the full potential of the biomass resources. Ideally, the biomass is regenerated over a short period of time when compared to fossil resources. This period can be few years for forest resources, seasonal for agricultural crops, and days for algae and cyanobacteria. Biomass is the source for natural atmospheric carbon dioxide fixation. Thus, with the use of biomass as feedstock for energy, fuels, and chemicals, the dependence on fossil resources can be reduced, and climate change issues related to resource utilization can be addressed.

3

Chemicals from Biomass

3.1 Introduction

Crude oil is the single largest source of energy for the United States, followed by natural gas and coal. Approximately 3% of the total crude oil is used as feedstock for the production of chemicals (Banholzer et al., 2008). Natural gas is used for the production of fertilizers and to supply energy to the production processes. Petroleum refineries extract and upgrade valuable components of crude oil using various physical and chemical methods into a large array of useful petroleum products. While the United States is one of the world's largest producers of crude oil, the country relies heavily on imports to meet the demand for petroleum products for consumers and industry. This reliance on international ties to petroleum trade has led to numerous upheavals in the industry over the last four decades, the most recent being when crude oil prices reached $134 per barrel in 2008 (EIA, 2010b), as shown in Figure 3.1. Natural disasters such as hurricanes in the Gulf Coast region (Katrina and Rita in 2005 and Gustav in 2008) caused major damages to offshore oil-drilling platforms and disruption of crude oil supply. Natural gas prices have also varied from $4 per cubic feet in 2001 to $13 per cubic feet in 2008 and back to $4 per cubic feet in 2011 with the development of shale deposits.

The consumption of energy resources in the world added 30.4 billion tons of carbon dioxide in 2008, an increase of approximately 12 billion tons higher than the 1980 figures (EIA, 2010c). The rate of carbon dioxide emissions are expected to go higher, unless alternate methods for obtaining energy, fuels, and chemicals are developed. Renewable resources are considered for supplementing and eventually substituting the dependence on oil and natural gas. These resources include biomass, wind, hydroelectric, and solar energy. They convert an alternate form of energy (different from fossil resource) into power, fuels, or chemicals. Some of these resources (wind, solar, and hydroelectric) do not emit carbon dioxide during resource utilization and, thus, are cleaner choices compared to fossil resources. They also reduce the dependence on foreign oil imports.

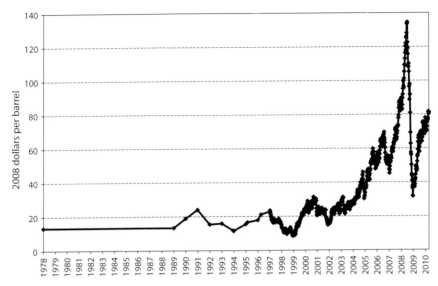

FIGURE 3.1
Historical crude oil prices. (From EIA, Weekly United States spot price FOB weighted by estimated import volume (Dollars per barrel), energy information administration, http://tonto.eia.doe.gov/dnav/pet/hist/LeafHandler.ashx?n=PET&s=WTOTUSA&f=W, accessed May 8, 2010, 2010b.)

The processes for the production of chemicals involve the conversion of traditional or conventional forms of energy (petroleum and natural gas) to materials by rearranging the atoms from the components, mainly carbon, hydrogen, and oxygen. The shift to renewable resources for the production of chemicals offers biomass as the only choice of raw material because only biomass can provide the necessary carbon, hydrogen, and oxygen atoms. The rest of the renewable resources can be used as supplement for energy requirements for conversion processes. The transition from fossil feedstock to biomass feedstock requires extensive process technology changes, market penetration of new chemicals from biomass replacing existing chemicals, and process energy requirements.

3.2 Chemicals from Nonrenewable Resources

The chemical industry in the United States is an integral part of the country's economy, producing more than 70,000 products each year. About 24% of the chemicals produced become raw materials for other products within the industry. For example, sulfuric acid is the second largest produced chemical in the United States, with 36 million short tons produced in 1997 (Energetics, 2000). Sulfuric acid is a key raw material for the fertilizer production process. The Department of Energy (DOE) gives an extensive list of chemicals and allied products manufactured in the United States, identified by SIC (Standard Industrial Classifications) codes. The major U.S. chemical industry SIC codes and their corresponding products are given in Table 3.1.

Based on the classifications of industrial chemicals in Table 3.1, they can be divided into five chains of chemicals. These include the ethylene chain, the propylene chain, the benzene–toluene–xylene (BTX) chain, the agricultural chemicals chain, and the chlor-alkali industry (Energetics, 2000). Among these, the production of ethylene, the building block for the ethylene chain

TABLE 3.1

Major U.S. Chemical Industry SIC Codes and Their Products

SIC	Major Products
281 Industrial Inorganic Chemicals	
2812 Alkalis and chlorine	Caustic soda (sodium hydroxide), chlorine, soda ash, potassium, and sodium carbonates
2813 Industrial gases	Inorganic and organic gases (acetylene, hydrogen, nitrogen, and oxygen)
2819 Industrial inorganic chemicals (not otherwise classified)	Compounds of aluminum, ammonium, chromium, magnesium, potassium, sodium, sulfur, and numerous other minerals; inorganic acids
282 Plastics and Rubbers	
2821 Plastic materials and resins	Synthetic resins, plastics, and elastomers (acrylic, polyamide, vinyl, polystyrene, polyester, nylon, and polyethylene)
2822 Synthetic rubber	Vulcanizable rubbers (acrylic, butadiene, neoprene, and silicone)
286 Industrial Organic Chemicals	
2865 Cyclic crudes and intermediates	Distilling coal tars; cyclic intermediates, that is, hydrocarbons and aromatics (benzene, aniline, toluene, and xylenes); and organic dyes and pigments
2869 Industrial organic chemicals, (not otherwise classified)	Aliphatic/acyclic organics (ethylene, butylene, and organic acids); solvents (alcohols, ethers, acetone, and chlorinated solvents); perfumes and flavorings; rubber processors and plasticizers
287 Agricultural Chemicals	
2873 Nitrogenous chemicals	Ammonia fertilizer compounds, anhydrous ammonia, nitric acid, urea, and natural organic fertilizers
2874 Phosphatic chemicals	Phosphatic materials and phosphatic fertilizers

Source: Adapted from Energetics, Energy and environmental profile of the U.S. chemical industry, Energy efficiency and renewable energy (U.S. DOE), http://www1.eere.energy.gov/industry/chemicals/pdfs/profile_chap1.pdf, accessed May 8, 2010, 2000.

of chemicals, depends on the availability of petroleum feedstock. Propylene, the building block for the propylene chain of chemicals, is almost entirely produced as a coproduct with ethylene in the steam cracking of hydrocarbons. The BTX chain of chemicals is coproduced by the catalytic reforming of naphtha. Agricultural chemicals, like ammonia, urea, ammonium phosphate, etc., are primarily dependent on natural gas for the production of hydrogen. Thus, the present chemical industry is almost entirely dependent on fossil resources for the production of chemicals. A significant amount of carbon dioxide and other greenhouse gases (GHGs) are also released during the production of these chemicals.

Historically, there had been no governmental regulations on carbon dioxide emissions by chemical industries. However, the increased concerns due to global warming, climate change, and pollution reduction programs have prompted the U.S. Government House of Representatives to pass the American Clean Energy and Security Act of 2009 (ACES, 2010). This bill, if passed, would introduce a cap-and-trade program aimed at reducing GHGs to address climate change. The Environmental Protection Agency (EPA) issued the Mandatory Reporting of Greenhouse Gases Rule in December 2009 (EPA, 2010). The rule requires the reporting of GHG emissions from large sources and suppliers in the United States and is intended to collect accurate and timely emissions data to inform future policy decisions. Under the rule, suppliers of fossil fuels or industrial GHGs, manufacturers of vehicles and engines, and facilities that emit 25,000 MT or more per year of GHG emissions are required to submit annual reports to the EPA.

With the government initiatives and increased global concerns for GHG emissions, alternate pathways for the production of chemicals from biomass are required. This chapter focuses on the use of biomass as feedstock for chemicals. This is an ongoing research area, and the chemicals discussed in this chapter are not an exhaustive list. However, an attempt is made to include the most promising chemicals from biomass that have the potential for commercialization and can replace the existing chain of chemicals from fossil resources.

3.3 Chemicals from Biomass as Feedstock

The world has a wide variety of biofeedstocks that can be used for the production of chemicals. Biomass includes plant materials such as trees, grasses, agricultural crops, and animal manure. The components of biomass are shown in Figure 3.2. As shown in the figure, all the biomass components are molecules of carbon, hydrogen, and oxygen atoms. Biomass can be divided into five major categories as shown in the figure: starch, cellulose, hemicellulose lignin, and oils. Cellulose, hemicellulose, and lignin are components

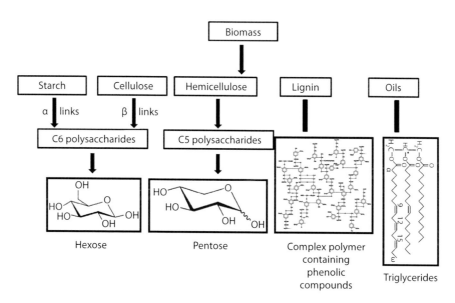

FIGURE 3.2
Biomass classifications and components.

of woody biomass, grasses, stalks, stover, etc. Starch and cellulose are both polymeric forms of hexose, a six-carbon sugar. Hemicellulose is a polymer of pentose. Lignin is composed of phenolic polymers, and oils are triglycerides. Starch is primarily found in corn, sweet sorghum, and other crops. Sugarcane contains sugar in monomeric form, but extraction of juice is required. Other biomass components, which are generally present in minor amounts, include sterols, alkaloids, resins, terpenes, terpenoids, and waxes.

The feedstock availability in the United States currently includes 142 million dry tons of forest biomass with a possibility of increasing it to 368 million dry tons (Perlack et al., 2005). The agricultural biomass currently available is 194 million dry tons with a possible increase to 998 million dry tons. Apart from forest and agricultural biomass, algae can be produced from power plant exhaust carbon dioxide and used for chemicals synthesis.

There are primarily two different platforms of conversion technologies for converting biomass feedstock to chemicals, the biochemical and the thermo-chemical (DOE, 2010c). The biochemical platform focuses on the conversion of carbohydrates (starch, cellulose, and hemicellulose) to sugars using bio-catalysts like enzymes and microorganisms and chemical catalysts. These sugars are then suitable for fermentation into a wide array of chemicals. Apart from this, chemical catalysis used in transesterification reaction can produce fatty acid methyl and ethyl esters and glycerol. The fermentation products such as ethanol and butanol can be starting material for numerous chemicals, for example, ethanol can be converted to ethylene and introduced to the ethylene chain of chemicals. The glycerol produced as by-product in the transesterification process can be converted to produce the propylene chain of chemicals. The thermochemical platform uses technology to convert biomass to fuels, chemicals, and power via thermal and chemical processes such as gasification and pyrolysis. Intermediate products in the thermochemical platform include clean synthesis gas or syngas (a mixture of primarily H_2 and CO) produced via gasification, bio-oil, and bio-char produced via pyrolysis. Synthesis gas is conventionally manufactured from natural gas, so the gasification procedure to produce synthesis gas from biomass is a possible replacement for the fossil resource.

Figure 3.3 shows the different routes for the production of chemicals from biomass. The feedstock base includes natural oils, sugars and starches as carbohydrates, and cellulose and hemicellulose. The main conversion technologies used are transesterification, fermentation, anaerobic digestion, acid dehydration, gasification, and pyrolysis. The primary products given in the figure are not an exhaustive list, but some representative chemicals.

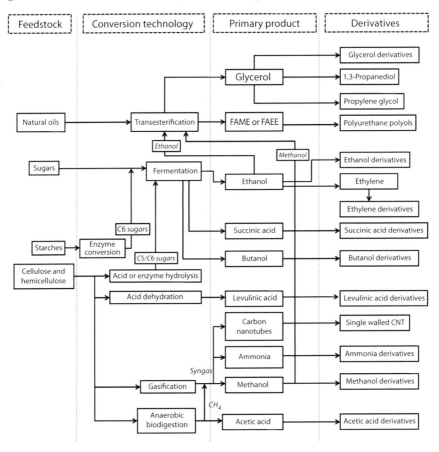

FIGURE 3.3
Biomass feedstock conversion routes to chemicals.

The various chemicals that can be manufactured from biomass are compiled based on carbon numbers and given in the following section. Some of these chemicals are presently made from nonrenewable feedstock like natural gas and petroleum, while others are new chemicals that have the potential to replace nonrenewable-feedstock-based chemicals. This description is not exhaustive but serves as a starting point for identifying the processes and feedstocks for conversion to chemicals.

3.4 Biomass Conversion Products (Chemicals)

Biomass can be converted to chemicals using the routes described in Section 3.3. The Biomass Research and Development Act of 2000 had set up a Biomass R&D Technical Advisory Committee, which has fixed a goal of supplying the United States with 25% of its chemicals from biomass by the year 2030 (Perlack, 2005). Bulk chemicals can be defined as those costing $1.00–$4.00 per kg and produced worldwide in volumes of more than 1 million MT/year (Short, 2007). The production cost of these chemicals can be reduced by 30% when petrochemical processes are replaced by bio-based processes. Some of these chemicals are discussed in the following sections.

3.4.1 Single-Carbon Compounds

3.4.1.1 Methane

Methane from natural gas is an important industrial raw material for the production of acetylene, synthesis gas, methanol, carbon black, etc. (Austin, 1984). Natural gas is a nonrenewable source, and ways to produce methane from biomass are needed.

Methane can be produced from the anaerobic digestion of biomass. Methanogenic bacteria are comprised of mesophilic and thermophilic species that convert biomass in the absence of oxygen. Anaerobic digestion of biomass is the treatment of biomass with a mixed culture of bacteria to produce methane (biogas) as a primary product. The four stages of anaerobic digestion are hydrolysis, acidogenesis, acetogenesis, and methanogenesis. In the first stage, hydrolysis, complex organic molecules are broken down into simple sugars, amino acids, and fatty acids with the addition of hydroxyl groups. In the second stage, acidogenesis, volatile fatty acids (e.g., acetic, propionic, butyric, and valeric) are formed along with ammonia, carbon dioxide, and hydrogen sulfide. In the third stage, acetogenesis, simple molecules from acidogenesis are further digested to produce carbon dioxide, hydrogen, and organic acids, mainly acetic acid. Then in the fourth stage, methanogenesis, the organic acids are converted to methane, carbon dioxide, and water. The last stage produces 65%–70% methane and 35%–30% carbon dioxide

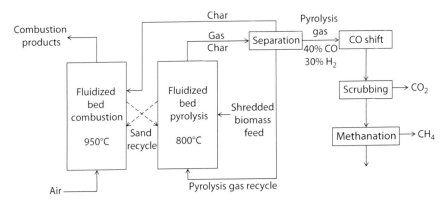

FIGURE 3.4
Pyrolytic gasification process using two fluidized bed reactors. (Adapted from Klass, D.L., *Biomass for Renewable Energy, Fuels and Chemicals*, Academic Press, San Francisco, CA, ISBN 0124109500, 1998.)

(Brown, 2003). Anaerobic digestion can be conducted either wet or dry, where dry digestion has a solid content of 30% or more and wet digestion has a solid content of 15% or less. Either batch or continuous digester operations can be used. In continuous operations, there is a constant production of biogas, while batch operations can be considered simpler and the production of biogas varies. Advantages of anaerobic digestion for processing biomass include the ability to use non-sterile reaction vessels, automatic product separation by outgassing, and relatively simpler equipment and operations. The primary disadvantages for the process are slow reaction rates and low methane yields.

An innovative process using pyrolytic gasification for methane production from biomass is given by Klass (1998), which is shown in Figure 3.4. Biomass is fed to the pyrolysis reactor operating at 800°C. The reactor temperature is maintained at this temperature by sand fed from the combustion reactor at 950°C. The biomass decomposes into pyrolysis gas (~40% CO, ~30% H_2 and others), which exits from the top of the reactor. Char is deposited on the sand, which is sent to the combustion reactor, and air is fed to this reactor to maintain the temperature at 950°C from combustion of the char. The pyrolysis gas can then be sent to a methanation reactor as shown in Figure 3.4.

3.4.1.2 Methanol

Methanol was historically produced by the destructive distillation of wood (Wells, 1999). Currently, 97% of methanol production is based on natural gas, naphtha, or refinery light gas. Large-scale methanol manufacture processes based on hydrogen–carbon oxide mixtures were introduced in the 1920s. In the 1970s, low-pressure processes replaced high-pressure routes for the product formation. Currently, methanol is produced using adiabatic route of ICI and isothermal route of Lurgi. Capacities of methanol plants range from 60,000 to 2,250,000 ton/year. Nearly 12.2 billion lb of methanol are produced annually in the United States and around 85% of it is converted to higher-value chemicals such as formaldehyde (37%), methyl tertiary butyl ether (28%), and acetic acid (8%) (Paster, 2003).

Synthesis gas, an intermediate in the conventional methanol process from natural gas, can be produced from gasification of biomass (Spath and Dayton, 2003). Details of the gasification process have been discussed in Section 2.4.5. The conventional process for methanol synthesis and the process modification for utilizing biomass as feedstock are given in Figure 3.5.

The chemical production complex in the lower Mississippi River corridor produces methanol from natural gas and carbon dioxide produced from ammonia plants. As ammonia plants are shut down due to rising natural gas prices, alternate methods for the production of methanol are needed. New processes for producing methanol in the chemical production complex using carbon dioxide as a feedstock are given by Xu (2004). If large-scale carbon nanotube processes in the order of 5000 MT/year are integrated into the complex (Agboola, 2005), comparable amounts of carbon dioxide will be produced, which can compensate for the carbon dioxide from the shut down plants.

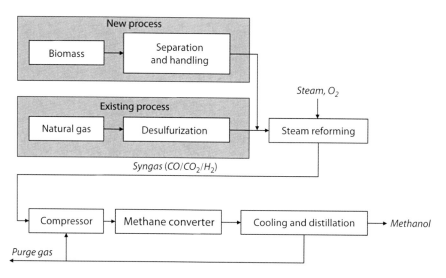

FIGURE 3.5
Conventional methanol process with modification for biomass-derived syngas. (Adapted from Spath, P.L. and Dayton, D.C., *Preliminary Screening—Technical and Economic Feasibility of Synthesis Gas to Fuels and Chemicals with the Emphasis on the Potential for Biomass-Derived Syngas*, National Renewable Energy Laboratory, Golden, CO, NREL/TP-510-34929, http://www.nrel.gov/docs/fy04osti/34929.pdf, 2003.)

3.4.2 Two-Carbon Compounds

3.4.2.1 Ethanol

Ethanol has been produced by the fermentation of carbohydrates for many thousands of years (Wells, 1999). Economic, industrial manufacture of ethanol began in the 1930s. Current processes to produce ethanol in the industry include direct and indirect hydration of ethylene and carbonylation of methyl alcohol and methyl acetate. Industrial uses of ethanol include use as solvents and in the synthesis of chemicals (Wells, 1999). About 45% of the total ethanol demand is for solvent applications. Ethanol is a chemical intermediate for the manufacture of esters, glycol ethers, acetic acid, acetaldehyde, and ethyl chloride; this demand as intermediate accounts for 35% of its production. Ethanol can also be converted to ethylene, which serves as a raw material for a wide range of chemicals that are presently produced from petroleum-based feedstock. Since ethylene is an important building block chemical and ethanol is its precursor, the processes for the manufacture of ethanol are discussed in detail in this section. There are four case studies presented for conversion of lignocellulosic biomass to ethanol.

Increasing prices of crude petroleum has prompted the research for manufacture of ethanol from biomass sources. Ethanol can be produced by the fermentation of starch (corn) sugar (sugarcane) or waste lignocellulosic biomass like corn stover or switch grass. The processes for conversion depend on the feedstock used. The reaction for the fermentation of glucose to ethanol is given by Equation 3.1:

$$C_6H_{12}O_6 \rightarrow 2C_2H_5OH + 2CO_2 \qquad (3.1)$$

Sugars can be directly converted to ethanol using *Saccharomyces cervisiae* without any pretreatment (Klass, 1998). For starch-containing grain feedstock, the cell walls must be disrupted to expose the starch polymers so that they can be hydrolyzed to free, fermentable sugars as yeast does not ferment polymers. The sugar polymers in grain starches contain about 10%–20% hot-water-soluble amylases and 80%–90% water-insoluble amylopectins. Both substances yield glucose or maltose on hydrolysis. Cellulosic or lignocellulosic biomass is mainly composed of crystalline and amorphous cellulose, amorphous hemicelluloses, and lignin as binder. The main problems associated with using this feedstock lie in the difficulty of hydrolyzing cellulosics to maximize glucose yields and the inability of yeasts to ferment the pentose sugars, which are the building blocks of the hemicelluloses.

Capacities of biomass-feedstock-based ethanol plants range from 1.5 to 420 million gal/year (EPM, 2010). Currently, 60% of the world's bio-based ethanol is obtained from sugarcane in Brazil. Sugar from sugarcane is used directly as a solution from the grinding of the cane and is sent directly to the fermenter rather than proceeding with clarification, evaporation, and crystallization to produce raw sugar that is sent to a sugar refinery.

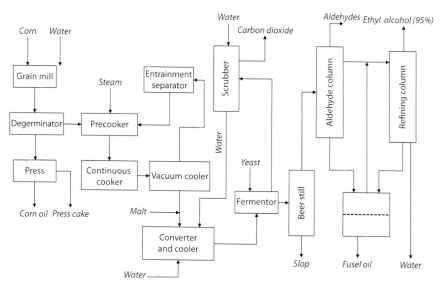

FIGURE 3.6
Corn dry-grind operation for the production of ethanol. (Adapted from Klass, D.L., *Biomass for Renewable Energy, Fuels and Chemicals*, Academic Press, San Francisco, CA, ISBN 0124109500, 1998.)

The corn dry-grind process for the production of ethanol is described by Klass (1998) and is shown in Figure 3.6. The production of ethanol in the United States increased from nearly 2 billion gal in 1999 to over 13 billion gal in 2010 (DOE, 2010c; EPM, 2010) as shown in Figure 3.7.

Cellulosic biomass refers to a wide variety of plentiful materials obtained from plants, including certain forest-related resources (mill residues, pre-commercial thinning, slash, and brush), many types of solid wood waste materials, and certain agricultural wastes (including corn stover, sugarcane bagasse), as well as plants that are specifically grown as fuel for generating electricity. These materials can be used to produce ethanol, which is referred to as "cellulosic ethanol." The cellulosic biomass contains cellulose, hemicellulose, and lignin. The cellulose and hemicellulose are converted to sugars using enzymes, which are then fermented to ethanol. Figure 3.8 gives the BCI process for the conversion of cellulosic biomass (sugarcane bagasse) to ethanol.

Six plants were selected by DOE to receive federal funding for cellulosic ethanol production (DOE, 2007). These plants received a sum of $385 million for biorefinery projects for producing more than 130 million gal of cellulosic ethanol per year. The Table 3.2 gives a list of these plants with their capacity of producing ethanol.

Four case studies are given in this section where biomass is converted to ethanol. The first two cases are the production of ethanol from cellulosic biomass, the third case is a fermentation process of glycerol to produce ethanol,

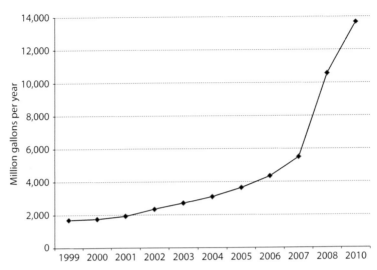

FIGURE 3.7
Production of ethanol in the United States from 1999 to 2010. (From DOE, Biomass energy databook, United States Department of Energy, http://cta.ornl.gov/bedb/biofuels.shtml, accessed May 8, 2010, 2010c; EPM, Plants list, *Ethanol Producers Magazine*, http://www.ethanolproducer.com/plant-list.jsp, accessed May 8, 2010, 2010.)

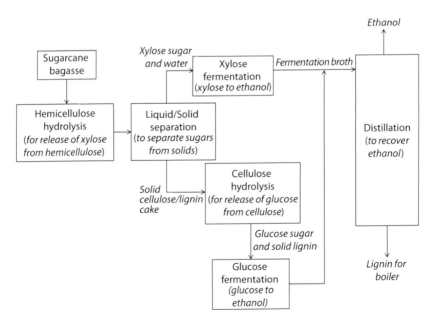

FIGURE 3.8
BCI process for converting sugarcane bagasse to ethanol. (Adapted from Smith, R.A., Analysis of a petrochemical and chemical industrial zone for the improvement of sustainability, MS thesis, Lamar University, Beaumont, TX, 2005.)

TABLE 3.2

DOE-Funded Cellulosic Ethanol Plants

Plant Name/ Location/ Start-Up Year	Feedstock	Feedstock Capacity (ton/day)	Products	Notes
Abengoa Bioenergy Biomass of Kansas LLC Colwich, Kansas, 2011	Corn stover Wheat straw Sorghum Stubble Switchgrass	700	Ethanol: 11.4 million gal/year Syngas	Thermochemical and biochemical processing
ALICO, Inc. LaBelle, Florida, 2010	Yard Wood Vegetative wastes (citrus peel)	770	Ethanol: 7 million gal/year (first unit) 13.9 million gal/year (second unit) Power: 6255 KW Hydrogen Ammonia	Gasification Fermentation of syngas to ethanol
BlueFire Ethanol, Inc. Southern California, 2009	Sorted green waste and wood waste from landfills	700	Ethanol: 19 million gal/year	Concentrated acid processing Fermentation
Broin Companies Emmetsburg, Palo Alto County, Iowa, 2010	Corn fiber Corn stover	842	Ethanol: 125 million gal/year Chemicals Animal feed	Fermentation of starch and lignocellulosic biomass (25%)
Iogen Biorefinery Partners, LLC Shelley, Idaho, 2010	Agricultural residues: wheat straw, barley straw, corn stover, switchgrass, and rice straw	700	Ethanol: 18 million gal/year (first plant) 250 million gal/year (future plants)	Enzymatic process converting cellulose to ethanol
Range Fuels, Inc. Near Soperton, Treutlen County, Georgia 2011	Unmerchantable timber and forest residues	1200	Ethanol: 10 million gal/year (first unit) ~40 million gal/year (commercial unit) Methanol: 9 million gal/year (commercial unit)	Thermochemical catalytic syngas conversion

Source: DOE, DOE selects six cellulosic ethanol plants for up to $385 million in federal funding, http://www.energy.gov/print/4827.htm, accessed October 2, 2007.

and the fourth case discusses fermentation of syngas to ethanol. There are several other methods to produce ethanol from biomass using corn, sugarcane, sugarcane bagasse, etc.

The fermentation of corn to ethanol is a well-established process (Klass, 1998) and detailed descriptions of corn wet milling and dry milling procedures have been given by Johnson (2006). Approximately, 93% of the ethanol currently produced in the United States comes from corn and 3% comes from sorghum (DOE, 2010c). Other feedstocks include molasses, cassava, rice, beets, and potatoes. However, these are primarily food and feed crops, and there is considerable debate on their usage, for example, the use of corn as feed versus feedstock. Cellulosic biomass to ethanol production is not yet fully developed for large-scale production, and some of these attempts are discussed in the following cases. The first two cases are discussed on the basis of selection of raw material and the optimum selection of plant size. These are currently the major concerns for a cellulosic-feedstock-based ethanol industry, and research is ongoing to reduce the cost of ethanol for these factors.

Case study 1: Iogen process for ethanol production from wheat straw and corn stover (Tolan, 2006)

Tolan (2006) discussed Iogen's process for the production of ethanol from cellulosic biomass. Iogen was one of the six companies identified by DOE to receive federal funding to produce ethanol from lignocellulosic feedstock. Iogen's facility produces 2,000 gal/day of ethanol from wheat straw in a pilot plant, with a proposal to scale up to 170,000 gal/day (60 million gal/year). The Iogen process uses steam explosion pretreatment for chopped, milled wheat straw mixed with corn stover. High-pressure steam and 0.5%–2% sulfuric acid are added to the feedstock at a temperature of 180°C–260°C. The acid hydrolysis releases the hemicellulose and converts it to xylose. The residence time in the pretreatment reactor is 0.5–5 min. The pressure is released rapidly to enable the steam explosion process. Hemicellulose reacts first in the process according to Equation 3.2. The dilute sulfuric acid produces xylose monomer, which dehydrates to furfural according to Equation 3.3 under further pretreatment conditions. Similar reactions occur for arabinose. Small amounts of cellulose react to glucose by Equation 3.4 and further degrade to hydroxymethylfurfural (HMF) according to Equation 3.5.

The lignin depolymerizes in this process but is insoluble in the acid or water.

$$\left(C_5H_8O_4\right)_n + H_2O \rightarrow \left(C_5H_8O_4\right)_{n-1} + C_5H_{10}O_5 \tag{3.2}$$

$$C_5H_{10}O_5 \rightarrow C_5H_4O_2 + 3H_2O \tag{3.3}$$

$$\left(C_6H_{10}O_5\right)_n + H_2O \rightarrow \left(C_6H_{10}O_5\right)_{n-1} + C_6H_{12}O_6 \tag{3.4}$$

$$C_6H_{12}O_6 \rightarrow C_6H_6O_3 + 3H_2O \tag{3.5}$$

The next step is the preparation of cellulase enzymes and cellulose hydrolysis. In the Iogen process, *Trichoderma*, a wood rotting fungus is used to produce cellulase enzymes. The cellulases are prepared in submerged liquid cultures in fermentation vessels of 50,000 gal. The liquid broth contains carbon source, salts, complex nutrients like corn steep liquor, and water. The carbon source is important and includes an inducing sugar (like cellobiose, lactose, sophorose, and other low-molecular-weight oligomers of glucose), promoting cellulase growth as opposed to glucose, which promotes growth of the organism. The nutrient broth is sterilized by heating with steam. The fermenter is inoculated with the enzyme production strain once the liquid broth cools down. The operating conditions of the fermenter are 30°C at pH 4–5. The temperature is maintained using cooling coils of water, and pH is maintained using alkali. A constant stream of air or oxygen is passed to maintain aerobic conditions required for *Trichoderma*. The cellulase enzyme production process requires about one week, and at the end of the run, the broth is filtered across a cloth to remove cells. The spent cell mass is disposed in landfills. Cellulase enzymes can be directly used at Iogen's ethanol manufacturing facility. The enzymes can also be stored provided that it is sterilized against microbial contamination by using sodium benzoate and protein denaturation by using glycerol. Iogen reduces the cost of their ethanol manufacture by having an on-site cellulase manufacture facility, reducing costs due to storage and transportation of enzymes. The cellulase enzymes are conveyed to hydrolysis tanks to convert cellulose to glucose. The slurry from pretreatment containing 5%–15% total solids is fed into hydrolysis tanks having a volume of 200,000 gal. Crude cellulase enzymes broth is added in dosages of 100 L/ton of cellulose. The contents are agitated to keep material dispersed in the tank. The hydrolysis proceeds for 5–7 days. The viscosity of the slurry decreases and lignin remains as insoluble particles. The cellulose hydrolysis process yields 90%–98% conversion of cellulose to glucose. Enzymatic hydrolysis of cellulose occurs according to Equations 3.6 and 3.7:

$$(C_6H_{10}O_5)_n + H_2O \rightarrow (C_6H_{10}O_5)_{n-2} + C_{12}H_{22}O_{11} \qquad (3.6)$$

$$C_{12}H_{22}O_{11} + H_2O \rightarrow 2C_6H_{12}O_6 \qquad (3.7)$$

The cellulose hydrolysis is followed by sugar separation and fermentation using recombinant yeast capable of fermenting both glucose and xylose. The hydrolysis slurry is separated from lignin and unreacted cellulose using a plate and frame filter. The filter plates are washed with water to ensure high sugar recovery. The sugar stream from pretreatment section is pumped to fermentation tanks. The lignin cakes can be used for power generation by combustion and excess electricity can be sold to neighboring plants. The sugar stream is fermented with genetically modified *Saccharomyces* yeast capable of fermenting both glucose and xylose.

The yeast is well developed for plant operations with good ethanol tolerance. The rates and yields of xylose fermentation are not high in the current process leaving scope for further improvement. The fermentation broth obtained after fermentation is pumped into a distillation column. Ethanol is distilled from the top and dehydrated. Yield of ethanol obtained in the process is 75 gal/ton of wheat straw.

The feedstock selection for the Iogen process depended on the following considerations:

- *Low cost*: Desired feedstock should be available and delivered to the plant at low cost. Primary and secondary tree growth, sawdust, and waste paper have existing markets and were not considered for the process.
- *Availability*: Feedstock availability should be consistent and in the order of 800,000 ton/year, which is not generally available from sugarcane bagasse.
- *Uniformity*: Feedstock available should be consistent, and, hence, municipal waste containing foreign matter was discarded.
- *Cleanliness*: High levels of silica can cause damage to equipment. Microbial contamination and toxic or inhibitory products should be prevented from the feedstock.
- *High potential ethanol yield*: Cellulose and hemicellulose should be present in high percentage in the feed to yield maximum ethanol by fermentation. Wood and forestry waste has high lignin content, which inhibits fermentation.
- *High efficiency of conversion*: The efficiency of conversion in the Iogen process depended on arabinan and xylan content in feedstock. These are constituent hemicelluloses and low content of these required high quantities of enzyme for conversion to cellulose, thereby increasing the process cost.

Case study 2: NREL process for conversion of 2000 MT/day of corn stover (Aden et al., 2002)

Aden et al. (2002) discuss the use of lignocellulosic biomass for the production of ethanol from corn stover. The plant size was such that 2000 MT/day of corn stover was processed in the facility. The cost estimate is based on the assumption that the plant developed is an "*n*th" plant of several plants that are already built using the same technology and are operating. The target selling price of ethanol is $1.07 per gallon with a start-up date for the plant in 2010. This cost was increased in an updated report (Aden, 2008) to $1.49 per gallon of ethanol. The conceptual design for this plant includes equipment design, corn stover handling, and purchase of enzymes from commercial facilities like Genencor International and

Novozymes Biotech. The design did not take into account the sale of by-products that are important commodity and specialty chemicals, but the report mentions that reduction of price of ethanol is possible with the sale of these chemicals. The design of the facility is divided into eight sections: feedstock storage and handling, pretreatment and hydrolyzate conditioning, saccharification and co-fermentation, product, solids and water recovery, wastewater treatment, product and feed chemical storage, combustor, boiler and turbo generator, and utilities. The process description for conversion of biomass is similar to the Iogen process for corn and wheat straw as raw material.

The NREL report gave the following considerations for the selection of plant size between 2000 and 4000 MT/day. These are listed as follows:

- *Economies of scale*: The plant size varies with capital cost according to Equation 3.8. If exponential, "exp," equals 1, linear scaling of plant size occurs. However, if the exponential value is less than 1, then the capital cost per unit size decreases as the equipment becomes larger. The NREL uses a cost scaling exponent of 0.7.

$$\text{New Cost} = \text{Original Cost} \left(\frac{\text{New Size}}{\text{Old Size}} \right)^{\text{exp}} \qquad (3.8)$$

- *Plant size and collection distance*: The distance traveled to collect corn stover increases as the plant size increases because more stover is required for feed. This collection distance is estimated as the radius of a circle around the plant within which the stover is purchased. This area around the plant is calculated using Equation 3.9:

$$\text{Area}_{\text{collection}} = (D_{\text{stover}} / (Y_{\text{stover}} * F_{\text{available acres}} * F_{\text{land in crops}})) \qquad (3.9)$$

where
  $\text{Area}_{\text{collection}}$ is the circle of collection around the plant
  D_{stover} is the annual demand for stover by an ethanol plant
  Y_{stover} is metric tons of stover collected per acre per year
  $F_{\text{available acres}}$ is the fraction of total farmland from which stover can be collected
  $F_{\text{land in crops}}$ is the fraction of surrounding farmland containing crops

The fraction of available acres takes into account the land use due to roads and buildings within the farm land. For example, if the farm area has 25% roads and other infrastructure, then the fraction of available land, $F_{\text{available acres}}$, is 0.75. The $F_{\text{land in crops}}$ is a variable parameter depending on the ability of farms around the ethanol plant to contribute to the corn stover demand. The parameter is used to vary the dependence of plant size on collection distance. The radius of collection is calculated from the $Area_{collection}$. The price of ethanol is also a function of plant size and percentage of available acres.

- *Corn stover cost*: The corn stover raw material cost depends on two direct costs: the cost of baling and staging stover at the edge of the field and the cost of transportation from the field to the plant gate. Apart from these, a farmer's premium and cost for fertilizers also add to the direct costs for corn stover as a raw material. A life-cycle analysis of the corn stover represents that 47% of cost was in the staging and baling process, 23% was for transport of stover to plant, 11% was farmer premium for taking the risk of added work of collecting and selling the residue and the remaining 12% was for fertilizer supplement for the land. This method of analysis gave a value of $62 per dry MT of corn stover. The report suggests that this cost will be reduced considerably over time, and an assumption of $33 per dry MT of corn stover was taken for further analysis. However, the update to the report in 2009 suggested that the cost for feedstock increased to $69.60 per dry ton of corn stover in 2007, which can be reduced to reach $ 50.90 per dry ton in 2012 (Aden, 2008).

- *Corn stover hauling cost*: The corn stover hauling cost (cost for farm to gate of plant) depended on distance from plant. The hauler cost is a function of radial distance from the plant. An increase in hauling cost shows the optimum plant size range to decrease. For a 50% increase in hauling costs per ton-mile, plant size range decreases from 2000–8000 MT/day to 2000–5000 MT/day. For a 100% increase, the optimal plant size is at around 3000 MT/day, and the price of ethanol increases drastically above or below this price.

- *Total cost of ethanol as a function of plant size*: The total cost of ethanol as a function of plant size was determined with the total feedstock and non-feedstock costs. The analysis was done with two plant sizes of 2,000 and 10,000 MT/day of stover. A net savings occurred for plant sizes between 6000 and 8000 MT/day of stover. Below 2000 MT/day, the selling price per gallon of ethanol increased rapidly. A minimum optimal plant size between 2000 and 4000 MT/day of corn stover was obtained for collection from 10% corn acres around a conversion plant.

Case study 3: Ethanol from fermentation of glycerol (Ito et al., 2005)

Ito et al. (2005) described a process where ethanol is produced from glycerol-containing waste discharged after transesterification process. *Enterobacter aerogenes* HU-101 microorganism is used to ferment the glycerol-rich waste and yields of 63 mmol/L/h of H_2 and 0.85 mol/mole glycerol of ethanol were reported using porous ceramics as support to fix cells in the reactor. There are no reports of scale-up of this process.

Case study 4: Ethanol from synthesis gas fermentation (Spath et al., 2003; Phillips et al., 2007; Snyder, 2007)

Synthesis gas can be used as feed to a fermenter that uses anaerobic bacteria to produce ethanol. Although it uses some of the oldest biological mechanisms in existence, technical barriers to be overcome include organism development, gas–liquid mass transfer, and product yield (Snyder, 2007).

Spath et al. (2003) give a detailed description of the process for the conversion of synthesis gas to ethanol. The first step in the process is to convert biomass synthesis gas and the syngas is then converted to ethanol using fermentation. The feedstock for this process was wood chips derived from forests. The overall schematic diagram is given in Figure 3.9.

The feed is received and placed in temporary storage on-site. It is then sent to the gasifier where it is converted into a raw syngas mixture rich in carbon monoxide and hydrogen. The indirect BCL/FERCO process gasifier was used for the production of syngas from biomass (Spath et al., 2003). The equipment includes an indirectly heated gasifier with operating temperatures at 700°C–850°C and pressures slightly greater than atmospheric. The biomass feed is dried and then fed to a fast fluidized bed where it is converted into a raw syngas. The resulting syngas contains significant amounts of methane, ethylene, and other light hydrocarbons and tars, which can be removed in the gas conditioning steps. The conditioned syngas is then fed to a fermentation reactor where it is converted to ethanol using bacteria. The resulting fermentation broth is dilute, typically containing 2% or less of ethanol. The ethanol can be recovered from the broth using recovery schemes used in the existing corn ethanol industry. The cell mass produced can be recycled as a portion of the feed to the gasifier. One advantage of the syngas fermentation route is that the chemical energy stored in all parts of the biomass, including

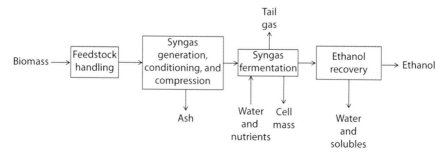

FIGURE 3.9
Synthesis gas to ethanol process. (Adapted from Spath, P.L. and Dayton, D.C., *Preliminary Screening—Technical and Economic Feasibility of Synthesis Gas to Fuels and Chemicals with the Emphasis on the Potential for Biomass-Derived Syngas*, National Renewable Energy Laboratory, Golden, CO, NREL/TP-510-34929, http://www.nrel.gov/docs/fy04osti/34929.pdf, 2003.)

the lignin fraction, contributes to the yield of ethanol. Equation 3.10 gives the method to calculate the capacity of ethanol produced by this process:

$$P = \frac{F \times HHV_F \times \eta_{Gas+Cond} \times X_{CO+H_2/EtOH}}{1.5 \times 10^5} \quad (3.10)$$

where
  P is the production of ethanol, million gallons per year
  F is the feed rate, tons per day (dry basis)
  HHV_F is the higher heating value of the feed in Btu per lb (dry)
  $\eta_{Gas+Cond}$ is the cold gas efficiency of gasifier + conditioning steps (a fraction less than 1)
  $X_{CO+H_2/EtOH}$ is the average conversion of carbon monoxide and hydrogen to ethanol, as a fraction of theoretical

Spath et al. (2003) give the overall reactions for the process as given in Equations 3.11 through 3.14. The microorganisms used for ethanol production from syngas mixtures are anaerobes that use a heterofermentative version of the acetyl-CoA pathway for acetogenesis. Acetyl-CoA is produced from carbon monoxide or hydrogen–carbon dioxide mixtures in this pathway. The acetyl-CoA intermediate is then converted into either acetic acid or ethanol as a primary metabolic product.

$$6CO + 3H_2O \rightarrow CH_3CH_2OH + 4CO_2 \quad \Delta G = -48.7 \, kcal/mol \quad (3.11)$$

$$2CO_2 + 6H_2 \rightarrow CH_3CH_2OH + 3H_2O \quad \Delta G = 28.7 \, kcal/mol \quad (3.12)$$

$$4CO + 2H_2O \rightarrow CH_3COOH + 2CO_2 \quad \Delta G = -39.2 \, kcal/mol \quad (3.13)$$

$$2CO_2 + 4H_2 \rightarrow CH_3COOH + 2H_2O \quad \Delta G = -25.8 \, kcal/mol \quad (3.14)$$

Spath et al. (2003) also report the cost analysis for the gasification process and fermentation. A facility for gasification processing 2000 ton (dry) per day of wood would produce 48.5 million gal/year of ethanol based on an ethanol yield of 71 gal/ton. Fixed capital was estimated at $153.6 million, or $3.17 per annual gallon of capacity. Cash costs were $0.697 per gallon with feedstock cost at $25 per ton. The price required for a zero net present value for the project with 100% financing and 10% real after-tax discounting, known as rational cost, was $1.33 per gallon.

  Phillips et al. (2007) described the feasibility of a forest-resources-based thermochemical pathway conversion to ethanol and mixed alcohols. Hybrid poplar was used as feed for the indirect gasification process. The detailed design included seven sections, namely, feed handling and drying, gasification, gas cleanup and conditioning, alcohol synthesis, alcohol separation, steam cycle, and cooling water. The syngas was heated to 300°C and 1000 psi pressure and converted to the alcohol mixture across a fixed bed catalyst. The minimum cost of ethanol, based on the operating cost, was $1.01 per gallon.

A similar study, with syngas from high-pressure oxygen-blown direct gasifiers gave a minimum cost of ethanol, based on the operating cost as $1.95 per gallon (Dutta and Phillips, 2009).

3.4.2.2 Acetic Acid

Acetic acid has a wide variety of uses. Currently, 44% of total acetic acid production is converted to vinyl acetate, which is used to form polyvinyl acetate and polyvinyl alcohols used for paints, adhesives, and plastics; 12% is converted to acetic anhydride, which is used to manufacture cellulose acetate, paper sizing agents, a bleach activator, and aspirin; 13% is used to produce acetates and esters used in solvents for coatings, inks, resins, gums, flavorings, and perfumes; and 12% is used in the production of terephthalic acid (TPA) used for polyethylene teraphthalate (PET) bottles and fibers.

Acetic acid was first made by the fermentation of ethyl alcohol and a very dilute solution of it is used as vinegar (Wells, 1999). Small quantities of acetic acid are recovered from pyroligneous acid liquor obtained from the destructive distillation of hardwood. The modern acetic acid industry began with the commercial availability of acetylene, which was converted to acetaldehyde and then oxidized to acetic acid. The three commercial processes for the manufacture of acetic acid are oxidation of acetaldehyde, liquid phase oxidation of n-butane or naphtha, and carbonylation of methyl alcohol. The carbonylation of methyl alcohol is the dominant technology because of low material and energy costs and the absence of significant by-products. Capacities of acetic acid plants range from 30,000 to 840,000 ton/year.

Synthesis gas is the raw material for the carbonylation process at low temperature and pressure using a proprietary catalyst, rhodium iodide, developed by BASF and Monsanto. The synthesis gas can be produced alternately from bio-based feedstock using gasification and pyrolysis as described in previous chapter. The fermentation of syngas can also be used to produce acetic acid, as shown in Equations 3.13 and 3.14.

Acetic acid can be produced by the anaerobic digestion of biomass. The four stages of anaerobic fermentation are given in the section for methane. The fourth stage of methane formation can be inhibited by the use of iodoform or bromoform, thus producing carboxylic acids, hydrogen, and carbon dioxide. Biomass is converted to acetic acid (CH_3COOH) under non-sterile anaerobic conditions according to Equation 3.15 (Holtzapple et al., 1999). Glucose ($C_6H_{12}O_6$) is used for the illustration of this reaction:

$$C_6H_{12}O_6 + 2H_2O + 4NAD^+ \rightarrow 2H_3CCOOH + 2CO_2 + 4NADH + 4H^+ \quad (3.15)$$

The reducing power of nicotinamide adenine dinucleotide (NADH) may be released as hydrogen using endogenous hydrogen dehydrogenase, as shown in Equation 3.16:

$$NADH + H^+ \rightarrow NAD^+ + H_2 \quad (3.16)$$

Methanogens are microorganisms that can produce methane by reacting carbon dioxide produced with hydrogen. The reaction is given in Equation 3.17:

$$CO_2 + 4H_2 \rightarrow CH_4 + 2H_2O \qquad (3.17)$$

Acetic acid can also be converted to methane in the presence of methanogens. So, the potential to convert all biomass to methane exists. The production of methane according to Equation 3.17 can be inhibited by the addition of iodoform or bromoform. Thus, combining Equations 3.15 and 3.16, Equation 3.18 is obtained where acetic acid is produced from glucose and the production of methane is inhibited.

$$C_6H_{12}O_6 + 2H_2O \rightarrow 2H_3CCOOH + 2CO_2 + 4H_2 \qquad (3.18)$$

The conversion of biomass mixtures of sugarcane bagasse/chicken manure (Thanakoses, 2003a), municipal solid waste/sewage sludge (Aiello-Mazzari et al., 2006), and corn stover/pig manure (Thanakoses, 2003b) to carboxylic acids has been reported.

   Cellulose acetate is a cellulose derivative prepared by acetylating cellulose with acetic anhydride (Wells, 1999). Fully acetylated cellulose is partially hydrolyzed to give an acetone-soluble product, which is usually between a di- and a tri-ester (Austin, 1984). The esters are mixed with plasticizers, dyes, and pigments and processed in different ways depending on the form of plastic desired. The important properties of cellulose acetate include mechanical strength, impact resistance, transparency, colorability, fabricating versatility, moldability, and high dielectric strength (Austin, 1984). Cellulose acetate is used to manufacture synthetic fibers like rayon, based on cotton or tree pulp cellulose.

   Research has been reported using waste cellulose from corn fiber, rice hulls, and wheat straw to produce cellulose acetate (Ondrey, 2007a). The raw materials are milled, slurried in dilute sulfuric acid, and pretreated in an autoclave at 121°C. This is followed by the acetylation to cellulose triacetate under ambient conditions at 80°C, using acetic acid, acetic anhydride, methylene chloride, and trace amounts of sulfuric acid. The cellulose acetate is soluble in methylene chloride and separated easily from the reaction medium. Conversions of cellulose to cellulose acetate have been 35%–40% in a laboratory study. The incentive to pursue this line of work was the price of cellulose acetate, approximately, $2.00 per lb—a more valuable product than ethanol.

3.4.2.3 Ethylene

Ethylene ranks fourth among chemicals produced in large volumes in the United States with about 48 billion lb produced in 1997 (Energetics, 2000). It is a principal building block for the petrochemicals industry, with almost

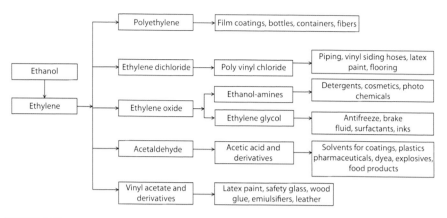

FIGURE 3.10
Ethylene product chain. (Adapted from Energetics, Energy and environmental profile of the U.S. chemical industry, energy efficiency and renewable energy (U.S. DOE), http://www1.eere.energy.gov/industry/chemicals/pdfs/profile_chap1.pdf, accessed May 8, 2010, 2000.)

all of the ethylene produced being used as a feedstock in the manufacture of plastics and chemicals.

Ethylene is used as a raw material in the production of a wide variety of chemicals and polymers as shown in Figure 3.10 (Energetics, 2000). Polyethylene (PE) is used in the manufacture of plastic films, packaging materials, moldings (e.g., toys, chairs, automotive parts, and beverage containers), wire and cable insulation, pipes, and coatings. Production of PE in United States in 1997 was about 27 billion lb (Energetics, 2000), which increased to 60 billion lb in 2008 (ICIS, 2009). Ethylene dichloride is used to manufacture poly vinyl chloride (PVC), which is used in drainage and sewer pipes, electrical conduits, industrial pipes, wire and cable coatings, wall panels, siding, doors, flooring, gutters, downspouts, and insulation. U.S. chemicals production of ethylene dichloride was over 20 billion lb in 1997. U.S. production of PVC was about 14 billion lb in 1997. Ethylene oxide is used for the production of ethylene glycol, which is commonly used antifreeze. Ethylene glycol also serves as a raw material in the production of polyester, used for manufacturing textiles. Ethylene oxide and ethylene glycol are both listed among the top 50 chemicals produced in the United States, with ethylene oxide ranking 27th (7.1 billion lb in 1997) and ethylene glycol ranking 29th (5.6 billion lb in 1997).

World demand for ethylene was about 180 billion lb in 1998, and was predicted to reach 250 billion lb by 2005 (Pellegrino, 2000). The PE industry was a 100 billion lb market with over 150 producers worldwide in 1998 (Energetics, 2000). The global market for PVC was estimated at about 7.5 billion lb capacity.

The petroleum refining industry is the major supplier of raw materials for ethylene production, and a large percentage of ethylene capacity is located at petroleum refineries that are in close proximity to petrochemical plants (Energetics, 2000). Currently, about 20% of ethylene is produced from naphtha (a light petroleum fraction) and 10% from gas oil from refinery processing units. In Western Europe and some Asian countries (South Korea, Taiwan, Japan), naphtha and gas oil account for 80%–100% of the feed to ethylene crackers. Overall, more than 50% of the ethylene production capacity is currently located at refineries. However, the current resources of petroleum are being depleted for use as fuels and the rising price of petroleum feedstock opens up new areas for research for the production of ethylene.

Ethanol can be used for the production of ethylene by dehydration. Ethanol, for the dehydration process to ethylene, can be produced from biomass feedstock as described in the earlier section. Ethanol is vaporized by preheating with high-pressure steam before passing over a fixed bed of activated alumina and phosphoric acid or alumina and zinc oxide contained in a reactor (Wells, 1999). The reactor can be isothermal or adiabatic, with the temperature maintained at 296°C–315°C. The reaction is endothermic, and the heat is supplied by condensing vapor latent heat. The temperature control in the reactor is important to prevent the formation of acetaldehyde or ethers as by-products. The gas is purified, dried, and compressed using conventional steps. A fluidized-bed modification of this process has been developed with efficient temperature controls and conversions up to 99%.

Takahara et al. (2005) have discussed the use of different catalysts for the dehydrogenation of ethanol into ethylene. The dehydration of ethanol into ethylene was investigated over various solid acid catalysts such as zeolites and silica–alumina at temperatures ranging from 453 to 573 K under atmospheric pressure. Ethylene was produced via diethyl ether during the dehydration process. H-mordenites were the most active for the dehydration.

Phillips and Datta (1997) reported the production of ethyl tert-butyl ether (ETBE) from biomass-derived hydrous ethanol dehydration over H-ZSM-5 catalyst. Temperatures between 413 and 493 K were studied for the process, at partial pressures of ethanol less than 0.7 atm and water feed molar ratio less than 0.25.

Varisli et al. (2007) reported the production of ethylene and diethyl-ether by dehydration of ethanol over heteropolyacid catalysts. The temperature range studied for this process was 413–523 K with three heteropolyacids, tungstophosphoricacid (TPA), silicotungsticacid (STA), and molybdophosphoricacid (MPA). Very high ethylene yields over 0.75 were obtained at 523 K with TPA. Among the three HPA catalysts, the activity trend was obtained as STA > TPA > MPA.

Tsao and Zasloff (1979) describe a detailed patented process for a fluidized-bed dehydration with over 99% yield of ethylene. Dow Chemical and Crystalsev, a Brazilian sugar and ethanol producer, announced the plans of a 300,000 MT/year ethylene plant in Brazil to manufacture 350,000 MT/year of low-density PE from sugarcane-derived ethanol. Braskem, a Brazilian petrochemical company announced their plans to produce 650,000 MT of ethylene from sugarcane-based ethanol, which will be converted to 200,000 MT/year of high-density PE (C&E News, 2007a).

3.4.3 Three-Carbon Compounds

3.4.3.1 *Glycerol*

Glycerol, also known as glycerine or glycerin, is a triol occurring in natural fats and oils. About 90% of glycerol is produced from natural sources by the transesterification process. The remaining 10% is commercially manufactured synthetically from propylene (Wells, 1999).

Glycerol is a major by-product in the transesterification process used to convert vegetable oils and other natural oils to fatty acid methyl and ethyl esters. Approximately, 10% by weight of glycerol is produced from the transesterification of soybean oil with an alcohol. Transesterification process is used to manufacture fatty acid methyl and ethyl esters, which can be blended in refinery diesel. As the production of fatty acid methyl and ethyl esters increases, the quantity of glycerol manufactured as a by-product also increases the need to explore cost-effective routes to convert glycerin to value-added products.

Glycerol currently has a global production of 500,000–750,000 ton/year (Werpy et al., 2004). The United States is one of the world's largest suppliers and consumers of refined glycerol. Referring to Figure 3.11, glycerin can

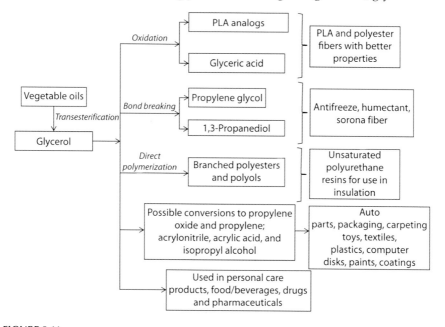

FIGURE 3.11
Production and derivatives of glycerol. (Adapted from Werpy, T. et al., Top value added chemicals from biomass: Volume 1 results of screening for potential candidates from sugars and synthesis gas, energy efficiency and renewable energy (U.S. DOE), http://www1.eere.energy.gov/biomass/pdfs/35523.pdf, accessed May 8, 2010, 2004; Energetics, Energy and environmental profile of the U.S. chemical industry, energy efficiency and renewable energy (U.S. DOE), http://www1.eere.energy.gov/industry/chemicals/pdfs/profile_chap1.pdf, accessed May 8, 2010, 2000.)

potentially be used in a number of paths for chemicals that are currently produced from petroleum-based feedstock. The products from glycerol are similar to the products currently obtained from the propylene chain. Uniqema, Procter & Gamble, and Stepan are some of the companies that currently produce derivatives of glycerol such as glycerol triacetate, glycerol stearate, and glycerol oleate. Glycerol prices are expected to drop if biodiesel production increases, enabling its availability as a cheap feedstock for conversion to chemicals. Small increases in fatty acid consumption for fuels and products can increase world glycerol production significantly. For example, if the United States displaced 2% of the on-road diesel with biodiesel by 2012, almost 800 million lb of new glycerol supplies would be produced.

Dasari et al. (2005) reported a low-pressure and low-temperature (200 psi and 200°C) catalytic process for the hydrogenolysis of glycerol to propylene glycol that is being commercialized and received the 2006 EPA Green Chemistry Award. Copper-chromite catalyst was identified as the most effective catalyst for the hydrogenolysis of glycerol to propylene glycol amongst nickel, palladium, platinum, copper, and copper-chromite catalysts. The low pressure and temperature are the advantages for the process when compared to a traditional process using severe conditions of temperature and pressure. The mechanism proposed forms an acetol intermediate in the production of propylene glycol. In a two-step reaction process, the first step of forming acetol can be performed at atmospheric pressure, while the second requires a hydrogen partial pressure. Propylene glycol yields >73% were achieved at moderate reaction conditions.

Karinen and Krause (2006) studied the etherification of glycerol with isobutene in liquid phase with acidic ion exchange resin catalyst. Five product ethers and a side reaction yielding C_8–C_{16} hydrocarbons from isobutene were reported. The optimal selectivity toward the ethers was discovered near the temperature of 80°C and isobutene/glycerol ratio of 3. The reactants for this process were isobutene (99% purity), glycerol (99% purity), and pressurized with nitrogen (99.5% purity). The five ether isomers formed in the reaction included two monosubstituted monoethers (3-tert-butoxy-1,2-propanediol and 2-tert-butoxy-1,3-propanediol), two disubstituted diethers (2,3-di-tert-butoxy-1-propanol and 1,3-di-tert-butoxy-2-propanol), and one trisubstituted triether (1,2,3-tri-tert-butoxy propane). *Tert*-butyl alcohol was added in some of the reactions to prevent oligomerization of isobutene and improve selectivity toward ethers.

Acrylic acid is a bulk chemical that can be produced from glycerol. Shima and Takahashi (2006) reported the production of acrylic acid involving steps of glycerol dehydration, in gas phase, followed by the application of a gas-phase oxidation reaction to a gaseous reaction product formed by the dehydration reaction. Dehydration of glycerol could lead to the commercially viable production of acrolein, an important intermediate for acrylic acid esters, superabsorber polymers, or detergents (Koutinas et al., 2008). Glycerol can also be converted to chlorinated compounds, such as dichloropropanol and epichlorohydrin. Dow and Solvay are developing a process to convert glycerol to epoxy resin raw material epichlorohydrin (Tullo, 2007a).

Several other methods for conversion of glycerol exist; however, commercial viability of these methods is still in the development stage. Some of these include catalytic conversion of glycerol to hydrogen and alkanes and microbial conversion of glycerol to succinic acid, polyhydroxyalkanoates (PHA), butanol, and propionic acid (Koutinas et al., 2008).

3.4.3.2 Lactic Acid

Lactic acid is a commonly occurring organic acid, which is valuable due to its wide use in food and food-related industries, and its potential for the production of biodegradable and biocompatible polylactate polymers. Lactic acid can be produced from biomass using various fungal species of the *Rhizopus* genus, which have advantages compared to the bacteria, including their amylolytic characteristics, low nutrient requirements, and valuable fermentation fungal biomass by-product (Zhang et al., 2007).

Lactic acid can be produced using bacteria also. Lactic acid producing bacteria (LAB) have high growth rate and product yield. However, LAB have complex nutrient requirements because of their limited ability to synthesize B-vitamins and amino acids. They need to be supplemented with sufficient nutrients such as yeast extracts to the media. This downstream process is expensive and increases the overall cost of production of lactic acid using bacteria.

An important derivative of lactic acid is polylactic acid (PLA). BASF uses 45% corn-based PLA for its product Ecovio®.

3.4.3.3 Propylene Glycol

Propylene glycol is industrially produced from the reaction of propylene oxide and water (Wells, 1999). Capacities of propylene glycol plants range from 15,000 to 250,000 ton/year. It is mainly used (around 40%) for the manufacture of polyester resins that are used in surface coatings and glass-fiber-reinforced resins. A growing market for propylene glycol is in the manufacture of nonionic detergents (around 7%) used in petroleum, sugar, and paper refining and also in the preparation of toiletries, antibiotics, etc. Around 5% of propylene glycol manufactured is used in antifreeze.

Propylene glycol can be produced from glycerol, a by-product of the transesterification process, by a low-pressure and low-temperature (200 psi and 200°C) catalytic process for the hydrogenolysis of glycerol to propylene glycol (Dasari et al., 2005) that is being commercialized and received the 2006 EPA Green Chemistry Award.

Ashland, Inc. and Cargill have a joint venture under way to produce propylene glycol in a 65,000 MT/year plant in Europe (Ondrey, 2007b,c). Davy Process Technology Ltd. (DPT) has developed the glycerin-to-propylene

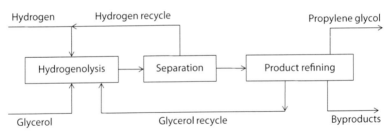

FIGURE 3.12
DPT process for the manufacture of propylene glycol from glycerol by hydrogenolysis. (From Ondrey, G., *Chem. Eng.*, 114(8), 12, 2007c.)

glycol process for this plant. The plant is expected to start up in 2009. The process is outlined in Figure 3.12. This is a two-step process where glycerin in the gas phase is first dehydrated into water and acetol over a heterogeneous catalyst bed, and then, propylene glycol is formed in situ in the reactor by the hydrogenation of acetol. The per pass glycerin conversion is 99% and by-products include ethylene glycol, ethanol, and propanols.

Huntsman Corporation plans to commercialize a process for propylene glycol from glycerin at their process development facility in Conroe, Texas (Tullo, 2007a). Dow and Solvay are planning to manufacture epoxy resin raw material epichlorohydrin from a glycerin-based route to propylene glycol.

3.4.3.4 1,3-Propanediol

1,3-Propanediol is a derivative that can be used as a diol component in the plastic polytrimethyleneterephthalate (PTT), a new polymer comparable to nylon (Wilke et al., 2006). Two methods to produce 1,3-propanediol exist, one from glycerol by bacterial treatment and another from glucose by mixed culture of genetically engineered microorganisms.

A detailed description of various pathways to microbial conversion of glycerol to 1,3-propanediol is given by Liu et al. (2010). Mu et al. (2006) give a process for conversion of crude glycerol to propanediol. They conclude that a microbial production of 1,3- propanediol by *Klebsiella pneumoniae* was feasible by fermentation using crude glycerol as the sole carbon source. Crude glycerol from the transesterification process could be used directly in fed-batch cultures of *K. pneumoniae* with results similar to those obtained with pure glycerol. The final 1,3-propanediol concentration on glycerol from lipase-catalyzed methanolysis of soybean oil was comparable to that on glycerol from alkali-catalyzed process. The high 1,3-propanediol concentration and volumetric productivity from crude glycerol suggested a low fermentation cost, an important factor for the bioconversion of such industrial by-products into valuable compounds. A microbial conversion process for propanediol from glycerol using *K. pneumoniae* ATCC 25955 was given by Cameron and

Koutsky (1994). A $0.20/lb of crude glycerol raw material, a product selling price of $1.10/lb of pure propanediol, and with a capital investment of $15 MM, a return on investment of 29% was obtained. Production trends in biodiesel suggest that the price of raw material (glycerol) is expected to go down considerably, and a higher return on investment can be expected for future propanediol manufacturing processes.

DuPont and Tate and Lyle Bio Products, LLC opened a $100 million facility in Loudon, Tennessee, to make 1,3-propanediol from corn (CEP, 2007). The company uses a proprietary fermentation process to convert corn to Bio-PDO, the commercial name of 1,3-propanediol used by the company. This process uses 40% less energy and reduces GHG emissions by 20% compared with petroleum-based propanediol. Shell produces propanediol from ethylene oxide and Degussa produces it from acroleine. It is used by Shell under the name Corterra to make carpets and DuPont under the name Sorona to make special textile fibers.

3.4.3.5 Acetone

Acetone is the simplest and most important ketone. It is colorless, flammable liquid miscible in water and many other organic solvents such as ether, methanol, and ethanol. Acetone is a chemical intermediate for the manufacture of methacrylates, methyl isobutyl ketone, bisphenol A, and methyl butynol, among others. It is also used as solvent for resins, paints, varnishes, lacquers, nitrocellulose, and cellulose acetate. Acetone can be produced from biomass by fermentation of starch or sugars via the acetone–butanol–ethanol (ABE) fermentation process (Moreira, 1983). This is discussed in detail in Section 3.4.4.1.

3.4.4 Four-Carbon Compounds

3.4.4.1 Butanol

Butanol or butyl alcohol can be produced by the fermentation of carbohydrates with bacteria, yielding a mixture of acetone and butyl alcohol (Wells, 1999). Synthetically, butyl alcohol can be produced by the hydroformylation of propylene, known as the oxo process, followed by the hydrogenation of the aldehydes formed yielding a mixture of n- and iso-butyl alcohol. The use of rhodium catalysts maximizes the yield of n-butyl alcohol. The principal use of n-butyl alcohol is as solvent. Butyl alcohol–butyl acetate mixtures are good solvents for nitrocellulose lacquers and coatings. Butyl glycol ethers formed by the reaction of butyl alcohol and ethylene oxide are used in vinyl and acrylic paints and lacquers, and to solubilize organic surfactants in surface cleaners. Butyl acrylate and methacrylate are important commercial derivatives that can be used in emulsion polymers for latex paints, in textile manufacturing, and in impact modifiers for rigid PVC. Butyl esters of acids like phthalic, adipic, and stearic acid can be used as plasticizers and surface coating additives.

The process for the fermentation of butanol is also known as Weizmann process or ABE fermentation. Butyric-acid-producing bacteria belong to the *Clostridium* genus. Two of the most common butyric-acid-producing bacteria are *Clostridium butylicum* and *C. acetobutylicum*. *C. butylicum* can produce acetic acid, butyric acid, 1-butanol, 2-propanol, H_2, and CO_2 from glucose, and *C. acetobutylicum* can produce acetic acid, butyric acid, 1-butanol, acetone, H_2, CO_2, and small amounts of ethanol from glucose (Klass, 1998). The acetone–butanol fermentation by *C. acetobutylicum* was the only commercial process for producing industrial chemicals by anaerobic bacteria that uses a monoculture. Acetone was produced from corn fermentation during World War I for the manufacture of cordite. This process for the fermentation of corn to butanol and acetone was discontinued in the 1960s for unfavorable economics due to the chemical synthesis of these products from petroleum feedstock.

The fermentation process involves conversion of glucose to pyruvate via the Embden-Meyerhof-Parnas (EMP) pathway; the pyruvate molecule is then broken to acetyl-CoA with the release of carbon dioxide and hydrogen (Moreira, 1983). Acetyl-CoA is a key intermediate in the process serving as a precursor to acetic acid, ethanol. The formation of butyric acid and neutral solvents (acetone and butanol) occurs in two steps. Initially, two acetyl CoA molecules combine to form acetoacetyl-CoA, thus initiating a cycle leading to the production of butyric acid. A reduction in the pH of the system occurs as a result of increased acidity. At this step in fermentation, a new enzyme system is activated, leading to the production of acetone and butanol. Acetoacetyl-CoA is diverted by a transferase system to the production of acetoacetate, which is then decarboxylated to acetone. Butanol is produced by reducing the butyric acid in three reactions. Detailed descriptions of batch fermentation, continuous fermentation, and extractive fermentation systems are given by Moreira (1983).

DuPont and BP are working with British Sugar to produce 30,000 MT/year of biobutanol using corn, sugarcane, or beet as feedstock (D'Aquino, 2007). U.K. biotechnology firm Green Biologics has demonstrated the conversion of cellulosic biomass to butanol, known as Butafuel. Butanol can also be used as a fuel additive instead of ethanol. Butanol is less volatile, not sensitive to water, less hazardous to handle, less flammable, has a higher octane number, and can be mixed with gasoline in any proportion when compared to ethanol. The production cost of butanol from bio-based feedstock is reported to be $3.75 per gal (D'Aquino, 2007).

3.4.4.2 Succinic Acid

Succinic acid, a DOE top-30 candidate, is an intermediate for the production of a wide variety of chemicals as shown in Figure 3.13. Succinic acid is produced biochemically from glucose using an engineered form of the organism *Anaerobiospirillum succiniciproducens* or an engineered *Eschericia coli* strain developed by DOE laboratories (Werpy et al., 2004).

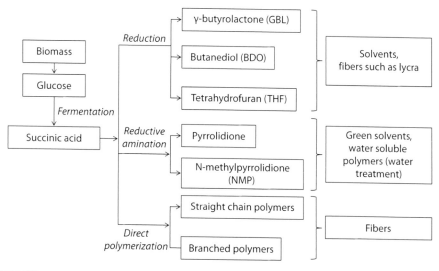

FIGURE 3.13

Succinic acid production and derivatives. (From Werpy, T. et al., Top value added chemicals from biomass: Volume 1 results of screening for potential candidates from sugars and synthesis gas, energy efficiency and renewable energy (U.S. DOE) http://www1.eere.energy.gov/biomass/pdfs/35523.pdf, accessed May 8, 2010, 2004.)

Zelder (2006) discusses BASF's efforts to develop bacteria that convert biomass to succinate and succinic acid. The bacteria convert the glucose and carbon dioxide with an almost 100% yield into the C4 compound succinate. BASF is also developing a chemistry that will convert the fermentation product into succinic acid derivatives, butanediol and tetrahydrofuran. Succinic acid can also be used as a monomeric component for polyesters.

Snyder (2007) reports the successful operation of a 150,000 L fermentation process that uses a licensed strain of *E. coli* at the Argonne National Laboratory. Opportunities for succinic acid derivatives include maleic anhydride, fumaric acid, dibase esters, and others in addition to the ones shown in Figure 3.13. The overall cost of fermentation is one of the major barriers to this process. Low-cost techniques are being developed to facilitate the economical production of succinic acid (Werpy et al., 2004).

BioAmber, a joint venture of Diversified Natural Products (DNP) and Agro Industries Recherche et Development will construct a plant that will produce 5000 MT/year of succinic acid from biomass in Pomacle, France (Ondrey, 2007d). The plant is scheduled for start-up in mid-2008. Succinic acid from BioAmber's industrial demonstration plant is made from sucrose or glucose fermentation using patented technology from the U.S. DOE in collaboration with Michigan State University. BioAmber will use patented technology developed by Guettler et al. (1996), for the production of succinic acid using biomass and carbon dioxide.

3.4.4.3 Aspartic Acid

Aspartic acid is an α-amino acid manufactured either chemically by the amination of fumaric acid with ammonia or the biotransformation of oxaloacetate in the Krebs cycle with fermentative or enzymatic conversion (Werpy et al., 2004). It is one of the chemicals identified in the DOE top-12 value-added chemicals from biomass list. Aspartic acid can be used as sweeteners and salts for chelating agents. The derivatives of aspartic acid include amine butanediol, amine tetrahydrofuran, aspartic anhydride, and polyaspartic with new potential uses as biodegradable plastics.

3.4.5 Five-Carbon Compounds

3.4.5.1 Levulinic Acid

Levulinic acid (LA) was first synthesized from fructose with hydrochloric acid by the Dutch scientist G.J. Mulder in 1840 (Kamm et al., 2006). It is also known as 4-oxopentanoic acid or γ-ketovaleric acid. In 1940, the first commercial-scale production of LA in an autoclave was started in the United States by A.E. Stanley, Decatur, Illinois. LA has been used in food, fragrance, and specialty chemicals. The derivatives have a wide range of applications like polycarbonate resins, graft copolymers, and biodegradable herbicide.

LA is formed by the treatment of six-carbon sugar carbohydrates from starch or lignocellulosics with acid (Figure 3.14). Five-carbon sugars derived from hemicelluloses like xylose and arabinose can also be converted to LA by the addition of a reduction step subsequent to acid treatment. The following steps are used for the production of LA from hemicellulose (Klass, 1998). Xylose from hemicelluloses is dehydrated by acid treatment to yield 64 wt% of furan-substituted aldehyde (furfural). Furfural undergoes catalytic decarbonylation to form furan. Furfuryl alcohol is formed by catalytic hydrogenation of the aldehyde group in furfural. Tetrahydrofurfuryl alcohol is formed after further catalytic hydrogenation of furfural. LA (γ-ketovaleric acid) is formed from tetrahydrofurfuryl alcohol on treatment with dilute acid. Werpy et al. (2004) report an overall yield of 70% for the production of LA.

A number of large-volume chemical markets can be addressed from the derivatives of LA (Werpy et al., 2004). Figure 3.14 gives the production of LA from hemicellulose and the derivatives of LA. In addition to the chemicals in the figure, the following derivative chemicals of LA also have a considerable market. Methyltetrahydrofuran and various levulinate esters can be used as gasoline and biodiesel additives, respectively. δ-aminolevulinic acid is a herbicide, and targets a market of 200–300 million lb per year at a projected cost of $2.00–$3.00 per lb. An intermediate in the production of δ-aminolevulinic acid is β-acetylacrylic acid. This material could be used in the production of new acrylate polymers, addressing a market of 2.3 billion lb per year with values of about $1.30 per lb. Diphenolic acid is of particular

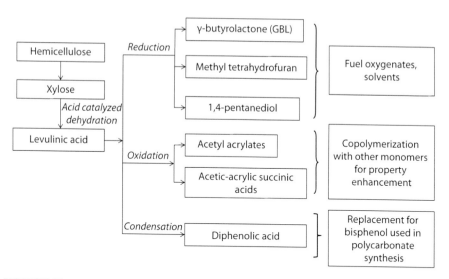

FIGURE 3.14
Production and derivatives of Levulinic acid. (Adapted from Werpy, T. et al., Top value added chemicals from biomass: Volume 1 results of screening for potential candidates from sugars and synthesis gas, energy efficiency and renewable energy (U.S. DOE) http://www1.eere.energy.gov/biomass/pdfs/35523.pdf, accessed May 8, 2010, 2004.)

interest because it can serve as a replacement for bisphenol A in the production of polycarbonates. The polycarbonate resin market is almost 4.0 billion lb/year, with product values of about $2.40 per lb. New technology also suggests that LA could be used for the production of acrylic acid via oxidative processes. LA is also a potential starting material for production of succinic acid. The production of LA-derived lactones offers the opportunity to enter a large solvent market, as these materials could be converted into analogs of N-methylpyrrolidinone. Complete reduction of LA leads to 1,4-pentanediol, which could be used for the production of new polyesters.

An LA production facility has been built in Caserta, Italy by Le Calorie, a subsidiary of Italian construction Immobilgi (Ritter, 2006). The plant is expected to produce 3000 ton/year of LA from local tobacco bagasse and paper mill sludge through a process developed by Biofine Renewables.

Hayes et al. (2006) give the details of the Biofine process for the production of LA. This process received the Presidential Green Chemistry Award in 1999. The Biofine process involves a two-step reaction in a two-reactor design scheme. The feedstock comprises of 0.5–1.0 cm biomass particles of cellulose and hemicellulose conveyed to a mixing tank by high-pressure air injection system. The feed is mixed with 2.5%–3% recycled sulfuric acid in the mixing tank. The feed is then transferred to the reactors. The first reactor is a plug flow reactor, where first-order acid hydrolysis of the carbohydrate polysaccharides occurs to soluble intermediates like HMF. The residence time in the reactor is 12 s at a temperature of 210°C–220°C and pressure of 25 bar.

The diameter of the reactor is small to enable the short residence time. The second reactor is a back mixed reactor operated at 190°C–200°C and 14 bar with a residence time of 20 min. LA is formed in this reactor favored by the completely mixed conditions of the reactor. Furfural and other volatile products are removed and the tarry mixture containing LA is passed to a gravity separator. The insoluble mixture from this unit goes to a dehydration unit where the water and volatiles are boiled off. The crude LA obtained is 75% and can be purified to 98% purity. The residue formed is a bone-dry powdery substance or char with calorific value comparable to bituminous coal and can be used in syngas production. Lignin is another by-product that can be converted to char and burned or gasified. The Biofine process uses polymerization inhibitors that convert around 50% of both five- and six-carbon sugars to LA.

3.4.5.2 Xylitol/Arabinitol

Xylitol and arabinitol are hydrogenation products from the corresponding sugars xylose and arabinose (Werpy et al., 2004). Currently, there is limited commercial production of xylitol and no commercial production of arabinitol. The technology required to convert the five-carbon sugars, xylose and arabinose, to xylitol and arabinitol, can be modeled based on the conversion of glucose to sorbitol. The hydrogenation of the five-carbon sugars to sugar alcohols occurs with one of many active hydrogenation catalysts such as nickel, ruthenium, and rhodium. The production of xylitol for use as a building block for derivatives essentially requires no technical development. Derivatives of xylitol and arabinitol are shown in Figure 3.15.

3.4.5.3 Itaconic Acid

Itaconic acid is a five-carbon dicarboxylic acid, also known as methyl succinic acid, and has the potential to be a key building block for deriving both commodity and specialty chemicals. The basic chemistry of itaconic acid is similar to that of the petrochemicals derived maleic acid/anhydride. The chemistry of itaconic acid to the derivatives is shown in Figure 3.16. Itaconic acid is currently produced via fungal fermentation and is used primarily as a specialty monomer.

The major applications include use as a copolymer with acrylic acid and in styrene–butadiene systems. The major technical hurdles for the development of itaconic acid as a building block for commodity chemicals include the development of very low cost fermentation routes. The primary elements of improved fermentation include increasing the fermentation rate, improving the final titer, and potentially increasing the yield from sugar. There could also be some cost advantages associated with an organism that could utilize both five-carbon and six-carbon sugars.

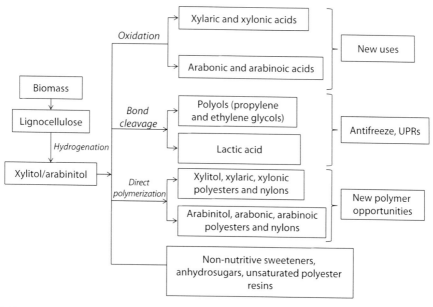

FIGURE 3.15
Production and derivatives of xylitol and arabinitol. (Adapted from Werpy, T. et al., Top value added chemicals from biomass: Volume 1 results of screening for potential candidates from sugars and synthesis gas, energy efficiency and renewable energy (U.S. DOE) http://www1.eere.energy.gov/biomass/pdfs/35523.pdf, accessed May 8, 2010, 2004.)

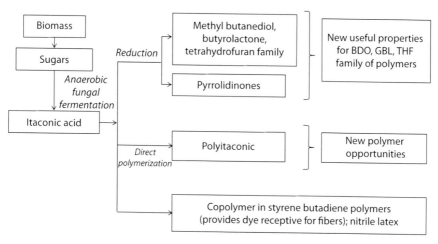

FIGURE 3.16
Production and derivatives of itaconic acid. (Adapted from Werpy, T. et al., Top value added chemicals from biomass: Volume 1 results of screening for potential candidates from sugars and synthesis gas, energy efficiency and renewable energy (U.S. DOE) http://www1.eere.energy.gov/biomass/pdfs/35523.pdf, accessed May 8, 2010, 2004.)

3.4.6 Six-Carbon Compounds

3.4.6.1 Sorbitol

Sorbitol is produced by the hydrogenation of glucose (Werpy et al., 2004). The production of sorbitol is practiced commercially by several companies and has a current production volume on the order of 200 million lb annually. The commercial processes for sorbitol production are based on batch technology and Raney nickel is used as the catalyst. The batch production ensures complete conversion of glucose.

Technology development is possible for conversion of glucose to sorbitol in a continuous process instead of a batch process. Engelhard (now a BASF-owned concern) has demonstrated that the continuous production of sorbitol from glucose can be done using a ruthenium on carbon catalyst (Werpy, 2004). The yields demonstrated were near 99% with very high weight hourly space velocity.

Derivatives of sorbitol include isosorbide, propylene glycol, ethylene glycol, glycerol, lactic acid, anhydrosugars, and branched polysaccharides (Werpy, 2004). The derivatives and their uses are described in Figure 3.17.

3.4.6.2 2,5-Furandicarboxylic Acid

FDCA is a member of the furan family, and is formed by an oxidative dehydration of glucose (Werpy, 2004). The production process uses oxygen, or electrochemistry. The conversion can also be carried out by oxidation of 5-hydroxymethylfurfural, which is an intermediate in the conversion of six-carbon sugars into LA. Figure 3.18 shows some of the potential uses of FDCA.

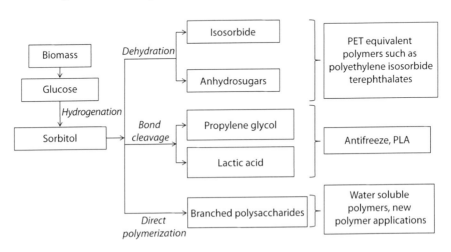

FIGURE 3.17
Production and derivatives of sorbitol. (Adapted from Werpy, T. et al., Top value added chemicals from biomass: Volume 1 results of screening for potential candidates from sugars and synthesis gas, energy efficiency and renewable energy (U.S. DOE) http://www1.eere.energy.gov/biomass/pdfs/35523.pdf, accessed May 8, 2010, 2004.)

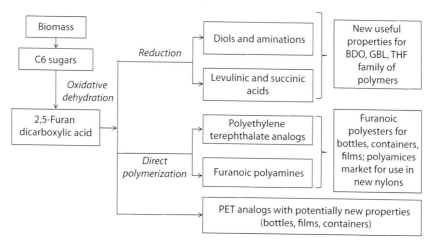

FIGURE 3.18

Production and derivatives of 2,5-FDCA. (From Werpy, T. et al., Top value added chemicals from biomass: Volume 1 results of screening for potential candidates from sugars and synthesis gas, energy efficiency and renewable energy (U.S. DOE) http://www1.eere.energy.gov/biomass/pdfs/35523.pdf, accessed May 8, 2010, 2004.)

FDCA resembles and can act as a replacement for TPA, a widely used component in various polyesters, such as PET and polybutylene terephthalate (PBT) (Werpy et al., 2004). PET has a market size approaching 4 billion lb per year, and PBT is almost a billion lb per year. The market value of PET polymers varies depending on the application but is in the range of $1.00–$3.00 per lb for use as films and thermoplastic engineering polymers. PET and PBT are manufactured industrially from TPA, which, in turn, is manufactured from toluene (Wells, 1999). Toluene is obtained industrially from the catalytic reforming of petroleum or from coal. Thus, FDCA derived from biomass can replace the present market for petroleum-based PET and PBT.

FDCA derivatives can be used for the production of new polyester, and their combination with FDCA would lead to a new family of completely biomass-derived products. New nylons can be obtained from FDCA, either through reaction of FDCA with diamines, or through the conversion of FDCA to 2,5-bis(aminomethyl)-tetrahydrofuran. The nylons have a market of almost 9 billion lb/year, with product values between $0.85 and $2.20 per lb, depending on the application.

3.5 Biopolymers and Biomaterials

Section 3.4 discussed the major industrial chemicals that can be produced from biomass. This section will focus on various biomaterials that can be produced from biomass. Around 13,000 million MT of polymers were made

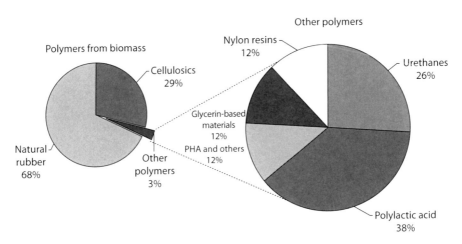

FIGURE 3.19
Production of polymers from biomass in 2007 (13,000 million MT) and breakdown of "other polymers." (From Tullo, A.H., *Chem. Eng. News*, 86(39), 21, 2008.)

from biomass in 2007, as shown in Figure 3.19, out of which 68% is natural rubber. New polymers from biomass, which attribute to a total of 3% of the present market share of bio-based polymers, consist of urethanes, glycerin-based materials, nylon resins, PHA, and PLA (Tullo, 2008).

A new product from a new chemical plant is expected to have a slow penetration (less than 10%) of the existing market for the chemical that it replaces. However, once the benefits of a new product is established, for example, replacing glass in soda bottles with petrochemical-based PET, the growth is rapid over a short period of time. Most renewable processes for making polymers have an inflection point at $70 per barrel of oil, above which the petroleum-based process costs more than the renewable process. For example, above $80 per barrel of oil, PLA is cheaper than PET (Tullo, 2008). In Table 3.2, a list of companies is given that have planned new chemical production based on biomass feedstock along with capacity and projected start-up date. Government subsidies and incentives tend to be of limited time and short-term value. Projected bulk chemicals from bio-based feedstocks are ethanol, butanol, and glycerin.

Some of these biomaterials have been discussed in association with their precursor chemicals in Section 3.4. The important biomaterials that can be produced from biomass, include wood and natural fibers, isolated and modified biopolymers, agromaterials, and biodegradable plastics (Vaca-Garcia, 2008). These are outlined in Figure 3.20. The production process for poly(3-hydroxybutyrate) is given by Rossell et al. (2006), and a detailed review for PHA as commercially viable replacement for petroleum-based plastics is given by Snell and Peoples (2009).

Lignin has a complex chemical structure and various aromatic compounds can be produced from lignin. Current technology is underdeveloped for the

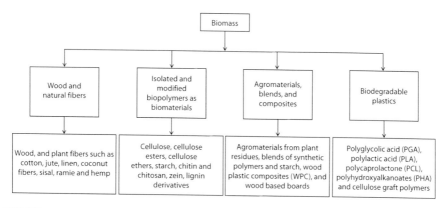

FIGURE 3.20
Biomaterials from biomass. (From Vaca-Garcia, C., Biomaterials, in: Clark, J.H. and Deswarte, F.E.I. (eds.), *Introduction to Chemicals from Biomass*, Wiley, West Sussex, U.K., ISBN 978-0-470-05805-3, 2008.)

industrial-scale production of lignin-based chemicals, but there is considerable potential to supplement the BTX chain of chemicals currently produced from fossil-based feedstock.

3.6 Natural-Oil-Based Polymers and Chemicals

Natural oils are mainly processed for chemical production by hydrolysis and or transesterification. Oil hydrolysis is carried out in pressurized water at 220°C, by which fatty acids and glycerol are produced. The main products that can be obtained from natural oils are shown in Figure 3.21. Transesterification is the acid catalyzed reaction in the presence of an alcohol to produce fatty acid alkyl esters and glycerol. Fatty acids can be used for the production of surfactants, resins, stabilizers, plasticizers, dicarboxylic acids, etc. Epoxidation, hydroformylation, and methesis are some of the other methods to convert oils to useful chemicals and materials. Sources of natural oil include soybean oil, lard, canola oil, algae oil, and waste grease, among others.

Soybean oil can be used to manufacture molecules with multiple hydroxyl groups known as polyols (Tullo, 2007b). Polyols can be reacted with isocyanates to make polyurethanes. Soybean oil can also be introduced in unsaturated polyester resins to make composite parts. Soybean oil-based polyols have the potential to replace petrochemical-based polyols derived from propylene oxide in polyurethane formulations (Tullo, 2007b). The annual market for conventional polyols is 3 billion lb in the United States and 9 billion lb globally.

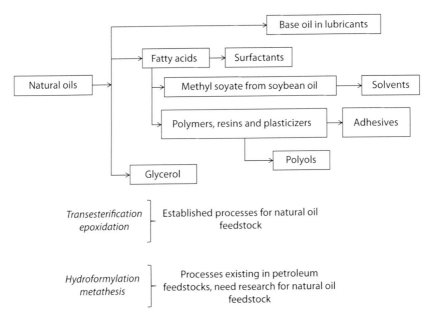

FIGURE 3.21
Natural-oil-based chemicals.

Dow Chemicals, the world's largest manufacturer of petrochemical polyols, also started the manufacture of soy-based polyols (Tullo, 2007b). Dow uses the following process for the manufacture of polyols. The transesterification of triglycerides gives methyl esters, which are then hydroformylated to add aldehyde groups to unsaturated bonds. This is followed by a hydrogenation step, which converts the aldehyde group to alcohols. The resultant molecule is used as a monomer with polyether polyols to build a new polyol. Urethane Soy Systems manufactures soy-based polyols at Volga, South Dakota with a capacity of 75 million lb per year and supplies them to Lear Corp., manufacturer of car seats for Ford Motor Company. The company uses two processes for the manufacture of polyols; an autoxidation process replacing unsaturated bonds in the triglycerides with hydroxyl groups and a transesterification process where rearranged chains of triglycerides are reacted with alcohols. Bio-Based Technologies supply soy polyols to Universal Textile Technologies for the manufacture of carpet backing and artificial turf. Johnson Controls uses their polyols to make 5% replaced foam automotive seats. The company has worked with BASF and Bayer Material Science for the conventional polyurethanes and now manufactures the polyols by oxidizing unsaturated bonds of triglycerides. The company has three families of products with 96%, 70%, and 60% of bio-based content.

Soybean oil can be epoxidized by a standard epoxidation reaction (Wool and Sun, 2005). The epoxidized soybean oil can then be reacted with acrylic acid to form acrylated epoxidized soybean oil (AESO). The acrylated epoxidized triglycerides can be used as alternative plasticizers in PVC as a replacement for phthalates.

Aydogan et al. (2006) give a method for the potential of using dense (sub/supercritical) CO_2 in the reaction medium for the addition of functional groups to soybean oil triglycerides (SOT) for the synthesis of rigid polymers. The reaction of SOT with $KMnO_4$ in the presence of water and dense CO_2 is presented in this paper. Dense CO_2 is utilized to bring the soybean oil and aqueous $KMnO_4$ solution into contact. Experiments are conducted to study the effects of temperature, pressure, $NaHCO_3$ addition, and $KMnO_4$ amount on the conversion (depletion by bond opening) of soybean-triglyceride double bonds (STDB). The highest STDB conversions, about 40%, are obtained at the near-critical conditions of CO_2. The addition of $NaHCO_3$ enhances the conversion; one mole of $NaHCO_3$ per mole of $KMnO_4$ gives the highest benefit. Increasing $KMnO_4$ up to 10% increases the conversion of STDB.

Holmgren et al. (2007) discuss the uses of vegetable oils as feedstock for refineries. Four processes are outlined as shown in Figure 3.22. The first process is the production of fatty acid methyl esters by transesterification process. The second process is the UOP/Eni Renewable Diesel Process that processes vegetable oils combined with crude diesel through a hydroprocessing unit. The third and fourth processes involve the

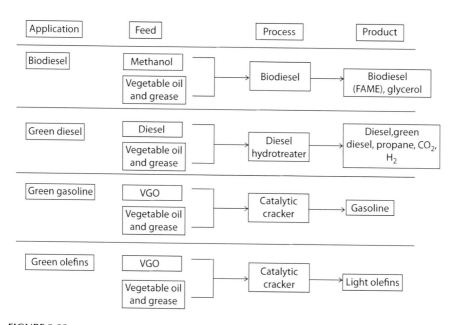

FIGURE 3.22
Processing routes for vegetable oils and grease. (From Holmgren, J. et al., *Petrol. Technol. Quart.*, 12(4), 119, 2007.)

catalytic cracking of pretreated vegetable oil mixed with virgin gas oil (VGO) to produce gasoline, olefins, light cycle oil, and clarified slurry oil. Petrobras has a comparable H-Bio process where vegetable oils can also be used directly with petroleum diesel fractions.

3.7 Summary

Various fractions in petroleum and natural gas are used for the manufacture of numerous chemicals. Biomass can be considered to have similar fractions. All types of biomass contain cellulose, hemicellulose, lignin, fats and lipids, and proteins as main constituents in varying ratios. Separate methods are used to convert these fractions into chemicals. Biomass containing mainly cellulose, hemicellulose, and lignin, referred to as lignocellulosics, can also undergo various pretreatment procedures to separate the components. Steam hydrolysis breaks some of the bonds in cellulose, hemicellulose, and lignin. Acid hydrolysis solubilizes the hemicellulose by depolymerizing hemicellulose to five-carbon sugars such as pentose, xylose, and arabinose. This can be separated for extracting the chemicals from five-carbon sugars. The cellulose and lignin stream is then subjected to enzymatic hydrolysis where cellulose is depolymerized to six-carbon glucose and other six-carbon polymers. This separates the cellulose stream from lignin. Thus, three separate streams can be obtained from biomass. The cellulose and hemicellulose monomers, glucose and pentose, can undergo fermentation to yield chemicals like ethanol, succinic acid, butanol, xylitol, arabinitol, itaconic acid, and sorbitol. The lignin stream is rich in phenolic compounds, which can be extracted, or the stream can be dried to form char and used for gasification to produce syngas.

Biomass-containing oils, lipids, and fats can be transesterified to produce fatty acid methyl and ethyl esters and glycerol. Vegetable oils can be directly blended in petroleum diesel fractions and catalytic cracking of these fractions produce biomass derived fuels. Algae have shown great potential for use as a source of biomass, and there have been algae strains that can secrete oil, reducing process costs for separation. Algae grow fast (compared to food crops), fix atmospheric and power-plant flue gas carbon sources, and do not require freshwater sources. However, algae production technology on an industrial scale for the production of chemicals and fuel is still in the research and development stage. Growth of algae for biomass is a promising field of research.

The glycerol from transesterification can be converted to propylene glycol, 1,3-propanediol, and other compounds that can replace current natural-gas-based chemicals. Vegetable oils, particularly soybean oil, have been considered for various polyols with a potential to replace propylene-oxide-based chemicals.

This chapter has outlined the various chemicals that are currently produced from petroleum-based feedstock that can be produced from biomass as feedstock. New polymers and composites from biomass are continually being developed, which can replace the needs of current fossil-feedstock-based chemicals.

4

Simulation for Bioprocesses

4.1 Introduction

This chapter describes the development of industrial-scale process designs for fermentation, anaerobic digestion, and transesterification processes for the production of chemicals from biomass. The chemicals produced from biomass were ethanol from corn and corn stover, fatty acid methyl ester (FAME) and glycerol from transesterification, acetic acid from anaerobic digestion, syngas from gasification of biomass, algae oil production, ethylene from ethanol, and propylene glycol from glycerol. The corn stover fermentation process, acetic acid process, FAME and glycerol process, propylene glycol process, and ethylene from ethanol process were designed in Aspen HYSYS®. The process cost estimation for these processes were made in Aspen ICARUS®. The corn ethanol process model was based on the USDA process for dry-grind ethanol, and the process model was obtained in SuperPro Designer® from Intelligen Inc. (Intelligen, 2009). The models for algae oil production and gasification of biomass processes were black box models since there was limited knowledge of processing details.

Figure 4.1 shows a conceptual design of the bioprocesses and the interconnections that were considered initially for inclusion in the chemical production complex. This conceptual design was developed from the processes identified in Chapters 2 and 3.

Bioprocess development and design is an ongoing field of research and is limited in information. The books by Petrides (2002) and Heinzle (2007) were helpful to gain insight in the development of bioprocess models. The detail of a process can be viewed either from a top-down or a bottom-up approach. The top-down and bottom-up are strategies of information processing and knowledge ordering mostly involving software, but they also involve other humanistic and scientific theories. These two approaches are discussed with respect to the research undertaken.

A top-down approach is essentially the breaking down of a system to gain insight into its compositional subsystems. In a top-down approach, the overview of a system is first formulated, without giving any details of the system.

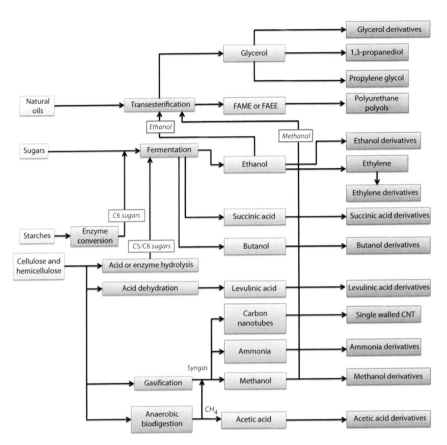

FIGURE 4.1
(See color insert.) Conceptual design of biomass-feedstock-based chemical production.

A bottom-up approach, on the other hand, is the piecing together of small systems to give rise to bigger systems. This makes the original system a collection of subsystems connected by detailed process knowledge of each of the subsystems.

Figure 4.1 can be considered as a top-down approach by looking at biomass feedstock for chemical products. Each of the boxes given in orange (transesterification, fermentation, acid hydrolysis, enzymatic hydrolysis, acid dehydration, gasification, and anaerobic digestion) is a black box model at the initial stage of research. This means that the raw materials going into the process and the products from the process are known, without definite knowledge of the processes that convert the raw materials to the products. The next step was a bottom-up approach, where each of these processes was modeled using a process simulator.

To convert the black box models in Figure 4.1 into process flow models (also called white box models) means developing detailed process knowledge of

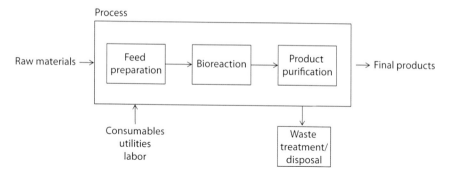

FIGURE 4.2
Process boundaries and material balance regions of a bioprocess. (Adapted from Heinzle, E. et al., *Development of Sustainable Bioprocesses: Modeling and Assessment*, John Wiley & Sons, England, U.K., ISBN 978-0-470-01559-9, 2007.)

the chemical reactions, mass flow rates, and energy requirements. Each of the biomass processes shown in Figure 4.1 was refined in greater detail to produce process flow models.

The different components of a bioprocess model can be outlined as given in Figure 4.2 (Heinzle et al., 2007). Raw materials enter a process and are converted to products through a series of reaction and purification steps. For bioprocesses, there is almost always a need to have a feed preparation, known as pretreatment, followed by the main biomass reaction, and then downstream processing for purification of products. Raw materials and additional materials like solvents and mineral salts are consumed in bioprocesses, and waste is generated from the processes (Heinzle et al., 2007).

Process modeling and simulation is used for optimization and identifying potential improvements in a process. The standard procedure to develop a detailed knowledge of the processes is to use process simulation software. Several process simulators are commercially available. These include Aspen Plus®, Aspen HYSYS, SuperPro Designer, PRO II®, PROSYS®, CHEMCAD®, among others. A choice of the software used for the process simulation is based on details required for process equipment, thermodynamic package, and cost of software. Three of the aforementioned tools, Aspen Plus, Aspen HYSYS, and SuperPro Designer, were compared for modeling bioprocesses. The main aim was to obtain plant models that predicted the flow rates of components and determined the energy requirements for the process. Aspen HYSYS was chosen among these three due to superior features for bioprocesses compared to Aspen Plus and SuperPro Designer. Aspen HYSYS had to advantage to export the process design to Aspen ICARUS for cost estimation. The cost estimation was based on equipment costs given by ICARUS, raw material costs provided by user, and utility costs from ICARUS database.

The bioprocess capacities developed in HYSYS used capacities of existing or proposed industrial-scale plants producing the same chemical. A difficulty in modeling the bioprocesses using conventional process simulation software was to obtain the thermodynamic package that incorporated biological materials in the design. The fermentation, anaerobic digestion, and transesterification processes had least thermodynamics properties available for modeling. Detailed discussions with professionals at Dechema (Sass and Meier, 2010), a leading source for thermodynamic databases, revealed that thermodynamic property estimations of biomass feedstock are difficult. There is almost no information available in their database on properties of cellulose, and no information was found for hemicellulose and lignin. The same applied to natural oils, for example, soybean oil.

The thermodynamic packages incorporated in Aspen HYSYS had limited thermodynamic data on the biomass components. Most of the biomass components were manually entered using the user-defined method and the structures of each compound were constructed using standard software (SYMYX Draw 3.2). Property estimation methods for HYSYS were used to predict the interaction parameters. The UNIQUAC thermodynamic package was used in all the processes, with UNIFAC methods for vapor–liquid equilibrium (VLE) estimations.

Input flow rates were specified, and suitable reactors and separation equipment were used wherever applicable. Conversion reactors were used for single reactions. Tanks were used for multiple reactions, as conversion reactors in HYSYS were incapable of handling multiple reactions. Adjusters were used to set the outlet temperature, reactor temperature, and separation extents required in separation equipment. These were specified by logical relations in the adjusters. Recycle for water, solvents, glycerol, and other components were used wherever necessary. Heat integration was used in the processes to minimize the energy requirements.

To do the initial cost estimations of the processes, the process model was exported to Aspen ICARUS. Aspen ICARUS is a sophisticated cost-estimation tool and widely used in the industry. Raw material costs were provided to the ICARUS cost-estimation model. The utility required by the processes was determined by Aspen ICARUS, and the cost for the process was calculated. The equipment costs were determined by Aspen ICARUS, and for special equipment like perfect separators, the equipment was chosen from ICARUS database, which resembled the equipment modeled. This was done for the centrifuges in the design, which were used for separation of solids.

Detailed description of the processes modeled in HYSYS is given in the following sections beginning with a brief discussion on the basis for design. This is followed by the literature sources for the design. Then the details of process flow are given. For the fermentation, anaerobic digestion, and transesterification processes, the process is divided into pretreatment, fermentation, and purification sections. Cost estimation using Aspen ICARUS was performed to determine the operating costs. These processes give generic plant designs with equipment and unit operations necessary to convert biomass feedstock to chemicals.

4.2 Ethanol Production from Corn Stover Fermentation

The process for fermentation of biomass to ethanol was designed in HYSYS based on the description given by the Department of Energy (Aden et al., 2002). The UNIQUAC thermodynamic model was used for estimating the interactions between reaction components. The biomass chosen as feed was corn stover. This design can use other feedstocks such as corn and sugarcane. Corn stover has a complex composition including cellulose, hemicellulose, and lignin. Corn is composed of starch, and sugarcane is composed of glucose. The corn stover has the highest complexity in composition, and the design can then be used for corn and sugarcane as feedstock.

The plant capacity was based on the processing of 2000 MT/day of corn stover (Aden et al., 2002), producing 54 MMgy (million gallons per year) of ethanol. Capacities of existing and under-construction ethanol plants in the United States range from 1.4 to 420 MMgy (EPM, 2009). Thus, the plant is a mid-sized ethanol plant in the United States.

Corn stover is composed of mainly cellulose, hemicellulose, and lignin. Cellulose and hemicellulose are organic compounds with the formula $(C_6H_{10}O_5)_n$ and $(C_5H_8O_4)_m$, respectively. $C_6H_{10}O_5$, also known as glucan, represents the monomer of cellulose, and $C_5H_8O_4$, also known as xylan, represents the monomer of hemicellulose. Aden et al. (2002) use the term glucan to represent cellulose and xylan to represent hemicellulose. The corn stover composition reported by Aden et al. (2002) is given in Table 4.1. The composition of corn stover for the design was adapted from Table 4.1 and is given in Table 4.2. Aden et al. (2002) calculated the unknown soluble solids with a mass balance closure. The acetate, protein, extractives, arabinan, galactan, and mannan are 18.3% of the corn stover and are not standard components in HYSYS. These components were considered as other solids for the HYSYS design. The dry biomass feed was adjusted to have 50% water going into the reactor, as given in Table 4.2.

Cellulose is the polymer of glucan, which, when hydrolyzed, produces glucose. Similarly, hemicellulose is a polymer of xylan, which, when hydrolyzed, gives xylose. The conversion of cellulose and hemicellulose to glucose (six-carbon sugar) and xylose (five-carbon sugar) are the main reactions in the pretreatment section. The reactions used in the design are given in Table 4.3 in the pretreatment section and Table 4.4 in the fermentation section. The conversion to oligomers of xylose and glucose and to furfural and other degradation products was small compared to the main reactions shown in Table 4.4 and, hence, not considered.

TABLE 4.1

Composition of Corn Stover on a Percent
Dry Basis

Component	% Dry Basis
Glucan	37.4
Xylan	21.1
Lignin	18.0
Ash	5.2
Acetate	2.9
Protein	3.1
Extractives	4.7
Arabinan	2.9
Galactan	2.0
Mannan	1.6
Unknown soluble solids	1.1
Moisture	15.0

Source: Aden, A. et al., *Lignocellulosic Biomass to Ethanol Process Design and Economics Utilizing Co-Current Dilute Acid Prehydrolysis and Enzymatic Hydrolysis for Corn Stover*, National Renewable Energy Laboratory, Golden, CO, NREL/TP-510-32438, 2002.

TABLE 4.2

Composition of Corn Stover Used
in HYSYS® Design

Component	% Mass Basis
Glucan	37.4
Xylan	21.1
Lignin	18.0
Ash	5.2
Other Solids	18.3
Mass percent of dry stover	100.00
Composition of feed into reactor	
Mass percent of dry stover	50.00
Water	50.00

4.2.1 Process Description for Ethanol Production from Corn Stover Fermentation

The HYSYS process flow diagram is shown in Figure 4.3. The design has three sections, a pretreatment section, a fermentation section, and a purification section. In the pretreatment section, the wet biomass is converted to

TABLE 4.3

Pretreatment Reactions Used in Corn Stover Fermentation

Vessel	Pretreatment Step	Reaction	Conversion (%)
V-100	Steam hydrolysis	$(Glucan)_n + nH_2O \rightarrow n\ Glucose$	7
V-100	Steam hydrolysis	$(Xylan)_n + nH_2O \rightarrow n\ Xylose$	70
V-102	Enzymatic hydrolysis	$(Glucan)_n + nH_2O \rightarrow n\ Glucose$	90

TABLE 4.4

Seed Production and Fermentation Reactions in Corn Stover Fermentation

Vessel	Step	Reaction	Conversion (%)
V-103 V-104	Seed fermentation (glucose and xylose used to grow the biocatalyst)	$0.56Glucose + 4.69O_2 \rightarrow 3.4CO_2 + 3.33H_2O + 0.23\ Z.\ mobilis$ $0.67Xylose + 4.69O_2 \rightarrow 3.52CO_2 + 3.33H_2O + 0.20\ Z.\ mobilis$	97 95
V-105	Fermentation (glucose and xylose are converted to ethanol)	$5Glucose \rightarrow 3\ Z.\ mobilis + 8.187CO_2$ $2Xylose \rightarrow Z.\ mobilis + 2.729CO_2$ $Glucose \rightarrow 2C_2H_5OH + 2CO_2$ $Xylose \rightarrow 1.68C_2H_5OH + 1.65CO_2$	1 1 99 99

Source: Petrides, D., Corn stover to EtOH-SuperPro Designer® Model, Private Communication, October 11, 2008.

digestible sugars using two pretreatment steps. The cellulose and hemicellulose in biomass are converted to glucose and xylose, respectively. The first pretreatment step, carried out in reactor V-100, was steam hydrolysis where 70% of the hemicellulose was converted to xylose (Petrides, 2008). This step was followed by a second pretreatment step in reactor V-102, known as saccharification (enzymatic hydrolysis) with cellulase enzymes to convert 90% of cellulose to glucose (Aden et al., 2002).

Cellulase enzymes are a collection of enzymes that attack different parts of the cellulose fibers. This collection contains endoglucanases, which attack randomly along cellulose fiber to reduce polymer size rapidly; exoglucanases, which attack the ends of cellulose fibers allowing it to hydrolyze the highly crystalline cellulose; and β-glucosidase, which hydrolyzes cellobiose, an intermediate polymer, to glucose. Several bacteria and fungi produce these enzymes naturally including bacteria in ruminant and termite guts and white rot fungus. The fungus, *Trichoderma reesei*, is used industrially to produce the cellulose enzymes. Genecor International and Novozymes Biotech are the two largest enzyme manufacturers in the world, and they have ongoing research for the production of cost-effective enzymes.

The fermentation section shown in Figure 4.3 followed the pretreatment section. A part of the digested biomass was used for seed production in V-103 and V-104 of the biocatalyst *Zymomonas mobilis* bacterium, which facilitated the fermentation (Aden et al., 2002). In the seed trains, the saccharified slurry

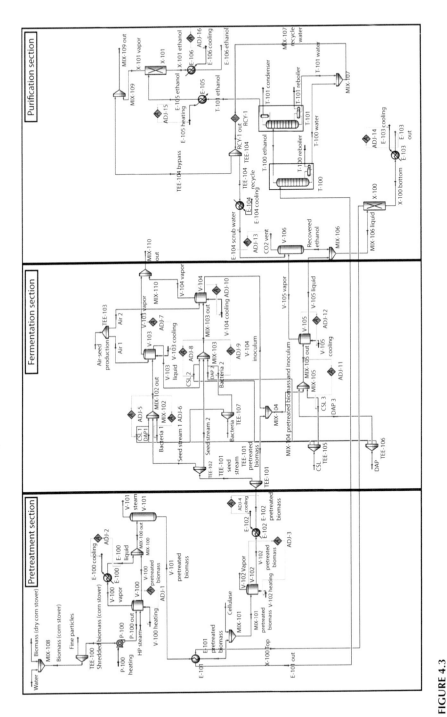

FIGURE 4.3
(See color insert.) Overall process design diagram for ethanol production from corn stover fermentation.

and nutrients were mixed with an initial seed inoculum in small vessels, V-103 and V-104. The result of each seed batch was used as the inoculum for the next seed size increment. This series of scale-ups was continued until the last step was large enough to produce enzymes to support the main ethanol production fermentation. A series of two seed fermentor trains were used for this design (Aden et al., 2002). The final seed was then combined in the mixer MIX-104 with the rest of the biomass for fermentation in V-105.

In the purification section that follows the fermentation section in Figure 4.3, the fermented broth was purified to separate the ethanol from the stream MIX-106 Liquid. Centrifuge X-100 was used to remove unreacted cellulose, hemicellulose, lignin, and other solids. The ethanol from the centrifuge contained mainly water, and this was removed in a reboiled absorption column T-100. The ethanol from the absorption column was transferred to a distillation column T-101 to get to the azeotropic composition of ethanol and water. The azeotrope from the distillation section required further drying, and this was conducted in a molecular sieve tray column X-101. The stillage bottoms from the centrifuge were considered to be a by-product of the process and assigned a cost of distiller's dry grain solids (DDGSs) in the cost analysis section. The simulation is explained in detail in the following three sections.

4.2.1.1 Pretreatment Section

In the pretreatment section shown in the Figure 4.4, water and dry biomass were mixed in MIX-108. The stream, Biomass (corn stover), at the rate of 166,700 kg/h was shredded to small pieces and passed through centrifuge TEE-100. In this design, the biomass comprised of 50% dry corn stover as given in Table 4.2 and 50% water. The fine particles, approximately 10% of the inlet stream, were removed in the centrifuge and the shredded biomass (corn stover) stream was pumped through P-100 to the first pretreatment reactor V-100. The pressure change across the pump was 900 kPa. The pretreatment reactor V-100 was designed for thermal hydrolysis of the corn stover. The design pressure was 1001.3 kPa in the vessel. Adjuster ADJ-1 was used to maintain the temperature at 190°C. High-pressure (HP) steam (1000 kPa and 200°C) was used for hydrolysis. The steam hydrolysis reactions and conversions used in the design were based on Petrides (2008) and are given in Table 4.3; 70% of the hemicellulose and 7% of the cellulose in P-100 Biomass stream were converted to xylose and glucose in the steam hydrolysis reactions.

The V-100 Vapor stream was condensed in E-100 using adjuster ADJ-2. The E-100 Liquid stream was mixed with V-100 Pretreated Biomass in MIX-100 and passed to the flash separator V-101. The MIX-100 Out stream was flash cooled to 101.3 kPa in V-101. Steam at 100°C and 101.3 kPa was recovered from the process. The biomass stream containing glucose, xylose, and unconverted cellulose, lignin, and hemicellulose was cooled in heat exchanger

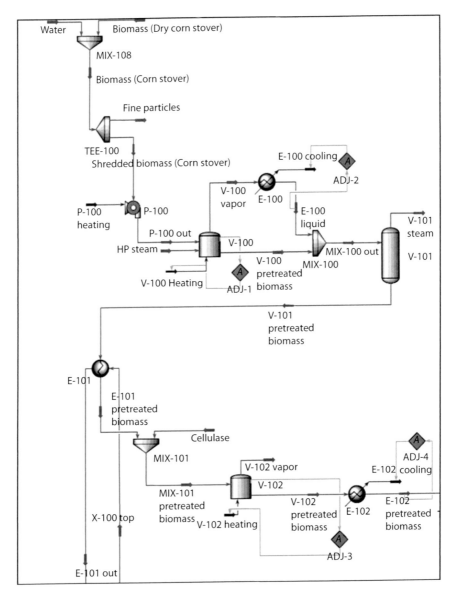

FIGURE 4.4
Pretreatment section for ethanol production from corn stover fermentation.

E-101 to 50°C. The energy from the hot biomass stream was transferred to the wet ethanol stream X-100 Top from the purification section.

The next pretreatment step, enzymatic hydrolysis, also known as sacchari-fication, was carried out in reactor V-102. Cellulase, a mixture of enzymes capable of converting cellulose to glucose, was added to the reactor at the rate of 2692 kg/h (Petrides, 2008). The reaction in V-102 is given in Table 4.3;

90% of the cellulose in MIX-101 Pretreated Biomass was converted to glucose in the saccharification step. The stream V-102 Pretreated Biomass was cooled in E-102 to 41°C. The saccharified slurry in stream E-102 Pretreated Biomass contained the sugars xylose (five-carbon sugar) and glucose (six-carbon sugar) in monomer form and was suitable for fermentation.

4.2.1.2 Fermentation Section

The fermentation section is shown in Figure 4.5. The saccharified slurry in E-102 Pretreated Biomass was split into two parts in TEE-101. The TEE-101 Seed stream, containing 10% of the pretreated biomass, was used for the seed

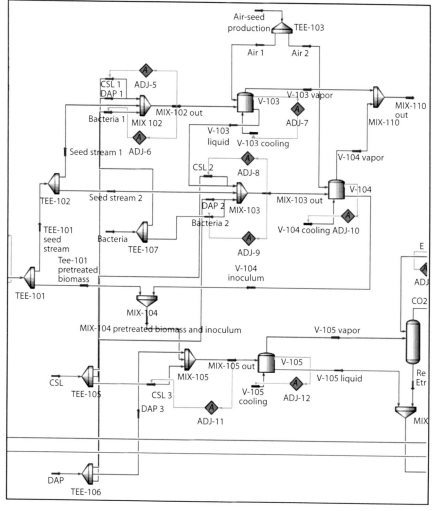

FIGURE 4.5
Fermentation section for ethanol production from corn stover fermentation.

production of bacteria required for fermentation. The recombinant bacterium *Z. mobilis* was used as the biocatalyst for producing ethanol from both glucose and xylose. In this design, two sequential seed fermentation train of vessels, V-103 and V-104, having five reactors in each train, were used for growing the bacteria. The TEE-101 Seed stream was split in TEE-102 with a 20% flow in seed stream 1 and the rest in seed stream 2. The seed reactors were large tanks with internal cooling coils and agitators. The overall conversion was given for the total volume of seed reactors, and this was incorporated in the design instead of five individual reactors for each train. Air, in stream Air-Seed Production, was used for the growth of bacteria. The air was split into two parts in TEE-103; 15% was sent to the reactor V-103 and the rest was sent to the reactor V-104.

An initial 10% volume of inoculum bacteria was fed to each train V-103 and V-104 (Aden et al., 2002). The stream Bacteria 1 constituted 10% standard ideal liquid volume of MIX-102 Out and the stream Bacteria 2 was 10% standard ideal liquid volume of MIX-103 Out. The adjusters ADJ-6 and ADJ-9 were used to modify the standard ideal liquid volume flow for stream Bacteria 1 and Bacteria 2, respectively. Corn steep liquor (CSL) and diammonium phosphate (DAP) were added as nutrients (nitrogen sources) for the growth of the bacteria (Aden et al., 2002). CSL in streams CSL 1 and CSL 2 was added at the rate of 0.5% standard ideal liquid volume of MIX-103 Out and MIX-104 Out, respectively. The adjusters ADJ-5 and ADJ-8 were used to modify the standard ideal liquid volume flow for streams CSL 1 and CSL 2, respectively. DAP addition rate was 0.67 g/L of fermentation broth; DAP 1 rate was 0.67 g/L of seed stream 1 and DAP 2 rate was 0.67 g/L of seed stream 2. The reactions occurring in the seed train are given in Table 4.4. The temperature in seed trains V-103 and V-104 was kept constant at 37°C using adjusters ADJ-7 and ADJ-10, respectively. The total vapor from the seed generation section is obtained in stream MIX-110 Out. The total DAP and CSL used in the model was given in streams DAP and CSL, respectively.

Liquid stream V-105 Liquid contained 11% (mass) ethanol, 63.5% (mass) water, and unreacted biomass and lignin. Vapor stream V-105 Vapor contained 4.2% (mass) ethanol, 93.3% (mass) carbon dioxide, and 2.5% (mass) water vapor. The ethanol was recovered from the vapor and liquid streams and purified as described in the following section.

4.2.1.3 Purification Section

The streams from the reactor V-105 containing ethanol were purified in this section, as shown in Figure 4.6. The fermentor vent, V-105 Vapor, containing carbon dioxide and ethanol was washed with water in a scrubber, V-106. The amount of scrub water, E-104, into the scrubber was determined using ADJ-13 to obtain a 100% recovery of ethanol in the stream Recovered Ethanol. The scrub water used in the scrubber was recovered water from the distillation sections described later. Carbon dioxide with trace amounts of water was vented in the CO_2 vent. The stream V-105 Liquid was mixed with the recovered

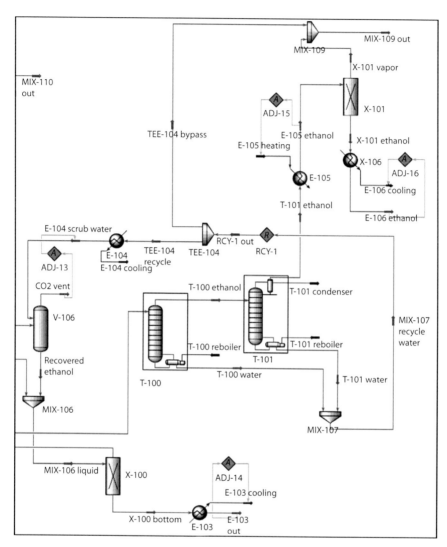

FIGURE 4.6
Purification section for ethanol production from corn stover fermentation.

ethanol stream in MIX-106. The stream MIX-106 Liquid was passed through the centrifuge X-100 to remove the unreacted lignin, cellulose, hemicellulose, ash, and other solids from the stream. For this design, all the solids and soluble impurities were removed in the centrifuge. In Petrides (2008), these impurities were removed as discussed in the following absorber and distillation sections.

The X-100 top stream contained 7% (mass) ethanol and 93% (mass) water. This stream was heated from 32°C to 56°C in heat exchanger E-101, as explained in Section 4.2.1.1. The stream E-101 Out, containing ethanol and water mixture, was transferred to a reboiled absorber, T-100. The absorber contained 10 trays, and the E-101 Out stream was introduced in the top stage. The ethanol–water mixture in the bottom stage was boiled in T-100 Reboiler, and the steam going up in the column helped in stripping the ethanol from the mixture. The pressure in the top stage was maintained at 70.93 kPa, and the pressure at the reboiler was 101.3 kPa. The specification for total recovery of ethanol in stream T-100 Ethanol was set in the absorber, and the required energy in the reboiler was calculated.

The T-100 Ethanol stream contained approximately 24% (mass) ethanol and was sent to distillation (rectification) column T-101. The column T-101 had 50 stages with feed introduced at Tray 25. A recovery rate of 100% and component fraction of 95% (mass) was set for ethanol in stream T-101 Ethanol. These specifications determined a reflux ratio of 10 in the column T-101. Water at 100°C was recovered from the columns in stream T-100 Water and T-101 Water. The water was recycled to the absorber V-106 for washing the carbon dioxide stream in recycle RCY-1. Excess water from the system was recovered in TEE-104 Bypass. The recycle water TEE-104 Recycle was cooled in E-104 to 25°C before sending it to V-106. The overhead vapor stream from distillation unit T-101 Ethanol was superheated to 116°C in E-105 and passed through Delta-T molecular sieve adsorption unit in X-101 (Aden et al., 2002). The adsorption setup was described in Aden et al. (2002), and a perfect separator was used in HYSYS to simulate the adsorption unit. The 99.5% pure ethanol vapor in X-101 Ethanol was condensed by heat exchange in E-106 and pumped to storage. The final output E-106 Ethanol was dehydrated in the adsorber to 99.5% purity. The water was regenerated from the adsorber in stream X-101 Vapor. The final ethanol stream was obtained in E-106 Ethanol at 30°C.

The overall mass balances in the inlet and outlet streams are given in Table 4.5. The detailed stream descriptions are given in Appendix F. Biomass (corn stover) at the rate of 83,300 kg/h was pretreated using HP

TABLE 4.5

Overall Mass Balance for Ethanol Production from Corn Stover Fermentation

Inlet Material Streams	Mass Flow (kg/h)	Outlet Material Streams	Mass Flow (kg/h)
Biomass (dry corn stover)	8.33E+04	Fine particles	8.33E+03
Water	8.33E+04	V-101 steam	2.54E+04
HP steam	6.00E+04	V-102 vapor	0.00E+00
Cellulase	2.69E+03	MIX-110 out	5.52E+04
Air-seed production	5.30E+04	CO_2 vent	1.90E+04
CSL	3.97E+02	E-103 out	1.20E+05
DAP	7.10E+01	MIX-109 out	3.72E+04
Bacteria	1.98E+03	E-106 ethanol	1.98E+04
Total flow of inlet streams	2.85E+05	Total flow of outlet streams	2.85E+05

TABLE 4.6

Overall Energy Balance for Ethanol Production from Corn Stover Fermentation

Inlet Streams	Energy Flow kJ/h	Outlet Streams	Energy Flow kJ/h
Biomass (dry corn stover)	−1.92E+08	Fine particles	−7.55E+07
Water	−1.32E+09	V-101 steam	−3.35E+08
HP steam	−7.85E+08	V-102 vapor	0.00E+00
Cellulase	−4.04E+07	MIX-110 out	−8.37E+07
Air-seed production	0.00E+00	CO2 vent	−1.71E+08
CSL	−8.50E+05	E-103 out	−1.28E+09
DAP	−1.52E+05	MIX-109 out	−5.74E+08
Bacteria	−4.53E+06	E-106 ethanol	−1.20E+08
Stream enthalpy in:	−2.34E+09	Stream enthalpy out:	−2.64E+09
V-100 heating	2.56E+08	V-103 cooling	1.80E+07
V-102 heating	9.83E+07	V-104 cooling	7.14E+07
P-100 heating	1.71E+05	V-105 cooling	2.49E+08
T-100 reboiler	1.78E+08	T-101 condenser	1.77E+08
T-101 reboiler	5.56E+07	E-100 cooling	2.82E+08
E-105 heating	1.31E+06	E-102 cooling	1.42E+07
		E-103 cooling	−1.31E+06
		E-104 cooling	5.88E+07
		E-106 cooling	2.12E+07
External energy in:	5.90E+08	External energy out:	8.90E+08
Total flow of inlet streams	−1.75E+09	Total flow of outlet streams	−1.75E+09

steam at the rate of 60,000 kg/h. Cellulase enzymes were used for enzymatic hydrolysis of cellulose in the biomass. Fermentation of the pretreated biomass was carried out with bacteria, and nutrient supplements for the process were CSL and DAP. Ethanol was produced in the process at the rate of 19,800 kg/h in stream E-106 Ethanol. Carbon dioxide was a byproduct in the process and vented out at the rate of 18,900 kg/h in stream CO_2 Vent. Impure carbon dioxide is vented out in the MIX-110 Out stream. The energy requirements for the inlet and outlet energy streams are given in Table 4.6. The total external energy required by the process was 5.90×10^8 kJ/h, and the total energy removed from the process (mainly cooling water) was 8.90×10^8 kJ/h.

4.2.2 Process Cost Estimation for Ethanol Production from Corn Stover Fermentation

The HYSYS flow sheet was exported to ICARUS Process Evaluator (IPE) using the embedded export tool in HYSYS. This tool can be accessed from Tools → Aspen Icarus → Export Case to IPE. The project results summary is given in Table 4.7. Table 4.8 gives the breakdown of the operating costs in percentage, and Figure 4.7 shows a pie chart of the distribution of operating

TABLE 4.7

Project Costs for Ethanol Production
from Corn Stover Fermentation

Cost	Amount	Unit
Yield of ethanol	53,000,000	gal/year
Total project capital cost	20,300,000	USD
Total operating cost	81,000,000	USD/year
Total raw materials cost	54,000,000	USD/year
Total utilities cost	17,000,000	USD/year
Total product sales	106,000,000	USD/year

TABLE 4.8

Operating Costs for Ethanol Production
from Corn Stover Fermentation

Operating Cost	Percentage
Total raw materials cost	67
Total operating labor and maintenance cost	3
Total utilities cost	21
Operating charges	0
Plant overhead	1
G and A cost	7

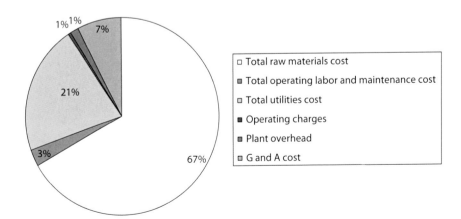

FIGURE 4.7
Operating cost breakdown for ethanol production from corn stover fermentation.

costs, which include the raw material and utilities costs. From Figure 4.7, it can be seen that the raw materials constitute 67% of the total operating costs, and 21% of the operating cost is for utilities. This is in accordance with high utility costs associated with the corn stover fermentation process. The equipment from the HYSYS case was mapped in IPE and is given in Appendix G with the respective costs of equipment as obtained from ICARUS.

The raw material and product unit costs used in ICARUS project basis are given in Table 4.9. The costs for biomass (corn stover) was given as $60/dry U.S. ton in Aden (2008); 2000 dry MT/day of corn stover was processed in the facility. The cost of DAP was reported as $142/ton in Aden et al. (2002), $249/ton in *ICIS Chemical Business* (2006), and $420/ton in *ICIS Chemical Business* (2007). This shows the sudden increase in the cost of the fertilizer in 2007. This increase in cost is attributed to the increase in demand of DAP as fertilizer for the growing biofuels business requiring agricultural products such as corn as feedstocks (ICIS Chemical Business, 2007). The maximum price reported till 2007 was used for cost estimation in IPE. This was calculated to $0.42/kg ($0.1906/lb) of DAP. The price for DAP is one of the costs included in sensitivity analysis. The raw material unit cost for CSL in the year 2000 was reported as $0.0804/lb in Aden et al. (2002) and total raw material cost for CSL was reported as $1.9 million/year. The raw material cost for CSL in 2007 was reported as $7.7 million/year (Aden, 2008). Using the same quantity usage of CSL in 2000 and 2007, the cost per unit of CSL was calculated for 2007 as $0.3258/lb or $0.72/kg.

The purchased cellulase enzyme unit cost was $0.010/gal of ethanol (equivalent to $0.0552/lb cellulase) and total cost for cellulase enzymes was $7 million/year in 2000 (Aden et al. 2002). The total cost for cellulase enzymes was $17.9 million/year in 2007 (Aden, 2008). Using the same quantity usage of cellulase in 2000 and 2007, the cost per unit of cellulase was calculated for 2007 as $0.1412/lb or $0.3112/kg.

TABLE 4.9

Raw Material and Product Unit Prices Used in ICARUS Cost Evaluation for Ethanol Production from Corn Stover Fermentation

Product/Raw Material	Flow Rate from HYSYS Simulation (kg/h)	Cost/Selling Price ($/kg)	Source
Corn stover	83,333	0.06	Aden (2008)
DAP	71	0.42	*ICIS Chemical Business* (2007)
Corn steep liquor	397	0.72	Aden et al. (2002) and Aden (2008)
Cellulase	2,690	0.31	Aden et al. (2002) and Aden (2008)
HP Steam at 165 psi	60,000	0.00983	ICARUS utility specification
Carbon dioxide	18,900	0.003	Indala (2004)
Ethanol	19,800	$1.517/gal	Minimum sale price based on operating cost

HP steam at 1000 kPa (165 psi) was used for steam hydrolysis. The cost for the steam used as material was similar to cost of steam at 165 psi as utility, which was $4.46/klb as described in ICARUS utility specification. The steam was specified in the raw materials section instead of the utilities section in IPE as it was used for prehydrolysis of corn stover reaction process.

The carbon dioxide obtained in the process was free from any impurities and contained trace amounts of water. The selling price of $0.003/kg for carbon dioxide was determined as the price at which it is available in the market from the pipeline (Indala, 2004).

The total operating cost was calculated in ICARUS with the aforementioned costs for raw materials. A minimum selling price for ethanol (Aden et al., 2002) was obtained by dividing the operating cost ($80,704,922/year) with the total gallons per year of ethanol produced (53,165,727 gal/year). The minimum sale price of $1.517/gal of ethanol was computed from the operating cost.

4.2.3 Summary of Ethanol Production from Corn Stover Fermentation

The design for fermentation of corn stover to ethanol including pretreatment of the stover was described. Two pretreatment steps, steam hydrolysis and enzymatic hydrolysis (using cellulase enzymes from *T. reesei*), were used to make the cellulose and hemicellulose in corn stover available for fermentation. The biocatalyst used in fermentation was *Z. mobilis* bacterium. The ethanol was purified to 99.5% purity in absorption, distillation, and molecular sieve separation columns.

About 53 MMgy (19,800 kg/h) of ethanol was produced in the process. This can be compared to a mid-sized ethanol production facility in the United States (*Ethanol Producer Magazine*, 2010). Carbon dioxide was a by-product in the process and was vented at the rate of 18,900 kg/h. The energy required by the process was 5.90×10^8 kJ/h and the energy liberated by the process was 8.90×10^8 kJ/h.

The economic analysis was performed in IPE. The total project capital cost was $20 million. The operating cost was $81 million/year, which included raw material costs of $54 million/year. A minimum product selling price computed from the operating cost was set at $1.52/gal for ethanol.

4.3 Ethylene Production from Dehydration of Ethanol

The standard industrial process for manufacturing ethylene is by steam cracking from a range of hydrocarbons including ethane, propane, butane, naphtha, liquefied petroleum gas (LPG), and gas oils. Refinery off-gases and light hydrocarbons recovered from natural gas are sources for ethane, propane, and butane.

Ethanol is readily converted to ethylene in a fluidized bed process with a 99% conversion (Wells, 1999). Tsao and Zasloff (1979) describe a process where ethanol is dehydrated to ethylene over a silica–alumina catalyst at 700°F–750°F (288°C–316°C) in a fluidized-bed reactor. Wells (1999) also describes a process where ethanol is converted to ethylene in a fixed-bed reactor with activated alumina and phosphoric acid or alumina and zinc oxide as catalysts. Takahara et al. (2005) described the dehydration of ethanol into ethylene over solid acid catalysts such as H-mordenites, zeolites, and silica–alumina at temperature ranges of 453–573 K (180°C–300°C) in a fixed-bed flow reactor. The conversion of ethanol to ethylene using H-mordenite (with SiO_2:Al_2O_3 ratio of 90%) gave a 99.9% yield of ethylene at 453 K (Takahara et al., 2005). In the fixed-bed reactors, the catalyst is regenerated every few weeks by passing air and steam over the bed to remove carbon deposits (Wells, 1999). Tsao and Zasloff (1979) describe the regeneration of catalyst in fluidized-bed reactor using a regeneration reactor. The chemical reaction that occurred is given by Equation 4.1:

$$C_2H_5OH \rightarrow C_2H_4 + H_2O \qquad (4.1)$$

The dehydration of ethanol to ethylene in a fluidized-bed reactor was simulated in HYSYS. The HYSYS flow sheet diagram for this process is shown in Figure 4.8. The process flow outlined in Wells (1999) was used to design the process. The plant capacity used for this simulation was 200,000 MT of ethylene production per year. This capacity was based on a Braskem-proposed ethanol to ethylene plant in Brazil (C&E News, 2007a). This amounts to 25,000 kg/h of ethylene production with 8000 h of plant operation per year. The simulated result gave a capacity of 24,970 kg/h production of ethylene. The UNIQUAC thermodynamic model was used for estimating the interactions between reaction components.

4.3.1 Process Description for Ethylene Production from Dehydration of Ethanol

The process for dehydration of ethanol to ethylene consists of two steps, as shown in Figure 4.8, a dehydration step for ethanol to ethylene and a purification step to remove water from ethylene (Wells, 1999). In the dehydration step, ethanol stream was vaporized by heating to 200°C in heater E-100. The heated ethanol stream was introduced to a fluidized-bed reactor CRV-100 with activated alumina catalyst. The catalyst was maintained in a fluidized state by gaseous ethanol introduced at the bottom of the reactor CRV-100 (Tsao and Zasloff 1979). The CRV-100 was a jacketed reactor maintained at a temperature of 300°C using ADJ-2. A 99% conversion of ethylene was obtained in the reactor. Ethylene and water vapor came out of the reactor in the stream CRV-100 top.

The purification step shown in Figure 4.8 involved the separation of ethylene from the vapor stream. CRV-100 Top containing ethylene, water, and

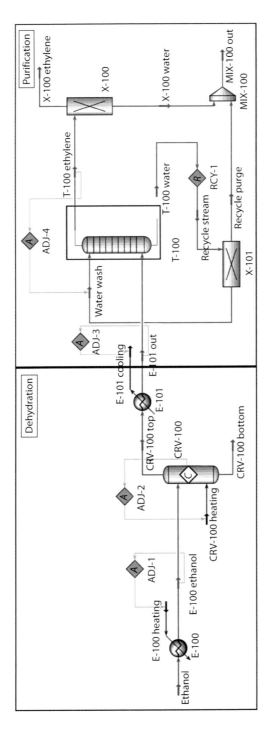

FIGURE 4.8
(See color insert.) Overall process design diagram for ethylene production from dehydration of ethanol.

TABLE 4.10

Overall Mass Balance for Ethylene Production from Dehydration of Ethanol

Inlet Material Streams	Mass Flow kg/h	Outlet Material Streams	Mass Flow kg/h
Ethanol	4.15E+04	X-100 ethylene	2.50E+04
		MIX-100 out	1.65E+04
Total flow of inlet streams	4.15E+04	Total flow of outlet streams	4.15E+04

TABLE 4.11

Overall Energy Balance for Ethylene Production from Dehydration of Ethanol

Inlet Streams	Energy Flow kJ/h	Outlet Streams	Energy Flow kJ/h
Ethanol	−2.50E+08	X-100 ethylene	4.69E+07
		MIX-100 out	−2.53E+08
Stream enthalpy in:	−2.50E+08	Stream enthalpy out:	−2.06E+08
CRV-100 heating	5.12E+07	E-101 cooling	5.84E+07
E-100 heating	5.13E+07		
External energy in:	1.03E+08	External energy out:	5.84E+07
Total flow of inlet streams	−1.48E+08	Total flow of outlet streams	−1.48E+08

residual ethanol was cooled to 35°C in cooler E-101. The cooled ethylene stream was separated in absorber T-100 with 20 stages. Water Wash stream at 25°C was introduced in the top stage of the absorber. The rate of wash water was determined using ADJ-4 to achieve 100% removal of residual ethanol in stream T-100 Ethylene. The T-100 Water stream containing trace amounts of ethylene and ethanol was separated in X-101. The water recovered from the separator was recycled to the absorber in stream Water Wash. The T-100 Ethylene stream contained 95% mole ethylene. This stream was passed through a drier unit X-100 to remove residual water. A 99.99% mole ethylene was obtained in the X-100 Ethylene stream. The waste water was collected in MIX-100 and obtained in MIX-100 Out stream from the process.

The overall mass balances and energy requirements for major inlet and outlet streams are given in Tables 4.10 and 4.11, respectively. From Table 4.10, it can be seen that 41,500 kg/h of ethanol was required to produce 25,000 kg/h of ethylene. The energy required by the process was 1.03×10^8 kJ/h and the energy removed from the process was 5.84×10^7 kJ/h. The detailed stream descriptions are given in Appendix F.

4.3.2 Process Cost Estimation for Ethylene Production from Dehydration of Ethanol

The HYSYS flow sheet was exported to IPE using the embedded export tool in HYSYS. This tool is accessed from Tools → Aspen Icarus → Export Case to IPE. The project result summary is given in Table 4.12. Table 4.13 gives the

TABLE 4.12

Project Costs for Ethylene Production
from Dehydration of Ethanol

Cost	Amount	Unit
Yield of ethylene	200,000,000	kg/year
Total project capital cost	3,100,000	USD
Total operating cost	186,500,000	USD/year
Total raw materials cost	168,800,000	USD/year
Total utilities cost	2,826,000	USD/year
Total product sales	186,500,000	USD/year

TABLE 4.13

Operating Costs for Ethylene Production
from Dehydration of Ethanol

Operating Cost	Percentage
Total raw materials cost	90
Total operating labor and maintenance cost	0
Total utilities cost	2
Operating charges	0
Plant overhead	0
G and A cost	7

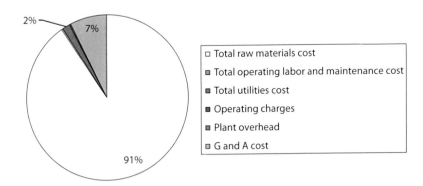

FIGURE 4.9
Operating cost breakdown for ethylene production from dehydration of ethanol.

breakdown of the operating costs in percentage, and Figure 4.9 shows a pie
chart of the distribution of operating costs, which include the raw material
and utilities costs. From Figure 4.9, it can be seen that the raw material, ethanol, constitute approximately 90% of the total operating costs. The equipment from the HYSYS case was mapped in IPE and is given in Appendix G
with the respective costs of equipment as obtained from ICARUS.

TABLE 4.14

Raw Material and Product Unit Prices Used in ICARUS Cost Evaluation for Ethylene Production from Dehydration of Ethanol

Product/Raw Material	Flow Rate from HYSYS Simulation (kg/h)	Cost/Selling Price ($/kg)	Source
Ethanol	41,500	$1.517/gal	Minimum selling price based on operating cost (Aden et al., 2002 and Aden, 2008)
Ethylene	25,000	0.93	Minimum selling price based on operating cost

The raw material and product unit costs used in project basis are given in Table 4.14. A minimum selling price for ethanol (Aden et al., 2002) was set by dividing the operating cost from the ethanol process ($80,704,922/year) with the total gallons per year of ethanol produced (53,165,727 gal/year) as explained in the ethanol production process from corn stover. The minimum sale price of $1.517/gal of ethanol was computed from the operating cost. This price was used as the raw material cost in the ethanol dehydration process to ethylene.

The U.S. market price of ethylene was reported as 33.5 cents/lb ($0.74/kg) (ICIS, 2009b). The Asian market price was reported as $1000–$1060/ton ($1.00–$1.06/kg) CFR (cost and freight) (ICIS, 2009b). The aforementioned prices for ethylene are based on petroleum-based feedstock as raw material. The minimum selling price of ethylene was obtained by dividing the total operating cost for the process ($186,500,000/year) by the quantity of ethylene produced per year (200,000 ton/year or 440,410,517 lb/year). This gave a value of $0.42/lb or $0.93/kg of ethylene. The value of $0.93/kg ethylene was used for calculations in IPE.

4.3.3 Summary of Ethylene Production from Dehydration of Ethanol

The design for ethylene production from dehydration of ethanol in a fluidized-bed reactor was described. The process consists of two steps, a dehydration process and a purification process. Ethylene of 99.99% purity was obtained in this process.

About 200,000 MT/year (25,000 kg/h) of ethanol was produced in the process. This can be compared to a Braskem-proposed ethanol to ethylene plant in Brazil (C&E News, 2007a). The energy required by the process was 1.03×10^8 kJ/h and the energy liberated by the process was 5.84×10^7 kJ/h.

The economic analysis was performed in IPE. The total project capital cost was $3 million. The operating cost was $187 million/year, which included raw material costs of $169 million/year. A minimum product selling price computed from the operating cost was $0.93/kg for ethylene.

4.4 Fatty Acid Methyl Ester and Glycerol from Transesterification of Soybean Oil

Transesterification is the reaction of fats and oils with an alcohol in the presence of a catalyst to produce a glycerol molecule and three fatty acid esters. Fats and oils are composed of triglycerides, a molecule containing a glycerol backbone attached to three fatty acid chains, as shown in Figure 4.10. The fatty acid chains can all be the same like linoleic acid shown in the Figure 4.10 or they can be different fatty acids. The common fatty acids present in oils and fats are oleic, linoleic, myristic, palmitic, stearic, linolenic, and lauric acids. The fatty acid content in oils are represented as percentages and given in Table 2.7 (Meher et al., 2006).

The design for transesterification for the production of methyl linoleate $(C_{19}H_{34}O_2)$ and glycerol $(C_3H_8O_3)$ from soybean oil and methanol (CH_3OH) using a sodium methoxide $(NaOCH_3)$ catalyst was developed in HYSYS. The chemical reaction that occurred is given in Equation 4.2 and shown in Figure 4.11. Soybean oil consists of 23% oleic acid and 54% linoleic acid along with myristic, palmitic, stearic, linolenic, and lauric acids (Meher et al., 2006). Trilinolein $(C_{57}H_{98}O_6)$, a triglyceride containing three linoleic acid chains, was used to simulate the soybean oil in the HYSYS design. Trilinolein was chosen for the design because soybean oil has the highest composition of linoleic acid. The other fatty acids have similar properties and differ in molecular weight.

$$C_{57}H_{98}O_6 + 3CH_3OH \rightarrow C_3H_8O_3 + 3C_{19}H_{34}O_2 \qquad (4.2)$$

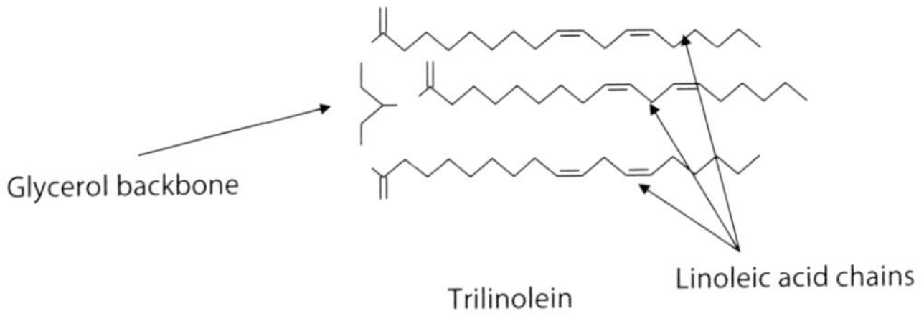

FIGURE 4.10
Molecular structure of trilinolein.

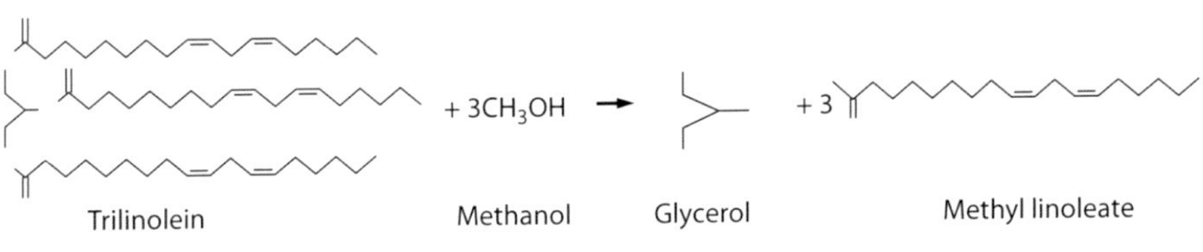

FIGURE 4.11
Transesterification reaction with trilinolein as representative triglyceride.

TABLE 4.15

Experimental Properties of Trilinolein

Property	Value	Condition	Note
Carbon-13 NMR spectrum	Spectrum given		Three references
Density	0.9334 g/cm³	Temp: 20°C	Four references
	0.9287 g/cm³	Temp: 18°C	
	0.9272 g/cm³	Temp: 20°C	
	0.9184 g/cm³	Temp: 40°C	
IR absorption spectrum	Spectrum given		One reference
IR spectrum			One reference
Mass spectrum	Spectrum given		Six references
Melting point	68°C to 69°C		Six references
	35°C to 37°C		
	13°C		
	−5°C to −4°C		
	−43°C to 44°C		
	−43.0°C to 42.5°C		
Proton NMR spectrum			One reference
Raman spectrum			One reference
Refractive index	1.4840	Wavelength: 589.3 nm	Six references
	1.4795	Temp: 20°C	
	1.4793	Wavelength: 589.3 nm	
	1.4719	Temp: 18°C	
	1.4709	Wavelength: 589.3 nm	
	1.4683	Temp: 20°C	
	1.46815	Wavelength: 589.3 nm	
		Temp: 50°C	
		Wavelength: 589.3 nm	
		Temp: 20°C	
		Wavelength: 589.3 nm	
		Temp: 50°C	
		Wavelength: 589.3 nm	
		Temp: 50°C	

Source: SciFinder Scholar, *SciFinder Research Tool,* American Chemical Society, Washington, DC, 2009.

A search for trilinolein was done in the substance identifier section of the SciFinder research tool from the American Chemical Society. There were a total of 16 references for experimental properties of trilinolein, and the results are given in Table 4.15. The results from the search showed that experimental measurements of thermophysical properties of trilinolein are very limited.

The calculation procedure for thermodynamic properties of liquids, gases, and vapors is explained in detail in the *Perry's Chemical Engineers' Handbook* (Perry and Green, 1997). The equation-of-state procedures can be used for calculation of liquid-phase and gas-phase properties. An alternate for liquid-phase property estimation is the application of excess properties. The excess property of importance for engineering applications is the excess Gibbs energy G^E. Several methods exist for the expression of G^E and the calculation of the properties, among which the most recent are given by the Wilson equation, NRTL equation, and UNIQUAC (UNIversal QUAsi-Chemical) equation. A development based on the UNIQUAC equation is the UNIFAC (UNIQUAC functional-group activity coefficients) method, which provides for the calculation of activity coefficients from contributions of the various groups making up the molecules of a solution. Thus, the group contribution method is a technique used to estimate and predict thermodynamic and other properties from molecular structures (Constantinou, 1994). The group contributions are obtained from known experimental data of well-defined pure components and mixtures. Databanks like Dortmund Databank, the Beilstein database, or the DIPPR data bank (from AIChE) are common sources of thermophysical data.

Trilinolein was created in HYSYS using the Hypo Manager tool. The UNIFAC functional groups were entered in the structure builder tool of Hypo Manager, and the properties of trilinolein were estimated by HYSYS proprietary method.

The properties of trilinolein obtained from the HYSYS estimation method using UNIFAC groups was verified with the online property estimation method available at the website for Dortmund Databank (DDBST, 2009). The structure of trilinolein was downloaded from the NIST Chemistry Webbook (NIST, 2009). The structure was stored in a *.mol file and uploaded to the "DDB Online Property Estimation by the Joback Method" tool. The properties were calculated and compared to the HYSYS property calculations, and comparisons are given in Table 4.16. The properties were similar obtained from the two sources and used in the design.

Methyl linoleate ($C_{19}H_{34}O_2$) was the FAME formed in reaction given by Equation 4.2 and was available in the traditional component list of HYSYS. Sodium methoxide and sodium chloride were created in HYSYS using the Hypo Manager tool by supplying data available on these molecules from SciFinder Scholar. The rest of the components used in the design were traditional components in HYSYS.

TABLE 4.16

Comparison of Property Estimation of Trilinolein in HYSYS and the Dortmund Databank Online Property Estimation by the Joback Method

Property	Value		Unit
	DDB Joback Method	HYSYS	
Heat of formation (ideal gas)	−1473.61	−1473.61	kJ/mol
Gibbs energy of formation	−23.17	−21.74	kJ/mol
Freezing point	813.66	Not available	K
Boiling point	1702.94	1702.74	K
Critical volume	3179.5	3179.5	cm3/mol
Critical pressure	247.61	247.610	kPa
Critical temperature	3665.67	3665.24	K
Enthalpy of fusion	154182	Not available	J/mol
Enthalpy of vaporization	164.057	Not available	kJ/mol

4.4.1 Process Description for Fatty Acid Methyl Ester and Glycerol from Transesterification of Soybean Oil

The plant capacity used for this simulation was 10 MMgy (Haas et al., 2006). This capacity was based on a mid-sized biodiesel manufacturing unit in the United States (NBB, 2008a). A flow rate of 1250 gal/h was required with 8000 h of plant operation in a year. The simulated result gave a capacity of 1260 gal/h production of methyl linoleate.

The UNIQUAC thermodynamic model was selected for estimating the interactions between reaction components. The overall HYSYS flow diagram for the process is shown in Figure 4.12. The design had three sections, the transesterification reaction section, the methyl ester purification section, and the glycerol recovery and purification section as described in the following.

The transesterification section is shown in Figure 4.13. Soybean oil was reacted with methanol and catalyst (sodium methoxide) according to Equation 4.2 in two sequential reactors, CRV-100 and CRV-101. Both the reactors were designed as conversion reactors with a 90% conversion of soybean oil reacting with methanol to methyl ester and glycerol (Freedman et al., 1984). The sequential reaction in two reactors ensured 99% overall conversion of the oil to ester.

The reaction yielded methyl linoleate esters and glycerol, which were separated in centrifuge X-101. The stream from the top of the centrifuge contained the impure methyl ester, and the bottom stream contained glycerol, free fatty acids, water, and residual methanol.

The methyl ester was purified as shown in Figure 4.14. The methyl ester stream was washed with water maintained at a pH of 4.5 using hydrochloric acid in the reactor CRV-102 to neutralize the catalyst and convert any soaps to free fatty acids. The product stream from CRV-102 was separated in centrifuge X-102, and the top stream contained methyl ester and water. The water was separated from the methyl ester stream in vacuum dryer unit X-103. U.S. biodiesel specifications require a maximum of 0.05% (v/v) of water and sediment in the product stream of methyl esters (Haas et al., 2006, NBB 2008b). A similar specification of methyl esters used as monomers for polymer manufacture was not available, so the standard specification of product esters outlined by the National Biodiesel Board was followed in this design.

The glycerol purification section is shown in Figure 4.15. The water and glycerol streams from the process were combined in CRV-103. Dilute hydrochloric acid was used to neutralize the catalyst sodium methoxide to sodium chloride and methanol, as shown in Equation 4.3. The water wash of the impure glycerol stream also ensured the conversion of soaps to free fatty acids. The reactions for soap formation, conversion of soap to free fatty acids with water, and the removal of the free fatty acids were simulated by removing a part of glycerol and soybean oil in the X-104 Top stream. The remaining acid was neutralized with sodium hydroxide in CRV-104. The glycerol stream containing methanol and water was separated in two distillation columns, T-100 and T-101. The processes are described in detail in the following three sections.

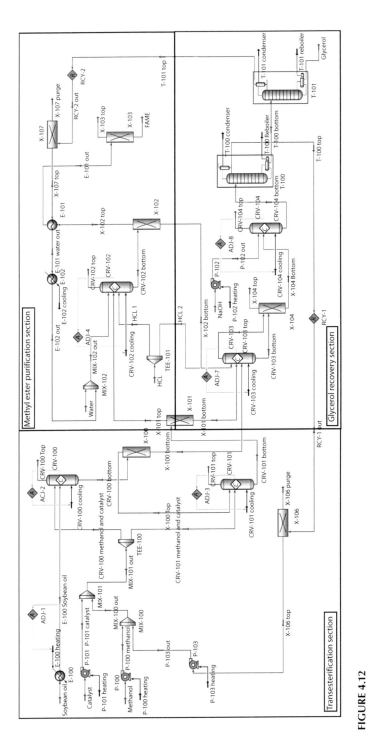

FIGURE 4.12
(See color insert.) Overall process design diagram for fatty acid methyl ester and glycerol from transesterification of soybean oil.

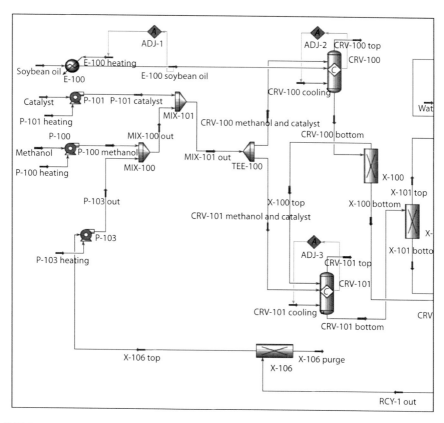

FIGURE 4.13
Transesterification section for fatty acid methyl ester and glycerol from transesterification of soybean oil.

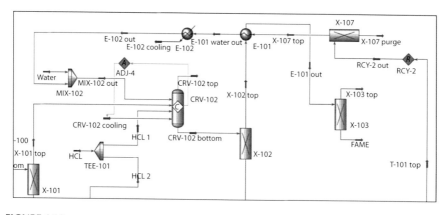

FIGURE 4.14
Methyl ester purification section for fatty acid methyl ester and glycerol from transesterification of soybean oil.

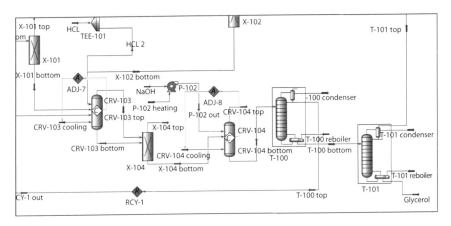

FIGURE 4.15
Glycerol recovery and purification section for fatty acid methyl ester and glycerol from transesterification of soybean oil.

4.4.1.1 Transesterification Section

The transesterification section shown in Figure 4.13 is described in the following. The stream, Soybean Oil, with a composition of 100% trilinolein was heated to 60°C in E-100. The stream, Methanol, with a composition of 100% methanol was pumped through P-100 and mixed with recycled methanol from P-103 Out in MIX-100. The stream, Catalyst, with a composition of 25% sodium methoxide and 75% methanol was pumped through P-101 and mixed in MIX-101 with methanol in MIX-100 Out stream. The composition of the stream MIX-101 Out stream was 1.78% (w/w) sodium methoxide in methanol.

The reaction was carried out in two sequential reactors, CRV-100 and CRV-101. Both the reactors were designed as conversion reactors with a 90% conversion of soybean oil reacting with methanol to methyl ester and glycerol as given by Equation 4.2 (Freedman et al., 1984). E-100 Soybean Oil was fed to the first reactor CRV-100. The 1.78% (w/w) sodium methoxide catalyst in methanol solution was split in TEE-100 in the ratio of 9:1, with 90% going to CRV-100 and the rest to CRV-101. The temperature of the reactors CRV-100 and CRV-101 was maintained at 60°C using ADJ-2 and ADJ-3, respectively. The pressure in the reactors was 446 kPa (Haas et al., 2006). The stream CRV-100 Bottom was separated in a centrifuge, X-100. The glycerol separated from the oil phase in X-100 and was removed in X-100 Bottom. The X-100 Top stream containing unreacted soybean oil and methanol was reacted with the CRV-101 Methanol and Catalyst streams in CRV-101.

The CRV-101 Bottom stream containing the methyl ester, unreacted methanol, glycerol, and soybean oil were separated in the centrifuge X-101. The glycerol separated from the oil and methyl ester and was recovered in X-101 Bottom. The X-101 Top stream contained the methyl ester, unreacted soybean oil, and catalyst. The methyl ester purification from stream X-101 Top and the glycerol recovery from X-101 Bottom are described in the following sections.

4.4.1.2 Methyl Ester Purification Section

The purification section of the methyl ester stream is shown in Figure 4.14. The crude methyl ester in X-101 Top was washed with water maintained at a pH of 4.5 in CRV-102. Water was supplied by makeup water in stream Water mixed with recycled water in stream E-102 Out. These two streams were mixed in MIX-102 and supplied to the CRV-102 in stream MIX-102 Out. Hydrochloric acid was used to maintain the pH at 4.5. The acid was supplied through HCL 1 stream to the reactor CRV-102. The acid neutralization reaction of the catalyst in CRV-102 is given in Equation 4.3. The temperature of the reactor CRV-102 was maintained at 25°C.

$$HCl + NaOCH_3 \rightarrow NaCl + CH_3OH \qquad (4.3)$$

The CRV-102 Bottom stream contained the methyl ester, with water, sodium chloride, methanol, and glycerol as impurities. The stream was separated in centrifuge X-102. The glycerol and water separated from the oil phase in X-102 Bottom and was sent to the glycerol recovery section. The methyl ester stream X-102 Top contained 2.2% water by volume and was heated in heat exchanger E-101 to 99°C before it was sent to vacuum dryer. The heat was supplied by water at 100°C from the distillation section. A vacuum dryer, X-103, was used to remove water from the methyl ester stream from an initial value of 2.2% (v/v) to a final value of 0.04% (v/v) (to conform to National Biodiesel Board Standard of water <0.05% (v/v) specification). The pure methyl ester at 50°C and 446.1 kPa was obtained in the FAME stream.

4.4.1.3 Glycerol Recovery and Purification

The glycerol recovery and purification section is shown in Figure 4.15. The impure, dilute, and aqueous glycerol streams from the system were collected in a glycerol pool in CRV-103. These three streams included X-100 Bottom (from centrifuge X-100) and X-101 Bottom (from centrifuge X-101) from the transesterification section and X-102 Bottom (from centrifuge X-102) from the methyl ester purification section. The impure glycerol stream was treated with dilute hydrochloric acid to neutralize remaining catalyst and convert soaps to free fatty acids. The neutralization reaction occurring in CRV-103 is given in Equation 4.3. The temperature in CRV-103 was maintained at 25°C using ADJ-7. A part of unconverted soybean oil and glycerol were removed in stream X-104 Top by the centrifuge X-104 to simulate the removal of free fatty acids.

The glycerol-rich stream in X-104 Bottom was treated with sodium hydroxide in CRV-104 to neutralize excess hydrochloric acid in the stream. Sodium hydroxide was pumped through P-102 to the reactor CRV-104.

The reaction occurring in CRV-104 is given in Equation 4.4. The CRV-104 Bottom containing glycerol, water, and methanol as main components was heated from 25°C to 68°C in heat exchanger E-103. The E-103 Out stream was passed to the first distillation column, T-100.

$$NaOH + HCl \rightarrow NaCl + H_2O \qquad (4.4)$$

Methanol having a lower boiling point than water and glycerol was removed first in T-100. The column, T-100, had 20 trays and feed was introduced in tray 10. The condenser and reboiler were operated at a pressure of 101.3 kPa. The condenser was operated as a total condenser to recover methanol in liquid phase. A methanol recovery rate of 99.99% and a methanol component recovery of 100% in T-100 Top were used as specifications for the column. The reflux ratio of 3 was given as an initial estimate for the column. The column converged at a reflux ratio of 20.

The methanol recovered in T-100 Top stream was recycled through RCY-1. The stream RCY-1 Out was passed through separator X-106 to separate the purge stream from the recycled methanol. X-106 Top containing 100% methanol was pumped through P-103 to the transesterification section.

The glycerol and water stream from T-100 Bottom was separated in distillation column T-101. The column T-101 had 10 trays with feed introduced in tray 5. The reboiler and condenser were operated at 101.3 kPa. The condenser was operated at full reflux to recover all water vapor at 100°C. A reflux ratio of 1 and a component fraction of 100% for glycerol in stream T-100 Bottom were used as specifications for running the column. The component recovery specification of glycerol in stream T-100 Bottom was monitored to ensure 100% recovery of glycerol in T-100 Bottom.

The T-100 Bottom stream exited the distillation column at 290°C. The heat exchanger E-103 was used to recover heat from T-101 Bottom and was used to raise the temperature of the stream CRV-104 Bottom. Glycerol was recovered from the process at 70°C and 101.3 kPa in stream Glycerol.

Water vapor was recovered at 100°C in T-101 Top. This stream was recycled through RCY-2. The RCY-2 Out stream was passed through separator X-107 to separate the purge stream from the recycled water vapor. The heat from X-107 Top was used to raise the temperature of the stream X-102 Top in heat exchanger E-101. The partially condensed water vapor stream E-101 Water Out was cooled in E-102 to 25°C. The water in stream E-102 Out was recycled to the methyl ester purification section through MIX-102.

The overall mass flow rates of the process inlet and outlet streams and the overall mass balance are shown in Table 4.17. Detailed stream descriptions are given in Appendix F. About 4260 kg/h of FAME was produced from this process and 413 kg/h of glycerol was the by-product. The reactants in this process were 4250 kg/h of soybean oil, 423 kg/h of fresh methanol, and 53 kg/h of catalyst containing 25% sodium methoxide in methanol (weight basis); 30 kg/h of diluted HCl acid containing 35% HCl and

TABLE 4.17

Overall Mass Balances for FAME and Glycerol
from Transesterification of Soybean Oil

Inlet Material Streams	Mass Flow (kg/h)	Outlet Material Streams	Mass Flow (kg/h)
Soybean oil	4.25E+03	FAME	4.26E+03
Methanol	4.23E+02	Glycerol	4.13E+02
Water	8.55E+01	X-103 top	1.22E+02
Catalyst	5.31E+01	X-104 top	4.87E+01
NaOH	1.91E+00		
HCL	3.03E+01		
Total flow of inlet streams	4.84E+03	Total flow of outlet streams	4.84E+03

65% water (weight basis) was required for purification of the methyl ester and glycerol; 86 kg/h of fresh water was required in the purification process; and 2 kg/h of caustic soda was required to neutralize excess HCl.

The energy requirements for the process are given in Table 4.18. Two heat exchangers, E-101 and E-103, were used in this design. The heat exchanger

TABLE 4.18

Overall Energy Balances for FAME and Glycerol
from Transesterification of Soybean Oil

Inlet Streams	Energy Flow (kJ/h)	Outlet Streams	Energy Flow (kJ/h)
Soybean oil	−7.12E+06	FAME	−8.36E+06
Methanol	−3.16E+06	Glycerol	−2.71E+06
Water	−1.35E+06	X-103 top	−1.66E+06
Catalyst	−3.39E+05	X-104 top	−3.32E+05
NaOH	−3.99E+03		
HCL	−3.38E+05		
Stream enthalpy in:	−1.23E+07	Stream enthalpy out:	−1.31E+07
P-100 heating	2.47E+02	CRV-100 cooling	1.13E+06
P-101 heating	3.20E+01	CRV-101 cooling	8.21E+04
P-102 heating	5.95E-01	CRV-102 cooling	8.77E+04
P-103 heating	1.66E+02	CRV-103 cooling	1.10E+05
T-100 reboiler	6.59E+06	CRV-104 cooling	2.26E+04
T-101 reboiler	4.60E+06	T-100 condenser	6.15E+06
E-100 heating	2.26E+05	T-101 condenser	2.19E+06
		E-102 cooling	2.39E+06
External energy in:	1.14E+07	External energy out:	1.22E+07
Total flow of inlet streams	−8.93E+05	Total flow of outlet streams	−8.95E+05

E-101 was used to recover heat from steam at 100°C from the distillation section. This heat was used to raise the temperature of methyl ester stream from 25°C to 99°C before it was dehydrated in vacuum dehydration unit. Heat exchanger E-103 was used to recover heat from glycerol at 290°C from the distillation section. This heat was used to raise the temperature of the glycerol–water–methanol mixture in stream CRV-104 Bottom from 25°C to 68°C before it was introduced to the distillation section. Using HYSYS flow sheet, the total energy required by the system in inlet energy streams was 1.14×10^7 kJ/h. The total energy removed in the outlet energy streams was 1.22×10^7 kJ/h.

4.4.2 Process Cost Estimation for Fatty Acid Methyl Ester and Glycerol from Transesterification of Soybean Oil

The HYSYS flow sheet was exported to IPE using the embedded export tool in HYSYS. This tool can be accessed from Tools → Aspen Icarus → Export Case to IPE. The project result summary is given in Table 4.19. Table 4.20 gives the breakdown of the operating costs in percentage, and Figure 4.16 shows a pie chart of the distribution of operating costs, which include the raw material and utilities costs. From Figure 4.16, it can be seen that the

TABLE 4.19

Project Costs for FAME and Glycerol
from Transesterification of Soybean Oil

Cost	Amount	Unit
Yield of FAME	10,363,000	gal/year
Total project capital cost	7,385,000	USD
Total operating cost	23,430,000	USD/year
Total raw materials cost	18,850,000	USD/year
Total utilities cost	301,000	USD/year
Total product sales	29,820,000	USD/year

TABLE 4.20

Operating Costs for FAME and Glycerol
from Transesterification of Soybean Oil

Operating Cost	Percentage
Total raw materials cost	80
Total operating labor and maintenance cost	6
Total utilities cost	1
Operating charges	7
Plant overhead	3
G and A cost	1

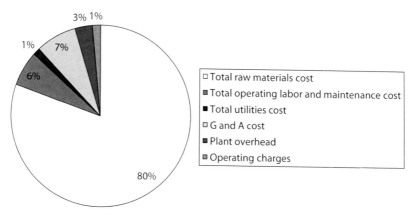

FIGURE 4.16

Operating cost breakdown for fatty acid methyl ester and glycerol from transesterification of soybean oil.

raw material, soybean oil, constitutes approximately 80% of the total operating costs. The equipment from the HYSYS case was mapped in IPE and is given in Appendix G with the respective costs of equipment as obtained from ICARUS.

The raw material and product unit costs used in project basis are given in Table 4.21. The costs for soybean oil, methanol hydrochloric acid, and sodium hydroxide as reported in Haas et al. (2006) were used for this design. The cost for 25% wt solution of sodium methylate ($NaOCH_3$ or NaOMe) catalyst in methanol was reported as $0.98/kg ($0.445/lb) (Haas et al., 2006). The price for sodium methylate is calculated in the following way (Seay, 2009). Sodium methylate is typically sold as a solution in methanol; the price of the solution is based on adding the cost of methanol to the solution according to Equation 4.5.

TABLE 4.21

Material and Product Unit Prices Used in ICARUS Cost Evaluation for FAME and Glycerol from Transesterification of Soybean Oil

Product/Raw Material	Cost/Selling Price ($/kg)	Source
Soybean oil	0.52	Haas et al. (2006)
Methanol	0.286	Haas et al. (2006)
Sodium methylate (25% w/w)	0.98	Haas et al. (2006)
HCl	0.132	Haas et al. (2006)
NaOH	0.617	Haas et al. (2006)
Water	$55/MM gal	ICARUS utility specification
Fatty acid methyl ester	$2.26/gal	Minimum selling price based on operating cost
Glycerol	1.94	*ICIS Chemical Business* (2008)

Methanol is also a raw material in the production process of sodium methylate, so the portion of the price based on sodium methylate is indexed to the price of methanol on a sliding scale as given in Equations 4.5 and 4.6. The index changes with changes in price of methanol. The index is not released by the companies, so the sodium methylate solution price available in Haas et al. (2006) was used for the raw material cost of the catalyst.

$$\text{Price of NaOMe Solution} = [\text{Methanol Price (\$/kg)} * \text{wt\% MeOH}]$$
$$+ [\text{NaOMe Price (\$/kg)} * \text{wt\% NaOMe}] \qquad (4.5)$$

$$\text{Price of NaOMe} = \text{Methanol Price} * \text{Index} \qquad (4.6)$$

Water at 25°C was used for washing the methyl ester in the methyl ester purification section. The properties of the water used were similar to the properties of cooling water in ICARUS utility specifications. The cost for water was included in the raw material specification instead of the utility section as it was used in the wash process of methyl ester.

The spot price of refined, pharmaceutical grade, 99.7% glycerol was reported as $0.88–$1.05/lb (ICIS Chemical Business, 2008). The selling price for 80% aqueous solution of crude glycerol was reported as $0.33/kg ($0.15/lb) (Haas et al., 2006). The glycerol obtained in the design case was 95% pure with sodium chloride as impurity. The lower range price for 99.7% glycerol ($0.88/lb) was used for computing the product sales in IPE.

The total operating cost was calculated in ICARUS with the aforementioned costs for raw materials. A minimum selling price for FAME (Aden et al., 2002) was set by dividing the operating cost ($23,435,000/year) with the total gallons per year of FAME produced (10,363,000 gal/year). The minimum sale price of $2.26/gal of ethanol was computed from the operating cost.

4.4.3 Summary of Fatty Acid Methyl Ester and Glycerol from Transesterification of Soybean Oil

The design of transesterification process for the production of FAME was developed for a medium-sized plant (10^6 MMgy production capacity) and was simulated in HYSYS. About 4250 kg/h of soybean oil, represented by trilinolein in HYSYS, was used as the triglyceride; 422 kg/h of methanol and 53 kg/h of sodium methylate (25% w/w solution in methanol) was used to convert 99% of the soybean oil in two sequential reactors.

The product, FAME, represented by methyl linoleate in HYSYS, was purified and obtained at the rate of 4260 kg/h. Crude glycerol stream was purified and obtained at the rate of 410 kg/h. About 270 kg/h of methanol and 970 kg/h of water were recycled in the process from the distillation section. The total energy required by the system in inlet energy streams was 1.14×10^7 kJ/h. The total energy removed in the outlet energy streams was 1.22×10^7 kJ/h.

The economic analysis was performed in IPE. The total project capital cost was \$7.4 million. The operating cost was \$23.4 million/year, which included raw material costs of \$18.9 million/year. A minimum product selling price computed from the operating cost was \$2.26/gal for the FAME.

4.5 Propylene Glycol Production from Hydrogenolysis of Glycerol

The standard industrial procedure to produce propylene glycol is from propylene oxide by hydration at a temperature of 200°C and pressure of 12 bar (Wells, 1999). Glycerol is a by-product of the transesterification process, and can be used to produce propylene glycol. The experimental study described by Dasari et al. (2005) was for the production of propylene glycol from glycerol and hydrogen using a copper chromite catalyst. This process is a low-temperature (200°C) and low-pressure (200 psi) process. Acetol is an intermediate in this reaction. The glycerol first is dehydrated to acetol and the acetol formed is hydrogenated to propylene glycol. The chemical reaction that occurred is given by Equation 4.7:

$$C_3H_8O_3 + H_2 \rightarrow C_3H_8O_2 + H_2O \qquad (4.7)$$

The aforementioned process was simulated in HYSYS and the process flow diagram is shown in Figure 4.17. The plant capacity used for this simulation was 65,000 MT/year of propylene glycol (~8125 kg/h with the plant operation for 8000 h per year). This capacity was based on a proposed Ashland/Cargill joint venture glycerol to propylene glycol plant in Europe (Ondrey, 2007b). The UNIQUAC thermodynamic model was selected for estimating the interactions between reaction components.

4.5.1 Process Description for Propylene Glycol Production from Hydrogenolysis of Glycerol

The reaction was carried out in two sequential reactors at 200°C and 200 psi hydrogen pressure, as shown in Figure 4.17. The conversion of glycerol was 54.8% in both the reactors (Dasari et al., 2005). Hydrogen was heated to 200°C and 200 psi (1379 kPa) pressure in E-100 and split into two streams in TEE-100. The hydrogen to reactor CRV-100 was CRV-100 Hydrogen and to CRV-101 was CRV-101 Hydrogen. Glycerol (80% wt. in water) at 25°C and atmospheric pressure was introduced in reactor CRV-100. The recycle stream E-101 Recycle Glycerol from the purification section was added to fresh Glycerol stream in MIX-100. The reactor CRV-100 was maintained at

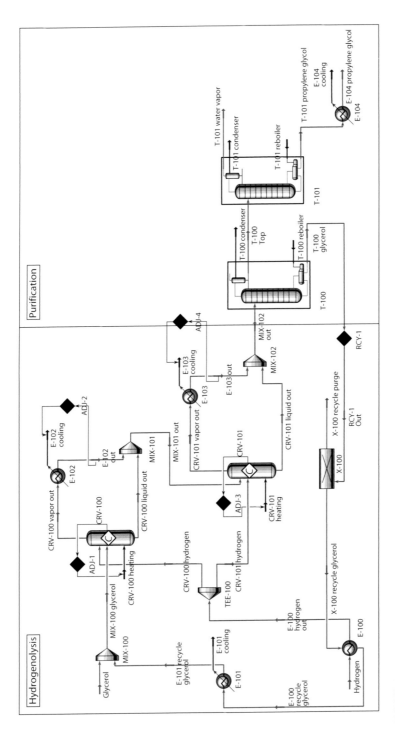

FIGURE 4.17
(See color insert.) Overall process design diagram for propylene glycol production from hydrogenolysis of glycerol.

200°C using adjuster ADJ-1. The vapor stream CRV-100 Vapor Out from the reactor was condensed in E-102. Adjuster ADJ-2 was set to completely condense the vapor by adjusting the energy stream, E-102 Cooling. The condensed stream was mixed with the reactor liquid stream CRV-100 Liquid Out in MIX-101. The Mix-101 Out stream contained unreacted glycerol, propylene glycol, and water. The stream was introduced into the second reactor CRV-101 with CRV-101 Hydrogen. The reactor CRV-101 was maintained at 200°C using adjuster ADJ-3. The vapor from the reactor, CRV-101 Vapor Out was condensed by cooling in E-103. Adjuster ADJ-4 was set to completely condense the vapor by adjusting the energy stream, E-103 Cooling. The condensed stream was mixed with the liquid stream from the reactor CRV-101 Liquid Out in mixer MIX-102. The MIX-102 Out stream was sent to the purification section.

The MIX-102 Out stream contained propylene glycol, unreacted glycerol, and water. The stream was separated into two consecutive distillation columns, T-100 and T-101. T-100 distillation column was used to separate glycerol from the propylene glycol and water stream. The distillation column had 10 stages with full reflux condenser and reboiler operating at 101.3 kPa. The reflux ratio 0.20 was used to achieve 99.99% separation of glycerol. The T-100 Glycerol stream was recycled to the hydrogenolysis section in RCY-1. The T-100 Top stream containing water and propylene glycol was separated in T-101. The column had 10 stages with full reflux condenser and reboiler operating at 101.3 kPa. A reflux ratio of 2 was required to obtain a 99.99% separation of propylene glycol in stream T-101 Propylene Glycol. Water vapor at 100°C was obtained in the process in T-101 Water Vapor. Propylene glycol was cooled to 25°C in cooler E-104 to obtain the propylene glycol in stream E-104 Propylene Glycol.

The overall mass balances and energy requirements for inlet and outlet streams are given in Tables 4.22 and 4.23, respectively. From Table 4.22, it can be seen that 14,800 kg/h of glycerol was required to produce 9,280 kg/h of propylene glycol. The energy required by the process was 1.02×10^8 kJ/h and the energy removed from the process was 1.00×10^8 kJ/h. The detailed stream descriptions are given in Appendix F.

TABLE 4.22

Overall Mass Balance for Propylene Glycol Production from Hydrogenolysis of Glycerol

Inlet Material Streams	Mass Flow (kg/h)	Outlet Material Streams	Mass Flow (kg/h)
Glycerol	1.48E+04	T-101 water vapor	5.74E+03
Hydrogen	2.46E+02	E-104 propylene glycol	9.28E+03
Total flow of inlet streams	1.50E+04	Total flow of outlet streams	1.50E+04

TABLE 4.23

Overall Energy Balance for Propylene Glycol Production
from Hydrogenolysis of Glycerol

Inlet Streams	Energy Flow (kJ/h)	Outlet Streams	Energy Flow (kJ/h)
Glycerol	−1.39E+08	T-101 water vapor	−7.59E+07
Hydrogen	0.00E+00	E-104 propylene glycol	−6.05E+07
Stream enthalpy in:	−1.39E+08	Stream enthalpy out:	−1.36E+08
CRV-100 heating	1.63E+07	E-101 cooling	1.60E+06
CRV-101 heating	1.99E+07	E-102 cooling	1.95E+07
T-100 reboiler	2.50E+07	E-103 cooling	2.44E+07
T-101 reboiler	4.11E+07	E-104 cooling	4.60E+06
		T-100 condenser	2.41E+07
		T-101 condenser	2.59E+07
External energy in:	1.02E+08	External energy out:	1.00E+08
Total flow of inlet streams	−3.64E+07	Total flow of outlet streams	−3.64E+07

4.5.2 Process Cost Estimation for Propylene Glycol Production from Hydrogenolysis of Glycerol

The HYSYS flow sheet was exported to IPE using the embedded export tool in HYSYS. This tool is accessed from Tools → Aspen Icarus → Export Case to IPE. The project result summary is given in Table 4.24. Table 4.25 gives the breakdown of the operating costs in percentage, and Figure 4.18 shows a pie chart of the distribution of operating costs, which include the raw material and utilities costs. From Figure 4.18, it can be seen that the raw material, glycerol, if bought at the current market price, constitutes 88% of the total operating costs. The equipment from the HYSYS case was mapped in IPE and is given in Appendix G with the respective costs of equipment as obtained from ICARUS.

The raw material and product unit costs used in the project basis are given in Table 4.26. The hydrogen price was computed from the price of natural

TABLE 4.24

Project Costs for Propylene Glycol Production
from Hydrogenolysis of Glycerol

Cost	Amount	Unit
Yield of propylene glycol	163,740,000	LB/year
Total project capital cost	6,600,000	USD
Total operating cost	83,400,000	USD/year
Total raw materials cost	73,300,000	USD/year
Total utilities cost	2,410,000	USD/year
Total product sales	133,000,000	USD/year

TABLE 4.25

Operating Costs for Propylene Glycol
Production from Hydrogenolysis of
Glycerol

Operating Cost	Percentage
Total raw materials cost	88
Total operating labor and maintenance cost	1
Total utilities cost	3
Operating charges	7
Plant overhead	1
G and A cost	0

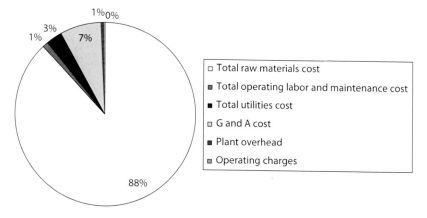

FIGURE 4.18
Operating cost breakdown for propylene glycol production from hydrogenolysis of glycerol.

TABLE 4.26

Raw Material and Product Unit Prices Used in ICARUS Cost Evaluation
for Propylene Glycol Production from Hydrogenolysis of Glycerol

Product/Raw Material	Flow Rate from HYSYS Simulation (kg/h)	Cost/Selling Price ($/kg)	Source
Hydrogen	246	1.50	Hydrogen price, Appendix C
Glycerol	14,774	0.60	*ICIS Chemical Business* (2008)
Copper chromite	884	0.55	Cost based on Dasari et al. (2005), used in capital cost
Propylene glycol	9,284	1.80	*ICIS Chemical Business* (2007)

gas as given in Appendix C. The price of glycerol and propylene glycol was obtained from *ICIS Chemical Business* (ICIS, 2007, 2008). The price of the catalyst was considered a one-time investment, and considered in the capital costs. The minimum selling price from operating costs ($83,400,000/year) per lb of propylene glycol produced (163,740,000 lb/year) was $0.53 cents per lb of propylene glycol.

4.5.3 Summary of Propylene Glycol Production from Hydrogenolysis of Glycerol

The design for propylene glycol production from hydrogenolysis of glycerol over copper chromite catalyst at 200°C and 200 psi is described in this section. The process consists of two steps, a hydrogenolysis reaction process and a purification process. Propylene glycol of 99.99% purity was obtained in this process.

About 65,000 MT/year (8125 kg/h) of propylene glycol was used as a design basis (based on a proposed plant by Ashland/Cargill joint venture), but the actual production rate was slightly higher (9300 kg/h). The energy required by the process was 1.02×10^8 kJ/h and the energy liberated by the process was 1.00×10^8 kJ/h.

The economic analysis was performed in IPE. The total project capital cost was $6.6 million. The operating cost was $83 million/year, which included raw material costs of $73 million/year. A minimum product selling price computed from the operating cost was $0.53/lb for propylene glycol.

4.6 Acetic Acid Production from Corn Stover Anaerobic Digestion

Anaerobic digestion of biomass is the treatment of biomass with a mixed culture of bacteria to produce methane (biogas) as a primary product. The four stages of anaerobic digestion are hydrolysis, acidogenesis, acetogenesis, and methanogenesis.

In the first stage, hydrolysis, complex organic molecules are broken down into simple sugars, amino acids, and fatty acids with the addition of hydroxyl groups. In the second stage, acidogenesis, volatile fatty acids (e.g., acetic, propionic, butyric, and valeric) are formed along with ammonia, carbon dioxide, and hydrogen sulfide. In the third stage, acetogenesis, simple molecules from acidogenesis are further digested to produce carbon dioxide, hydrogen, and organic acids, mainly acetic acid. Then in the fourth stage, methanogenesis, the organic acids are converted to methane, carbon dioxide, and water. The fourth stage of methane formation can be inhibited by the use of iodoform or bromoform, thus producing carboxylic acids, hydrogen, and carbon dioxide.

Anaerobic digestion can be conducted either wet or dry, where dry digestion has a solids content of 30% or greater and wet digestion has a solids content of 15% or less. Either batch or continuous digester operations can be used. In continuous operations, there is a constant production of biogas while batch operations can be considered simpler and the production of biogas varies.

The standard process for anaerobic digestion of cellulose waste to biogas (65% methane–35% carbon dioxide) uses a mixed culture of mesophilic or thermophilic bacteria. Mixed cultures of mesophilic bacteria function best at 37°C–41°C and thermophilic cultures function best at 50°C–52°C for the production of biogas (Kebanli et al., 1981).

Thanakoses et al. (2003a) describe a modification of the anaerobic digestion process, the MixAlco process, where corn stover and pig manure are converted to carboxylic acids. In the MixAlco process, anaerobic digestion is used to produce mixed alcohols by inhibiting the fourth stage, methanogenesis.

The process described by Thanakoses et al. (2003a) for the conversion of 80% corn stover and 20% pig manure mixture to carboxylic acid was used for the HYSYS design case. Other raw materials that can be used include municipal solid waste/sewage sludge mixture (Aiello-Mazzarri et al., 2006) and sugarcane bagasse/chicken manure mixture (Thanakoses, 2003b). The compositions of the other raw materials were not readily available, so the corn stover/pig manure conversion to acetic acid was designed in HYSYS.

The composition of corn stover was obtained from the Department of Energy (Aden et al., 2002) and is given in Table 4.1. The composition of corn stover used in this design is given in Table 4.2. The composition of pig manure used in this design is given in Table 4.27. The flow rate for pig manure was computed from the flow rate of corn stover to form the 80% corn stover and 20% pig manure mixture.

The compositions of corn stover and pig manure are as follows. Corn stover is composed of mainly cellulose, hemicellulose, and lignin. Cellulose and hemicellulose are organic compounds with the formula $(C_6H_{10}O_5)_n$ and $(C_5H_8O_4)_m$, respectively. $C_6H_{10}O_5$, also known as glucan, represents the

TABLE 4.27

Composition of Pig Manure Used in Design

Component	% Mass Basis
Glucan	52.5
Ash	30.0
Other solids	17.5
Mass percent of dry pig manure	100.00
Composition of feed into reactor	
Mass percent of dry pig manure	50.00
Water	50.00

monomer of cellulose and $C_5H_8O_4$, also known as xylan, represents the monomer of hemicellulose. Aden et al. (2002) use the terms glucan to represent
cellulose and xylan to represent hemicellulose. Aden et al. (2002) calculate
the unknown soluble solids as the mass balance closure. The acetate, protein,
extractives, arabinan, galactan, and mannan are not standard components in
HYSYS. So these components are considered as other solids for the HYSYS
design. The dry biomass feed was adjusted to have 50% water going into the
reactor, as given in Table 4.2.

The composition of dry pig manure was obtained from Thanakoses et al.
(2003a). The composition of dry manure after pretreatment as given in the
paper was 54.3% cellulose (all the carbohydrates in pig manure was cellulose), 28.7% ash, and 17% other solids (proteins). Using these calculations, the
inlet composition of the pig manure was determined using the pretreatment
reaction conditions. The inlet composition of pig manure is given in Table
4.27. The flow rate of the pig manure was computed on the basis of the flow
rate of corn stover to make an 80% corn stover–20% pig manure mixture; 50%
water was added to the dry pig manure to make wet pig manure. The bacteria contained in pig manure was not mentioned to be either thermophilic or
mesophilic. Considering the reaction condition of 77% conversion at 40°C, it
is assumed that the mixed culture of bacteria was mesophilic.

4.6.1 Process Description for Acetic Acid Production from Corn Stover Anaerobic Digestion

The anaerobic digestion plant was designed for processing of 2000 MT/day
(dry basis). This amounts to 83,333 kg/h conversion of dry corn stover. The
UNIQUAC thermodynamic model was used for estimating the interaction
between reaction components. Acetic acid was the representative carboxylic
acid for the HYSYS design.

The HYSYS process flow diagram is given in Figure 4.19. The design is
described in the following three parts, the pretreatment section (Figure 4.20),
the anaerobic digestion section (Figure 4.21), and the purification and recovery section (Figure 4.22). In these sections, a brief overview of the whole process is given first, followed by a detailed description of the streams.

4.6.1.1 Pretreatment Section

The pretreatment section is shown in Figure 4.20. The 80%–20% corn
stover/pig manure mixture was pretreated with lime and steam in V-100.
Steam converted 20% of the cellulose and hemicellulose in biomass to the
monomers, glucose and xylose, respectively. Lime pretreatment was used
to facilitate the enzymatic hydrolysis of corn stover (Kaar and Holtzapple,
2000). Lime addition rate was given as 0.05–0.15 g/g biomass at a temperature
between 70°C and 130°C in Wyman et al. (2005). The concentration of solids after lime pretreatment was 5%–20%. So, considering 10% of remaining

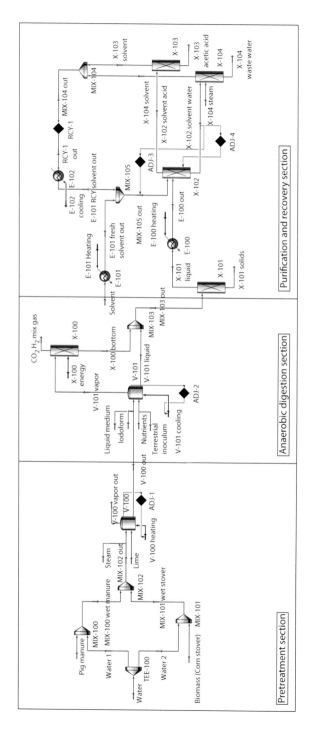

FIGURE 4.19
(See color insert.) Overall process design diagram for acetic acid production from corn stover anaerobic digestion.

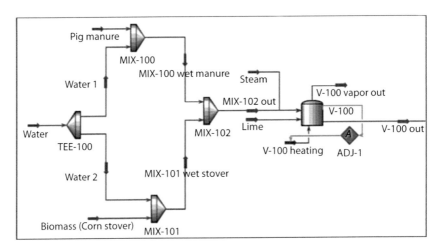

FIGURE 4.20
Pretreatment section for acetic acid production from corn stover anaerobic digestion.

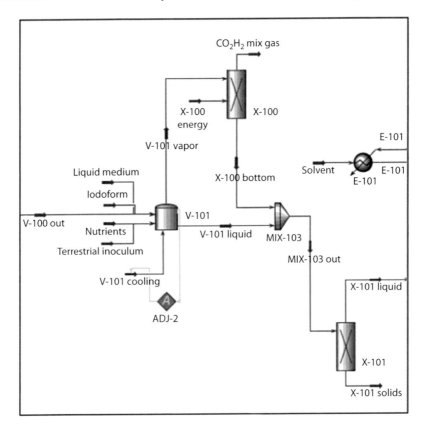

FIGURE 4.21
Anaerobic digestion section for acetic acid production from corn stover anaerobic digestion.

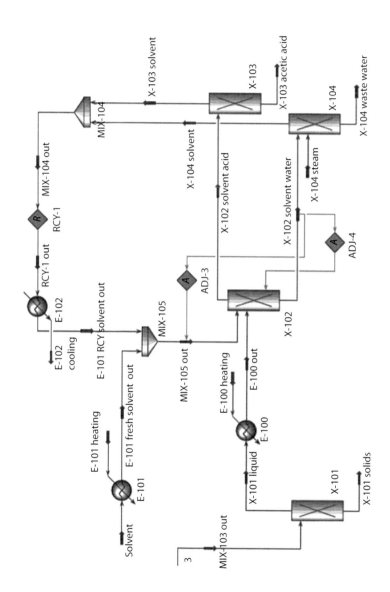

FIGURE 4.22
Purification and recovery section for acetic acid production from corn stover anaerobic digestion.

solids, the lime pretreatment converted 92% of the biomass remaining after steam hydrolysis. A conversion of 92% for the cellulose and hemicellulose to glucose and xylose was used in the pretreatment reactor.

Referring to Figure 4.20, Biomass (corn stover) stream containing dry biomass (composition given in Table 4.2) was mixed with equal mass of water (Water 2) in MIX-101 and sent to MIX-102. Dry pig manure rate was computed from the dry biomass rate and water was added to make an 80% corn stover–20% pig manure mixture. Pig manure and water 1 (equal mass of pig manure) were combined in MIX-100 and then mixed with the biomass in MIX-102.

The corn stover–pig manure mixture is henceforth referred to as biomass. The biomass in stream MIX-102 Out was pretreated in V-100 with steam and lime. The pretreatment reactor was maintained at 100°C using ADJ-1. The steam pretreatment converted 20% of the biomass from stream MIX-102 Out. The conversion obtained by lime pretreatment was 90% of the remaining biomass after steam hydrolysis. Since both the pretreatment reactions were carried out in the same vessel, an overall conversion of 92% was obtained for the biomass in the pretreatment reactor V-100. The stream, V-100 Out, was sent to the anaerobic digester, V-101.

4.6.1.2 Anaerobic Digestion Section

The anaerobic digestion section is shown in Figure 4.21. The pretreated biomass is converted to acetic acid in V-101. A liquid medium, iodoform, nutrients, and terrestrial inoculum were necessary to convert the pretreated biomass to acetic acid.

Biomass is converted to acetic acid (CH_3COOH) under non-sterile anaerobic conditions according to the Equation 4.8 (Holtzapple et al., 1999). Glucose ($C_6H_{12}O_6$) is used for illustration of this reaction.

$$C_6H_{12}O_6 + 2H_2O + 4NAD^+ \rightarrow 2H_3CCOOH + 2CO_2 + 4NADH + 4H^+ \quad (4.8)$$

The reducing power of nicotinamide adenine dinucleotide (NADH) may be released as hydrogen using endogenous hydrogen dehydrogenase, as shown in Equation 4.9:

$$NADH + H^+ \rightarrow NAD^+ + H_2 \quad (4.9)$$

Methanogens are microorganisms that can produce methane by reacting carbon dioxide produced with hydrogen. The reaction is given in Equation 4.10:

$$CO_2 + 4H_2 \rightarrow CH_4 + 2H_2O \quad (4.10)$$

Acetic acid can also be converted to methane in the presence of methano-gens. So, the potential to convert all biomass to methane exists. The pro-duction of methane, according to Equation 4.10, can be inhibited by the addition of iodoform or bromoform. Thus, combining Equations 4.8 and 4.9, Equation 4.11 is obtained where acetic acid is produced from glucose and the production of methane is inhibited:

$$C_6H_{12}O_6 + 2H_2O \rightarrow 2H_3CCOOH + 2CO_2 + 4H_2 \tag{4.11}$$

The reaction for xylose is similar to glucose and can be represented by Equation 4.12:

$$C_5H_{10}O_5 + 1.67H_2O \rightarrow 1.67H_3CCOOH + 1.67CO_2 + 3.33H_2 \tag{4.12}$$

In the anaerobic digestion section shown in Figure 4.21, the pretreated bio-mass was anaerobically fermented in V-101. Iodoform, inoculum, nutrients, and a liquid medium were used in the reactor. These components were added as described in Thanakoses et al. (2003a). Acetic acid, carbon dioxide, and hydrogen were formed in the reactor V-101 according to Equations 4.11 and 4.12.

The reaction conversions and temperatures for the process were 41% at 40°C and 80% at 55°C, respectively, for acetic acid (Granda, 2007) in a mix-ture of carboxylic acids. The reaction conversion for acetic acid mentioned in Thanakoses et al.(2003a) was 77% at 40°C. The reaction conversion of 77% at 40°C for acetic acid was used in V-101. The unreacted biomass was removed in centrifuge X-101. The acetic acid and water mixture was separated in a liquid–liquid extraction process.

The liquid medium described in Thanakoses et al. (2003a) was water. The rate of addition of the liquid medium was not mentioned in the paper. So, the rate of the inlet water stream in the process, Water, was used as the basis for the liquid medium addition rate.

The anaerobic digestion of biomass to methane was inhibited by the addi-tion of iodoform. The addition rate of iodoform was determined on the basis of Thanakoses et al. (2003a). The iodoform solution was made with 20 mg/L ethanol. The iodoform solution addition rate to the reactor was 12 mg/L liq-uid medium.

Thanakoses et al. (2003a) reported the nutrient mixture addition rate to the reactor as 1.0 g/L of liquid medium. The rate of nutrient addition was deter-mined as 1.0 g/L of the stream, Water. The flow rate for terrestrial inoculum was not mentioned in Thanakoses et al. (2003a). There were no costs associ-ated with the collection of the inoculum as given in Holtzapple et al. (1999). The inoculum flow rate of 1.0 g/L of liquid medium, equal to the nutrient flow rate, was used.

Referring to Figure 4.21, the pretreated biomass V-100 Out was sent to the reactor V-101. Iodoform, nutrients, liquid medium, and terrestrial inoculum were added at the flow rates mentioned earlier. The temperature in the reactor was maintained at 40°C using ADJ-2. The glucose and xylose in the biomass were converted to acetic acid, carbon dioxide, and hydrogen according to the reactions given by Equations 4.11 and 4.12.

The top stream from the reactor, V-101 Vapor contained carbon dioxide, hydrogen, and acetic acid. The vapor stream was cooled in X-100. The top stream was CO_2 H_2 Mix Gas and the bottom stream was X-100 Bottom, containing the condensed acetic acid. The X-100 Bottom stream was mixed with V-101 Liquid in MIX-103 and sent to the purification and recovery section.

4.6.1.3 Purification and Recovery Section

The separation of acetic acid included a liquid–liquid extraction process for the separation of acetic acid and water. This separation using rectification is difficult (De Dietrich, 2010). Different methods are used to separate acetic acid from water, depending on the concentration of acetic acid present in feed. Between 50% and 70% w/w acetic acid, extractive distillation is used. A third component is added to increase the volatility of water and achieve separation with less energy. For mixtures with less than 40% (w/w) acetic acid, liquid–liquid extraction process is appropriate. Acetic acid is extracted from water using a suitable solvent in order to obtain pure acetic acid. Liquid–liquid extraction is also useful when other contaminants such as salts interfere with direct distillation.

The acetic acid concentration after centrifugal separation in X-101 was 17% (w/w) acetic acid in water. The concentration was less than 40% (w/w), so acetic acid was removed from the mixture of acetic acid and water using liquid–liquid extraction process. Methyl isobutyl ketone or ethyl acetate is the solvent for the process (De Dietrich, 2010). Methyl isobutyl ketone is a standard component in HYSYS and was used for the design. Methyl isobutyl ketone can be used for mixtures having up to 50% acetic acid concentration, giving greater flexibility to the system compared to using ethyl acetate.

Figure 4.22 shows the extraction process carried out in a liquid–liquid extraction column. The process described by De Dietrich (2010) for acetic acid separation was used in this design. HYSYS has a liquid–liquid extraction column among the equipment available. However, the available information on liquid–liquid extraction process was best suited for a perfect separator. So, a perfect separator, X-102, represented the liquid–liquid extraction column. The top stream from the extraction column contained the acetic acid and solvent mixture. It was sent to the solvent rectification column X-103 where acetic acid was separated from the solvent and the solvent was recycled; 100% acetic acid was recovered from the rectification column. The bottom stream from the liquid extraction column was sent to a stripping section X-104.

Steam was used to separate the solvent from water. The recovered solvent was recycled in the system.

In the purification and recovery section shown in Figure 4.22, the MIX-103 Out stream contained acetic acid, water, and unreacted biomass. The unreacted biomass was separated in centrifuge X-101. The acetic acid–water mixture in X-101 Liquid contained 17% acetic acid. It was separated in the perfect separator, simulated as liquid–liquid extractor, X-102. The stream, X-101 Liquid was heated to the boiling point of water in E-100 before passing the stream to the extractor. Fresh solvent, methyl isobutyl ketone, was introduced in stream Solvent, mixed with recycled solvent from the process in MIX-105, and introduced at the bottom of the extractor. The acetic acid–water mixture, having a higher density compared to the solvent, was introduced at the top of X-102 extractor. The top stream from the liquid–liquid extraction column contained the acetic acid and solvent mixture in stream X-102 Solvent Acid. It was sent to the solvent rectification column X-103 where acetic acid was separated from the solvent and the solvent was recycled in stream X-103 Solvent; 100% acetic acid was recovered from the rectification column. The bottom stream from the extraction column contained the water and solvent mixture in stream X-102 Solvent Water. The water and solvent were separated in a stripping column, X-104. Steam in stream X-104 Steam was used to separate the solvent from water. The solvent was recovered in stream X-104 Solvent.

The ADJ-3 was used to determine the flow rate of the solvent required in the liquid–liquid extraction column to maintain a fraction of 25% (w/w) water in the solvent/water azeotrope in stream X-102 Solvent Water. The ADJ-4 was used to determine the overhead fraction for acetic acid required to maintain a 0.05% mole acetic acid concentration in stream X-102 Solvent Water. The recovered solvent from the X-103 Solvent and X-104 Solvent was mixed in MIX-104 and recycled. The solvent was cooled in E-102 to 100°C and mixed with fresh solvent in MIX-105.

The overall mass balances in the inlet and outlet streams are given in Table 4.28. The detailed stream descriptions are given in Appendix F. Biomass (corn stover) (dry) at the rate of 83,000 kg/h and pig manure (dry) at the rate of 20,800 kg/h was pretreated using HP steam at the rate of 2,140 kg/h and lime at the rate of 8,300 kg/h. Anaerobic digestion of the pretreated biomass was carried out with terrestrial inoculum at the rate of 191 kg/h, nutrients at the rate of 191 kg/h, and iodoform at the rate of 2.3 kg/h. Carbon dioxide and hydrogen gas mixture was obtained from the system at the rate of 2520 kg/h. The gas mixture was in the ratio of 1:2 (mole ratio, according to stoichiometry). Methyl isobutyl ketone was used as solvent for the separation of acetic acid and water and was required at the rate of 13,400 kg/h in the Solvent stream. Acetic acid was obtained in the process at the rate of 29,200 kg/h in the stream X-103 Acetic Acid.

The energy requirement for the inlet and outlet energy streams is given in Table 4.29. The total energy required by the process was 4.59×10^8 kJ/h, and the total energy removed from the process was 3.87×10^8 kJ/h.

TABLE 4.28

Overall Mass Balances for Acetic Acid Production from Corn
Stover Anaerobic Digestion

Inlet Material Streams	Mass Flow (kg/h)	Outlet Material Streams	Mass Flow (kg/h)
Biomass (Corn stover)	8.33E+04	CO_2 H_2 mix gas	2.52E+04
Pig manure	2.08E+04	X-101 solids	7.20E+04
Water	1.04E+05	X-103 acetic acid	2.92E+04
Steam	2.14E+03	X-104 waste water	1.08E+05
Lime	8.33E+03		
Iodoform	2.29E+00		
Nutrients	1.91E+02		
Terrestrial inoculum	1.91E+02		
Solvent	1.34E+04		
X-104 steam	1.80E+03		
Total flow of inlet streams	2.34E+05	Total flow of outlet streams	2.34E+05

TABLE 4.29

Overall Energy Balances for Acetic Acid Production from Corn Stover
Anaerobic Digestion

Inlet Streams	Energy Flow (kJ/h)	Outlet Streams	Energy Flow (kJ/h)
Biomass (corn stover)	−1.92E+08	CO_2 H_2 mix gas	−2.06E+08
Pig manure	−4.78E+07	X-101 solids	−1.83E+08
Water	−1.65E+09	X-103 acetic acid	−2.20E+08
Steam	−2.84E+07	X-104 waste water	−1.36E+09
Lime	−6.13E+07		
Iodoform	−1.35E+04		
Nutrients	−1.80E+06		
Terrestrial inoculum	−4.37E+05		
Solvent	−4.37E+07		
X-104 steam	−2.39E+07		
Stream enthalpy in:	−2.05E+09	Stream enthalpy out:	−1.97E+09
V-100 heating	2.09E+08	E-102 cooling	9.98E+07
E-100 heating	5.79E+06	V-101 cooling	2.81E+08
E-101 heating	2.44E+08	X-100 energy	5.28E+06
External energy in:	4.59E+08	External energy out:	3.87E+08
Total flow of inlet streams	−1.59E+09	Total flow of outlet streams	−1.59E+09

4.6.2 Process Cost Estimation for Acetic Acid Production from Corn Stover Anaerobic Digestion

The HYSYS flow sheet was exported to IPE using the embedded export tool in HYSYS. This tool is accessed from Tools → Aspen Icarus → Export Case to IPE. The project results summary is given in Table 4.30. Table 4.31 gives the breakdown of the operating costs in percentage, and Figure 4.23 shows a pie chart of the distribution of operating costs, which include the raw material and utilities costs. From Figure 4.23, it can be seen that the raw material, corn stover, constitutes approximately 76% of the total operating costs. The equipment from the HYSYS case was mapped in IPE and is given in Appendix G with the respective costs of equipment as obtained from ICARUS.

The raw material and product unit costs used in project basis are given in Table 4.32. The costs for biomass (corn stover) was given as $60/dry U.S. ton in Aden (2008); 2000 dry MT/day of corn stover was processed in the facility. The pig manure is a waste product, so a cost is considered small enough to not be included for that raw material.

The cost of lime used in pretreatment section was reported as $42/ton and inhibitor (iodoform) used in the reaction section was reported as $3.30/kg

TABLE 4.30

Project Costs for Acetic Acid Production from Corn Stover Anaerobic Digestion

Cost	Amount	Unit
Yield of acetic acid	515,000,000	LB/year
Total project capital cost	6,090,250	USD
Total operating cost	56,666,300	USD/year
Total raw materials cost	42,902,500	USD/year
Total utilities cost	8,360,290	USD/year
Total product sales	117,181,000	USD/year

TABLE 4.31

Operating Costs for Acetic Acid Production from Corn Stover Anaerobic Digestion

Operating Cost	Percentage
Total raw materials cost	76
Total operating labor and maintenance cost	1
Total utilities cost	15
Operating charges	0
Plant overhead	1
G and A cost	7

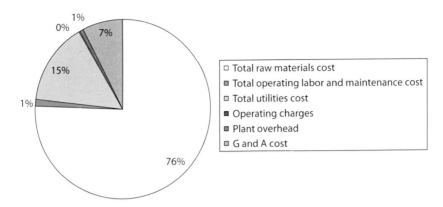

FIGURE 4.23
Operating cost breakdown for acetic acid production from corn stover anaerobic digestion.

TABLE 4.32

Raw Material and Product Unit Prices Used in ICARUS Cost Evaluation
for Acetic Acid Production from Corn Stover Anaerobic Digestion

Product/Raw Material	Flow Rate from HYSYS Simulation (kg/h)	Cost/Selling Price ($/kg)	Source
Corn stover	83,333	0.06	Aden (2007)
Lime	8,333	0.043	Holtzapple et al. (1999)
Inhibitor (iodoform)	2.29	3.3	Holtzapple et al. (1999)
HP steam (hydrolysis)	2,140	0.00983	ICARUS utility specification
HP steam (solvent recovery)	1,800	0.00983	ICARUS utility specification
Solvent (MIBK)	1,340	1.29	*ICIS Chemical Business* (2008)
Acetic acid	29,200	0.40	*ICIS Chemical Business* (2009)
$CO_2 + H_2$	25,200	0.123	Indala (2004) and Appendix [hydrogen price]

(Holtzapple et al., 1999). HP steam at 1000 kPa (145 psi) was used for steam
hydrolysis. The cost for the steam used was similar to cost of Steam at 165 psi,
which was $4.46/klb as described in ICARUS utility specification. The steam
was specified in the raw materials section instead of the utilities section in
IPE as it was used for prehydrolysis of the corn stover reaction process. Steam
was also used in the stripping column to recover solvent from water and solvent solution. The cost of solvent, methyl isobutyl ketone, was reported as
€950–€1000/MT ($1290–$1360/ton) (ICIS Chemical Business, 2008).

The carbon dioxide and hydrogen mixture obtained in the process was in
the ratio of 1:2 (molar ratio). This was equivalent to 92% CO_2 and 8% H_2 mass
ratio. The carbon dioxide price of $0.003/kg was determined as the price at
which it is available in the market from pipeline (Indala, 2004). The price of

hydrogen was determined as given in Appendix C. The unit cost of $1.001/kg of the mixture of CO_2 and H_2 obtained in the process was calculated from the unit costs of the individual gases, as shown in the following:

Unit cost of CO_2 = $0.003/kg

Unit cost of H_2 = $1.50/kg

CO_2 and H_2 gas mixture: 92% CO_2, 8% H_2

Unit cost of CO_2 + H_2 mixture = (0.003*92+1.5*8)/100 = 0.123$/kg

Acetic acid selling price in the market was reported as $400/ton (ICIS, 2009a). This price was used to compute the operating costs and product sales in ICARUS. The minimum selling price computed based on operating costs from the ICARUS results was $0.24/kg for acetic acid.

4.6.3 Summary of Acetic Acid Production from Corn Stover Anaerobic Digestion

The design of a process for anaerobic digestion of 80% corn stover and 20% pig manure mixture to produce acetic acid, and mixture of carbon dioxide and hydrogen gas was described. Steam hydrolysis with lime addition was used as the pretreatment step in the process. The anaerobic digestion of the pretreated biomass in the presence of a terrestrial inoculum (mixed culture) and nutrients gave acetic acid as a product. Further degradation of acetic acid into methane was inhibited by using iodoform. Unreacted solids and acetic acid and water mixture were separated using a centrifugal separator followed by a liquid–liquid extraction process with methyl isobutyl ketone as solvent; 100% pure acetic acid was obtained from the process. The carbon dioxide and hydrogen mixture can be used as a fuel or for the manufacture of chemicals.

About 83,333 kg/h of biomass was converted in the process, with the production of 29,200 kg/h acetic acid and 25,200 kg/h of CO_2 and H_2 mix gas. The energy required by the process was 4.59×10^8 kJ/h, and the total energy removed from the process was 3.87×10^8 kJ/h.

The economic analysis was performed in IPE. The total project capital cost was $6 million. The operating cost was $57 million/year, which included raw material costs of $42 million/year. A minimum product selling price computed from the operating cost was $0.24/kg for acetic acid.

4.7 Ethanol Production from Corn Dry-Grind Fermentation

A process and cost model for a conventional corn dry-grind processing facility producing 119 million kg/year (40MMgy) of ethanol was developed for use in evaluating new processing technologies and products from starch-based commodities by the USDA (Kwiatkowski et al., 2006). The capacity of the corn ethanol plant was comparable to any mid-sized corn ethanol plant existing in the United States (*Ethanol Producer Magazine*, 2010). The model was developed using SuperPro Designer software, and they include the composition of raw materials and products, sizing of unit operations, utility consumptions, estimation of capital and operating costs, and the revenues from products and coproducts (Intelligen, 2009). The model was based on data gathered from ethanol producers, technology suppliers, equipment manufacturers, and engineers working in the industry. This model was available for educational uses from the USDA and used in the analysis for corn ethanol production. The overall process flow diagram is shown in Figure 4.24.

In the paper by Kwiatkowski et al. (2006), the process simulator (SuperPro Designer) was used to calculate the processing characteristics, energy requirements, and equipment parameters of each major piece of equipment for the specified operating scenario. Volumes, composition, and other physical characteristics of input and output streams for each equipment item were identified. This information became the basis of utility consumptions and purchased equipment costs for each equipment item.

The design details are available in the paper, and the information used for the process model formulation in the next chapter is explained in this section. The composition for corn was used from this paper and given in Table 4.33. The components of corn include corn starch, water, non-starch polysaccharides (denoted by NSP), soluble and insoluble proteins, oil, and other solids (denoted by NFDS). The design can be divided into three sections, pretreatment, fermentation, and purification, as shown in Figure 4.24.

The pretreatment section, as shown in Figure 4.24, had a grain receiving unit, followed by liquefaction and saccharification. Liquefaction is the process step where starch is hydrolyzed (broken down) with thermostable alpha-amylase into oligosaccharides also known as dextrins. The conversion of the oligosaccharides by glucoamylase to glucose is referred to as saccharification. Process water, thermostable alpha-amylase, ammonia, and lime were mixed with corn in the liquefaction tank. Alpha-amylase was added at 0.082% (db) of corn brought to the slurry, while ammonia and lime were added at 90 and 54 kg/h, respectively. In the saccharification tank, sulfuric acid was used to lower the pH to 4.5. Glucoamylase was added at 0.11% (db) during the saccharification step, and the starch was further hydrolyzed from dextrins into glucose at a temperature of 61°C.

Starch is not a pure (standard) component in SuperPro Designer and was user defined. The molecular weight of starch used in the design was 18.20 g/gmol. The pretreatment reaction in the saccharification reactor converted 99% of starch to glucose, as shown in Equation 4.13:

$$8.9 \text{Starch} + H_2O \rightarrow C_6H_{12}O_6 \qquad (4.13)$$

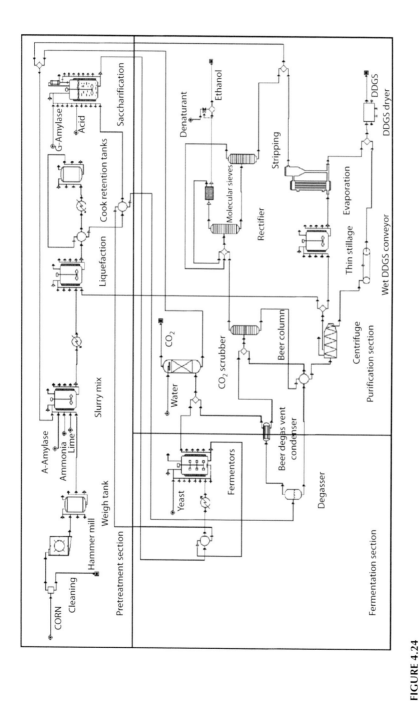

FIGURE 4.24

Overall SuperPro Designer® process design diagram for ethanol production from corn dry-grind fermentation. (From *Ind. Crop. Prod.*, 23(3), Kwiatkowski, J.R., McAloon, A.J., Taylor, F., and Johnston, D.B., Modeling the process and costs of fuel ethanol production by the corn dry-grind process, 288–296, Copyright 2006, with permission from Elsevier.)

TABLE 4.33

Composition of Corn

Component	Mass Percent
Starch	59.5
Water	15.0
Non-starch polysaccharides	7.0
Other solids	6.7
Protein—insoluble	6.0
Protein—soluble	2.4
Oil	3.4
Total	100

Source: Kwiatkowski, J.R. et al., *Ind. Crops Prod.*, 23(3), 288, 2006.

The fermentation section, as shown in Figure 4.24, converted the glucose obtained from the pretreatment section to ethanol. Fermentation is the conversion of glucose to ethanol and carbon dioxide using yeast. The fermentation simulated in the process model was a batch process with six fermentors of approximately 1.9 million l (504,000 gal) each. The reactions occurring in the reactor are given in Equations 4.14 and 4.15. The conversion in Equation 4.14 was 100% for glucose. The conversion in Equation 4.15 for NFDS was 6.8%. The term NFDS was used in the SuperPro design to signify other solids in the process. The molecular weight of NFDS and protein used in the design was 180.16 g/gmol.

$$C_6H_{12}O_6 \rightarrow 1.9C_2H_5OH + 1.9CO_2 + 0.05NFDS \qquad (4.14)$$

$$NFDS \rightarrow Protein \qquad (4.15)$$

The purification section, as shown in Figure 4.24, had beer from the fermentation heated using the process stream inlet to the saccharification tank, and then sent through a degasser drum to flash off the vapor. The vapor stream contained ethanol and water with some residual carbon dioxide. The ethanol and water vapors were condensed and recombined with the liquid stream prior to distillation. Any uncondensed vapor was combined with the carbon dioxide produced during fermentation and sent through the carbon dioxide scrubber prior to venting or recovery. Water was used in the carbon dioxide scrubbing process.

The ethanol recovery section consisted of multiple steps. In the first step, the beer column captured nearly all of the ethanol produced during fermentation. In the second step, water was removed from the process by rectification/stripping. The third step was a complete removal of water in molecular sieves. The detailed explanation is available in the paper. The ethanol recovered after molecular sieve adsorption was of 99.6% pure.

TABLE 4.34

Overall Mass Balance for Ethanol Production from Corn
Dry-Grind Fermentation

Inlet Material Streams	Mass Flow (kg/h)	Outlet Material Streams	Mass Flow (kg/h)
Corn	4.52E+04	Ethyl alcohol	1.44E+04
Lime	5.36E+01	PC	5.55E+02
Liq. ammonia	8.97E+01	Exhaust	4.19E+04
Alpha-amylase	3.15E+01	DDGS	1.50E+04
Gluco-amylase	4.54E+01	CO_2	1.38E+04
Sulfuric acid	8.97E+01		
Caustic	2.26E+03		
Yeast	1.09E+01		
Water	1.34E+04		
Air	2.45E+04		
Total flow of inlet streams	8.57E+04	Total flow of outlet streams	8.57E+04

The stillage bottoms from the beer column contained 15% solids and remaining water. About 83% of the water present was recovered during centrifugation producing wet distiller's grains at 37% solids. Processing steps were applied to recover the DDGS from the process. The DDGS was sold as an animal feed with its values based on the protein content.

The overall mass balance for the process is given in Table 4.34. The corn flow rate is 45,200 kg/h, and the ethanol obtained from the process was 14,400 kg/h. The energy balance for the process was not given in the paper or the design. Instead, the utilities were specified and the total utility cost was given. This is shown in Table 4.35.

TABLE 4.35

Utility Costs for Ethanol Production from Corn
Dry-Grind Fermentation

Utility	Annual Amount	Reference Units	Annual Cost ($)
Electricity	1.70E+07	kWh	8.51E+05
Natural Gas	7.40E+06	Kg	2.14E+06
CT Water	1.30E+10	Kg	9.13E+05
CT Water 35Cout	3.47E+08	Kg	2.43E+04
CT Water 31Cout	8.26E+08	Kg	5.78E+04
Steam 50 PSI	9.57E+07	Kg	1.63E+06
Steam 6258 BTU	3.15E+07	Kg	5.38E+05
Steam 556 BTU	1.64E+08	Kg	2.80E+06
Total			8.96E+06

4.8 Summary

This chapter described the process simulation models developed for fermentation, anaerobic digestion, and transesterification processes for the production of chemicals from biomass. The chemicals produced from the biomass were ethanol from corn and corn stover, FAME and glycerol from transesterification, acetic acid from anaerobic digestion, ethylene from ethanol, and propylene glycol from glycerol. The corn stover fermentation process, acetic acid process, FAME and glycerol process, propylene glycol process, and ethylene from ethanol process were designed in Aspen HYSYS. The process cost estimation for these processes were made in Aspen ICARUS. The corn ethanol process model was based on the USDA process for dry-grind ethanol and the relevant details from that process were discussed in this chapter.

Chapter 5 formulates the models for optimization. The process flow models from this chapter are converted to input–output block models, and the mass and energy balance equations describing the system are constructed. These equations are used to formulate the superstructure for optimization. The models for syngas from gasification of biomass and algae oil production were black box models described by a conversion equation and plant capacity information. The description of these process models are given in Chapter 5.

5

Bioprocesses Plant Model Formulation

5.1 Introduction

This chapter describes the development of input–output block models for the biomass processes to be used in the determination of the optimal structure. The overall diagram of the bioprocesses is shown in Figure 5.1. In this chapter, the model formulation of the individual processes shown in green is described. These processes were simulated in HYSYS as shown in Chapter 4.

The process design for ethanol production from corn stover fermentation was converted to the block flow diagram as shown in Figure 5.1. The block diagram for ethanol production from corn stover had three units: pretreatment (corn stover), fermentation (corn stover), and purification (corn stover EtOH). The total number of corn stover ethanol plants required to meet the capacity for ethylene is designated by EP1 on the diagram. The solid boundary in green around the corn stover ethanol fermentation blocks denotes that every stream within the boundary is multiplied by EP1 to get the flow rates into and out of the system.

The process design for ethanol production from dry-grind corn ethanol fermentation was converted to the block flow diagram as shown in Figure 5.1. The block diagram for ethanol production from corn contains three units: pretreatment (corn), fermentation (corn), and purification (corn EtOH). The total number of corn ethanol plants required to meet the capacity for ethylene is designated by EP2 on the diagram. The solid boundary in green around the corn ethanol fermentation blocks denotes that every stream within the boundary is multiplied by EP2 to get the flow rates into and out of the system.

The process design for ethylene production from the dehydration of ethanol was converted to the block flow diagram as shown in Figure 5.1. The block diagram for ethylene production from ethanol contains one unit, ethylene. Ethanol from the corn stover fermentation and corn fermentation sections were combined and this was the feed to the ethylene plant.

The process design for acetic acid production from corn stover anaerobic digestion was converted to the block flow diagram as shown in Figure 5.1.

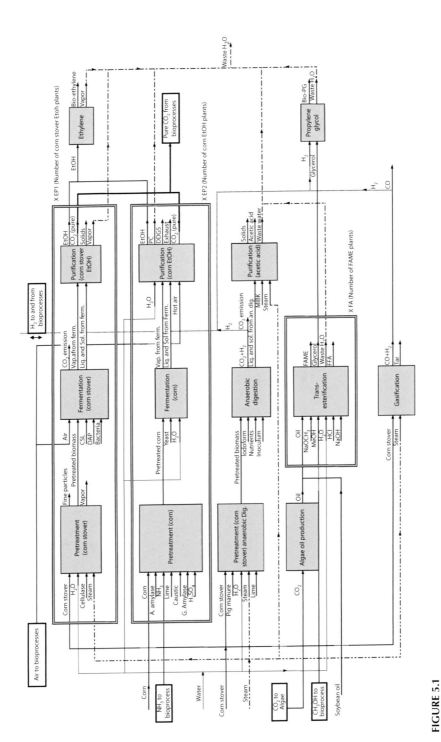

FIGURE 5.1
(See color insert.) Overall biochemical processes block flow diagram.

The block diagram for acetic acid production from corn stover contains three units: pretreatment (Corn Stover), anaerobic digestion, and purification (acetic acid).

The process design for fatty acid methyl ester (FAME) and glycerol from transesterification was converted to the block flow diagram as shown in Figure 5.1. The block diagram for FAME and glycerol production from natural oils contains one unit, transesterification. The number of FAME plants required to produce glycerol to meet the capacity of the propylene glycol plant is designated by FA. The solid boundary around the transesterification block denotes that every stream within the boundary is multiplied by FA to get the flow rates into and out of the system.

The process design for propylene glycol production from hydrogenolysis of glycerol was converted to the block flow diagram as shown in Figure 5.1. Glycerol from the total number of FAME plants (FA) was the feed to the propylene glycol plant.

As shown in Figure 5.1, two other process blocks, one for algae oil production and another for syngas gasification, were added to the biomass-based production complex. A black box model for algae oil production from carbon dioxide was included. The algae oil produced was combined with purchased soybean oil, and this was the feed to the transesterification process. A black box model for syngas production from gasification of biomass (corn stover) was included in the model.

The development of the block flow models using the chemical complex analysis system (Appendix E) for the aforementioned processes is given in Section 5.2. Each model includes material and energy balances, rate equation, and equilibrium relations. The organization of the sections is done in the following way. The reactions that describe the processes are given first, with the name of the HYSYS (or SuperPro Designer) model from where these relations were obtained. This is followed by the block flow diagram for the process. The models for fermentation and anaerobic digestion were divided into three sections due to the complexity of the HYSYS models, and to differentiate between the distinct boundaries within each process. The block flow models for transesterification, ethylene production, and propylene glycol production converted from HYSYS contained one unit each. The production of algae oil and the syngas gasification from biomass were also single-unit models created with the input and output information available for those processes.

The variables used in the optimization model are described in a table for each model. The model formulation equations are explained after the definition of the streams. The parameters used in the model are explained with respect to each block where it was used. Then the results from the optimization model validation with HYSYS or SuperPro results are presented in a table for each block. After this, all the equations for the material and energy balances for each block are given in a table.

5.2 Ethanol Production from Corn Stover Fermentation

The process design for ethanol production from corn stover fermentation from Chapter 4 was converted to the block flow diagram as shown in Figure 5.2. The block diagram for ethanol production from corn stover had three units: pretreatment (corn stover), fermentation (corn stover), and purification (corn stover EtOH). These denote the pretreatment section, fermentation section, and purification section from the HYSYS model. These three processes are separately represented in three blocks in the optimization model. The reactions occurring in the process are given in Table 5.1. The streams are shown in Figure 5.2 and the stream descriptions are given in Table 5.2. The parameters for the process are given in Table 5.3. The overall balance for each section and the individual species mass balance and energy balance equations are given in Tables 5.7 and 5.8, respectively. The inlet and outlet stream flow rates for the blocks from the HYSYS design corresponding to the streams in Table 5.2 are given in Appendix F.

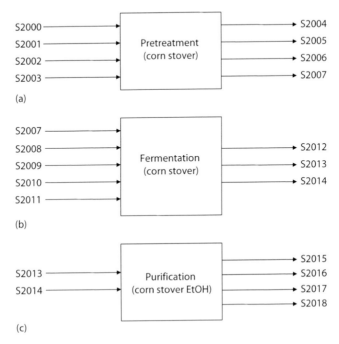

FIGURE 5.2
Block flow diagram of ethanol production from corn stover fermentation: (a) pretreatment (corn stover), (b) fermentation (corn stover), and (c) purification (corn stover EtOH).

TABLE 5.1

Reactions for Ethanol Production from Corn Stover Fermentation

Step	Reaction	Conversion (%)
Steam hydrolysis	$(Glucan)_n + nH_2O \rightarrow nC_6H_{12}O_6$	7
Steam hydrolysis	$(Xylan)_n + nH_2O \rightarrow nC_5H_{10}O_5$	70
Enzymatic hydrolysis	$(Glucan)_n + nH_2O \rightarrow nC_6H_{12}O_6$	90
Seed fermentation	$0.56C_6H_{12}O_6 + 4.69O_2 \rightarrow 3.4CO_2$ $+ 3.33H_2O + 0.23$ *Z. mobilis*	97
	$0.67C_5H_{10}O_5 + 4.69O_2 \rightarrow 3.52CO_2$ $+ 3.33H_2O + 0.20$ *Z. mobilis*	95
Fermentation	5 Glucose $\rightarrow$ 3 *Z. Mobilis* + 8.187CO_2	1
	2 Xylose $\rightarrow$ *Z. Mobilis* + 2.729CO_2	1
	Glucose $\rightarrow 2C_2H_5OH + 2CO_2$	99
	Xylose $\rightarrow 1.654C_2H_5OH + 1.68CO_2$	99

The mass balance for the streams is according to the equation

$$F_{in}^{(i)} - F_{out}^{(i)} + F_{gen}^{(i)} - F_{cons}^{(i)} = 0$$

where
  F^i is the flow rate of ith component in the system
  the subscripts in, out, gen, and cons denote the flow rates into, out of, and generation of and consumption of component i in the system

The energy balance for the process is according to the equation

$$Q_{in} - Q_{out} \times F_p + \sum_j H_j F_{j_{in}} - \sum_j H_j F_{j_{out}} = 0$$

H_j is the enthalpy per unit mass of jth stream computed from HYSYS. Q_{out} is the energy removed from the system per unit mass of product p having a flow rate of F_p. Q_{in} is the energy required by the system and calculated in the model using the aforementioned equation with the values for the other terms specified from HYSYS.

5.2.1 Pretreatment (Corn Stover)

The block for pretreatment was created in the chemical complex analysis system, with S2000–S2003 as inlet streams and S2004–S2007 as outlet streams. The parameters for biomass composition and cellulase enzyme composition were entered in the system as given in Table 5.3. The flow rate of S2001 was fixed with the capacity constraint for 2000 MT/day of dry corn stover. The flow rate of the remaining inlet streams were obtained as fractions of the dry biomass stream and its components. The inlet water stream, S2000, was equal

TABLE 5.2

Description of Process Streams in Ethanol Production
from Corn Stover Fermentation

Name of Streams	Description
Pretreatment (Corn Stover)	
Input streams	
S2000	Water added to dry biomass
S2001	Dry biomass (corn stover)
S2002	HP steam to steam hydrolysis reactor V-100
S2003	Cellulase to enzyme hydrolysis reactor V-102
Output streams	
S2004	Fine particles from centrifuge TEE-100
S2005	Steam from flash separator V-101
S2006	Vapor from reactor V-102
S2007	Pretreated biomass stream from V-102
Energy streams	
QFEPRO	Heat removed by cooling water in pretreatment section
QFEPRI	Heat required from steam in pretreatment section
Fermentation (Corn Stover)	
Input streams	
S2007	Pretreated biomass stream from V-102
S2008	Air–seed production
S2009	Corn steep liquor to fermentation section
S2010	DAP to fermentation section
S2011	Bacteria to seed fermentors
Output streams	
S2012	Vapor from seed reactors in MIX-110
S2013	Vapor from fermentation reactor V-105
S2014	Liquid from fermentation reactor V-105
Energy streams	
QFEFEO	Heat removed by cooling water in fermentation section
QFEFEI	Heat required from steam in fermentation section
Purification (Corn Stover)	
Input streams	
S2013	Vapor from fermentation reactor V-105
S2014	Liquid from fermentation reactor V-105

TABLE 5.2 (continued)

Description of Process Streams in Ethanol Production
from Corn Stover Fermentation

Name of Streams	Description
Output streams	
S2015	Ethanol from the process in stream E-106
S2016	CO_2 liberated in the process from flash separator V-106
S2017	Residual solids from centrifuge X-100
S2018	Vapor from adsorption and distillation section in MIX-109
Energy streams	
QFEPUO	Heat removed by cooling water in purification section
QFEPUI	Heat required from steam in purification section
Overall energy stream: QFE = QFEPRI + QFEFEI + QFEPUI	

to the flow rate of dry biomass, S2001. The steam for steam hydrolysis was used for the conversion of hemicellulose in the biomass; hence, the flow rate of steam, S2002, was computed from the fraction of steam required per unit mass of the hemicellulose in the biomass stream S2001H, added to the Scalar4 set as HPSTFRAC. The cellulase enzyme for enzymatic hydrolysis targeted the conversion of cellulose in the biomass. The flow rate of cellulase, S2003, was computed from the fraction of cellulase required per unit mass of the cellulose in the biomass stream, S2001C, added to the Scalar4 set as CELLFRAC.

The steam hydrolysis and enzymatic hydrolysis occurred in two reactors. The reactions are given in Table 5.1 for the cellulose and hemicellulose conversion to glucose and xylose, respectively. The steam hydrolysis converted 7% of the cellulose and 70% of the hemicellulose. The enzymatic hydrolysis converted 90% of the cellulose. These conversion factors, STHYCELCONV, STHYHECELCONV, ENHYCELCONV, and ENHYHECELCONV are given in Table 5.3. The parameter for enzymatic hemicellulose conversion, ENHYHECELCONV, was also included in the model equations for possible modifications to the model based on simultaneous enzymatic hydrolysis of hemicellulose and cellulose.

The outlet stream, S2004, was the fine particles removed from the centrifuge, and it was a fraction of inlet biomass and water stream. This fraction was specified in the Scalar4 list as FPFRAC. The stream, S2005 is the steam that comes out of the flash separation after steam hydrolysis. This is a fraction of the inlet stream for steam, S2002. This fraction is specified in the Scalar4 list as STOUTFRAC.

There were 32 variables in the pretreatment section and 33 equations, including two overall equations. Therefore, the degree of freedom for the pretreatment section was $32 - (33 - 2) = 1$. The constraint for capacity of

TABLE 5.3

Parameters in Ethanol Production from Corn Stover Fermentation

Name	Meaning	Value
Biomass composition		
MFCELP	Mass fraction of cellulose in S2001 corn stover	0.374
MFHEMP	Mass fraction of hemicellulose in S2001 corn stover	0.211
MFLIGP	Mass fraction of lignin in S2001 corn stover	0.180
MFASHP	Mass fraction of ash in S2001 corn stover	0.052
MFOTHP	Mass fraction of other solids in S2001 corn stover	0.183
Cellulase composition		
MFCELL	Mass fraction of cellulase in S2003	0.10
MFWATER	Mass fraction of water in S2003	0.90
Conversion in reactors		
STHYCELCONV	Steam hydrolysis conversion of cellulose to glucose	0.07
STHYHECELCONV	Steam hydrolysis conversion of hemicellulose to xylose	0.70
ENHYCELCONV	Enzymatic hydrolysis conversion of cellulose to glucose	0.90
ENHYHECELCONV	Enzymatic hydrolysis conversion of hemicellulose to xylose	0.00
SFEGLCONV	Seed fermentation conversion of glucose to bacteria	0.97
SFEXYCONV	Seed fermentation conversion of xylose to bacteria	0.95
FEGLBCONV	Fermentation conversion of glucose to bacteria	0.01
FEXYBCONV	Fermentation conversion of xylose to bacteria	0.01
FEGLECONV	Fermentation conversion of glucose to ethanol	0.99
FEXYECONV	Fermentation conversion of xylose to ethanol	0.99
Stream fractions		
FPFRAC	Fine particles fraction from centrifuge TEE-100	0.05
HPSTFRAC	High pressure steam fraction with respect to hemicellulose flow rate in biomass stream S2001	3.412
CELLFRAC	Cellulase enzyme fraction with respect to cellulose flow rate in biomass stream S2001	0.086
STOUTFRAC	Stream fraction out of flash drum V-101 with respect to HP Steam into the pretreatment reactor	0.423
SSTFRAC	Seed stream fraction in Tee-101	0.10
SFSTFRAC	Seed fermentor Seed Stream 1 fraction in Tee-102	0.20
AIRFRAC	Fraction of air with respect to pretreated biomass flow rate in stream S2007	0.27
CSLFRAC	Fraction of corn steep liquor with respect to pretreated biomass flow rate in stream S2007	0.00203
DAPFRAC	Fraction of diammonium phosphate with respect to pretreated biomass flow rate in stream S2007	0.000363
SBACFRAC	Fraction of bacteria with respect to pretreated biomass flow rate in stream S2007	0.01

TABLE 5.4

Pretreatment (Corn Stover) Section Optimization
Model Results Validated with Data from HYSYS (kg/h)

Stream Name	Data from HYSYS	Data from the System	Percent Difference (%)
S2004	8.33E+03	8.33E+03	0
S2004A	2.17E+02	2.17E+02	0
S2004C	1.56E+03	1.56E+03	0
S2004H	8.79E+02	8.79E+02	0
S2004H$_2$O	4.17E+03	4.17E+03	0
S2004L	7.50E+02	7.50E+02	0
S2004OS	7.62E+02	7.62E+02	0
S2005	2.54E+04	2.54E+04	0
S2005A	6.42E+01	6.34E+01	−1
S2005H$_2$O	2.53E+04	2.53E+04	0
S2007	1.96E+05	1.96E+05	0
S2007A	4.05E+03	4.05E+03	0
S2007C	2.75E+03	2.75E+03	0
S2007CA	2.69E+02	2.68E+02	0
S2007G	2.98E+04	2.98E+04	0
S2007H	5.01E+03	5.01E+03	0
S2007H$_2$O	1.12E+05	1.12E+05	0
S2007L	1.42E+04	1.42E+04	0
S2007OS	1.45E+04	1.45E+04	0
S2007X	1.33E+04	1.33E+04	0

processing 2000 MT/day of corn stover in the plant in stream S2001 specified the degree of freedom. The stream flow rates for S2004, S2005, and S2007 are compared with HYSYS results to check the validity of the model and are given in Table 5.4.

The energy balance equations are given in Table 5.8. The variables for the energy balance equation were H2001–H2007. A set with elements 2000 × 2050 was created in set "setbio." The mass enthalpy per unit mass of corresponding streams for corn stover fermentation process extension in the complex was included in the list H2, defined on "setbio," with the description "Enthalpy of biomass streams in complex."

The external energy variable for this process was QFEPRI, where "I" denotes the input coefficient for heat supplied by steam. The heat removed from this process was QFEPRO, where "O" denotes the output coefficient for heat removed. The value for QFEPRO and the enthalpy of the individual streams were specified from HYSYS. QFEPRI was calculated from the overall energy balance equation as given in Table 5.8. There were eight unknown variables in the energy balance for the pretreatment section and eight equations. So the degree of freedom was $8 - 8 = 0$.

The total degree of freedom for the pretreatment section was $(32 + 8) - (33 - 2 + 8) = 1$.

5.2.2 Fermentation (Corn Stover)

The input stream for this section was the S2007 pretreated biomass stream. The component flow rates for this stream were calculated in the pretreatment section. The variables in this section were S2007–S2014. The rest of the input streams to this section, S2008–S2011, were fractions of the pretreated biomass stream, and the parameters, AIRFRAC, CSLFRAC, DAPFRAC, and SBACFRAC, were used to give the stream relations.

The fermentation process involved seed generation of bacteria and fermentation reaction. The reactions given in Table 5.1 give the relations used for conversion reactions for glucose and xylose to bacteria and ethanol. SFEGLCONV, SFEXYCONV, FEGLBCONV, FEXYBCONV, FEGLECONV, and FEXYECONV are the conversion parameters in the process, and explained in Table 5.2.

The outlet streams in this section were S2012–S2014. The S2012 stream was the vapor from the seed reaction section, containing nitrogen, oxygen, carbon dioxide, and water vapor. The water vapor in the stream, S2012H2O, was a fraction of the inlet pretreated biomass stream, and this relation was computed from HYSYS and used in the model. The outlet vapor and the liquid/solids stream from the fermentation section, S2013 and S2014, respectively, were the inputs to the purification section. The components in these streams were calculated from species material balances.

There were 44 variables in the fermentation section and 37 equations, with three dependent equations. Therefore, the degree of freedom for the fermentation section was $44 - (37 - 3) = 10$. The flow rate variables for individual components in stream S2007 was computed from the previous section, and these ten mass flow rate variables were used to reduce the degree of freedom. The stream flow rates for S2012–S2014 are compared with HYSYS results to check the validity of the model and are given in Table 5.5.

The variables for the energy balance equation were H2007–H2014. The enthalpy of biomass streams for fermentation process extension in the complex was included in the list H2 with the description "Enthalpy of biomass streams in complex." The enthalpy per unit mass was entered in the list for the corresponding streams.

The external energy variable for this process was QFEFEI, where "I" denotes the input coefficient for heat supplied by steam. The heat removed from this process was QFEFEO, where "O" denotes the output coefficient for heat removed by cooling water. The value for QFEFEO and the enthalpy of

TABLE 5.5

Fermentation (Corn Stover) Section Optimization Model
Results Validated with Data from HYSYS (kg/h)

Stream Name	Data from HYSYS	Data from the System	Percent Difference (%)
S2012	5.52E+04	5.52E+04	0
S2012CO_2	6.28E+03	6.28E+03	0
S2012H_2O	2.11E+03	2.11E+03	0
S2012N_2	3.94E+04	3.94E+04	0
S2012O_2	7.41E+03	7.42E+03	0
S2013	2.03E+04	2.03E+04	0
S2013CO_2	1.89E+04	1.89E+04	0
S2013ETOH	8.41E+02	8.41E+02	0
S2013H_2O	5.09E+02	5.09E+02	0
S2014	1.76E+05	1.76E+05	0
S2014A	4.05E+03	4.05E+03	0
S2014B	3.86E+03	3.83E+03	−1
S2014C	2.75E+03	2.75E+03	0
S2014CA	2.69E+02	2.68E+02	0
S2014CO_2	1.29E+02	1.28E+02	0
S2014CS	3.97E+02	3.97E+02	0
S2014DA	7.08E+01	7.10E+01	0
S2014ETOH	1.88E+04	1.88E+04	0
S2014H	5.01E+03	5.01E+03	0
S2014H_2O	1.12E+05	1.12E+05	0
S2014L	1.42E+04	1.42E+04	0
S2014OS	1.45E+04	1.45E+04	0

the individual streams were specified from HYSYS. QFEFEI was calculated from the overall energy balance equation as given in Table 5.8. There were 10 unknown variables in the fermentation section and 10 equations. So the degree of freedom was 10 − 10 = 0.

The total degree of freedom for the fermentation section was (44 + 10) − (37 − 3 + 10) = 10.

5.2.3 Purification Section (Corn Stover EtOH)

The input stream for the purification section was the S2013 and S2014 stream containing ethanol, water, solids, and carbon dioxide. The variables in this section were S2013–S2018. The processes involved in this section included centrifugation to remove the solids from the stream, followed by distillation and molecular sieve adsorption separation of ethanol from water. The carbon dioxide was removed from the process from the vapor stream, S2013. The ethanol from the process was obtained in S2015. The carbon dioxide was liberated in S2016 along with some ethanol vapor. The ethanol split in the S2015ETOH stream was determined from the vapor split in the S2013 and S2014 streams from HYSYS. The carbon dioxide in the vapor stream was determined from the amount of dissolved CO_2 in stream S2017 from HYSYS. The water in streams S2017 and S2018 was determined from fractions computed from the inlet S2013 and S2014 streams from HYSYS. These relations were used as constraints in the model.

There were 37 variables in the purification section and 25 equations, with two dependent equations. Therefore, the degree of freedom for the purification section was $37 - (25 - 2) = 14$. The flow rate variables for individual components in stream S2013 and S2014 were computed from the previous section, and the 14 mass flow rate variables in those streams solved the degree of freedom. The stream flow rates for S2015–S2018 were compared with HYSYS results to check the validity of the model and are given in Table 5.6.

The variables for the energy balance equation were H2013–H2018. The enthalpy of biomass streams for fermentation process extension in the complex was included in the list H2 with the description "Enthalpy of biomass streams in complex." The mass enthalpy per unit mass was entered in the list for the corresponding streams.

The external energy variable for this section was QFEPUI, where "I" denotes the input coefficient for heat supplied by steam. The heat removed from this process was QFEPUO, where "O" denotes the output coefficient for heat removed by cooling water. The value for QFEPUO and the enthalpy of the individual streams were specified from HYSYS. QFEPUI was calculated from the overall energy balance equation as given in Table 5.8. There were nine unknown variables in the purification section and nine equations. So the degree of freedom was $9 - 9 = 0$.

The total degree of freedom for the purification section was $(37 + 9) - (25 - 2 + 9) = 14$.

The overall energy required from steam by the ethanol from the corn stover process was QFE, which was equal to the sum of QFEPRI, QFEFEI, and QFEPUI. The value for QFE was also validated in this section, given by QFE in Table 5.6.

The mass balance equations for overall and species balances for the corn stover pretreatment section, fermentation section, and purification section are given in Table 5.7. The conversion terms for the fermentation section are specified in separate variables in the block in chemical complex analysis system, CON1–CON6, but the complete equation is given in Table 5.7.

TABLE 5.6

Purification (Corn Stover EtOH) Section Optimization
Model Results Validated with Data from HYSYS (kg/h)

Stream Name	Data from HYSYS	Data from the System	Percent Difference (%)
S2015	1.98E+04	1.98E+04	0
S2015ETOH	1.96E+04	1.96E+04	0
S2015H$_2$O	1.03E+02	1.02E+02	−1
S2016	1.90E+04	1.90E+04	0
S2016CO$_2$	1.87E+04	1.87E+04	0
S2016ETOH	2.23E+01	2.23E+01	0
S2016H$_2$O	2.79E+02	2.79E+02	0
S2017	1.20E+05	1.20E+05	0
S2017A	4.05E+03	4.05E+03	0
S2017B	3.86E+03	3.83E+03	−1
S2017C	2.75E+03	2.75E+03	0
S2017CA	2.69E+02	2.68E+02	0
S2017CO$_2$	3.46E+02	3.46E+02	0
S2017CS	3.97E+02	3.97E+02	0
S2017DA	7.10E+01	7.10E+01	0
S2017H	5.01E+03	5.01E+03	0
S2017H$_2$O	7.45E+04	7.45E+04	0
S2017L	1.42E+04	1.42E+04	0
S2017OS	1.45E+04	1.45E+04	0
S2018	3.72E+04	3.72E+04	0
S2015	1.98E+04	1.98E+04	0
S2015ETOH	1.96E+04	1.96E+04	0
S2015H$_2$O	1.03E+02	1.02E+02	−1
S2016	1.90E+04	1.90E+04	0
S2016CO$_2$	1.87E+04	1.87E+04	0
S2016ETOH	2.23E+01	2.23E+01	0
S2016H$_2$O	2.79E+02	2.79E+02	0
S2017	1.20E+05	1.20E+05	0
S2017A	4.05E+03	4.05E+03	0
S2017B	3.86E+03	3.83E+03	−1
S2017C	2.75E+03	2.75E+03	0
S2017CA	2.69E+02	2.68E+02	0
S2017CO$_2$	3.46E+02	3.46E+02	0
S2017CS	3.97E+02	3.97E+02	0
S2017DA	7.10E+01	7.10E+01	0
S2017H	5.01E+03	5.01E+03	0
S2017H$_2$O	7.45E+04	7.45E+04	0
S2017L	1.42E+04	1.42E+04	0
S2017OS	1.45E+04	1.45E+04	0
S2018	3.72E+04	3.72E+04	0
QFE (kJ/h)	5.90E+08	5.90E+08	0

TABLE 5.7

Mass Balance Equations for Ethanol Production from Corn Stover Fermentation

Material Balance	IN−OUT + GENERATION−CONSUMPTION = 0

Pretreatment (Corn StoverR)

Overall

$$(F_{2000} + F_{2001} + F_{2002} + F_{2003}) - (F_{2004} + F_{2005} + F_{2006} + F_{2007}) = 0$$

where

$$F_{2001} = F_{2001}^{(Cellulose)} + F_{2001}^{(Hemicellulose)} + F_{2001}^{(Lignin)} + F_{2001}^{(Ash)} + F_{2001}^{(Other\ Solids)}$$

$$F_{2003} = F_{2003}^{(H_2O)} + F_{2003}^{(Cellulase)}$$

$$F_{2004} = F_{2004}^{(H_2O)} + F_{2004}^{(Cellulose)} + F_{2004}^{(Hemicellulose)} + F_{2004}^{(Lignin)} + F_{2004}^{(Ash)} + F_{2004}^{(Other\ Solids)}$$

$$F_{2005} = F_{2005}^{(H_2O)} + F_{2005}^{(Ash)}$$

$$F_{2007} = F_{2007}^{(H_2O)} + F_{2007}^{(Cellulose)} + F_{2007}^{(Hemicellulose)} + F_{2007}^{(Lignin)} + F_{2007}^{(Ash)} + F_{2007}^{(Other\ Solids)} + F_{2007}^{(Glucose)} + F_{2007}^{(Xylose)} + F_{2007}^{Cellulose}$$

Species:

Cellulose

$$F_{2001}^{(Cellulose)} - F_{2004}^{(Cellulose)} - F_{2007}^{(Cellulose)}$$

$$- \left(STHYCELCONV \times \left(F_{2001}^{(Cellulose)} - F_{2004}^{(Cellulose)} \right) \times \frac{mw(Cellulose)}{mw(Cellulose)} \right)$$

$$- \left(ENHYCELCONV \times (1-STHYCELCONV) \times \left(F_{2001}^{(Cellulose)} - F_{2004}^{(Cellulose)} \right) \times \frac{mw(Cellulose)}{mw(Cellulose)} \right) = 0$$

Hemicellulose

$$F_{2001}^{(\text{Hemicellulose})} - F_{2004}^{(\text{Hemicellulose})} - F_{2007}^{(\text{Hemicellulose})}$$

$$-\left(\text{STHYHECELCONV} \times \left(F_{2001}^{(\text{Hemicellulose})} - F_{2004}^{(\text{Hemicellulose})}\right) \times \frac{\text{mw(Hemicellulose)}}{\text{mw(Hemicellulose)}}\right)$$

$$-\left(\text{ENHYHECELCONV} \times (1-\text{STHYHECELCONV}) \times \left(F_{2001}^{(\text{Hemicellulose})} - F_{2004}^{(\text{Hemicellulose})}\right) \times \frac{\text{mw(Hemicellulose)}}{\text{mw(Hemicellulose)}}\right) = 0$$

H$_2$O

$$\left(F_{2000} + F_{2002} + F_{2003}^{(\text{H}_2\text{O})}\right) - \left(F_{2004}^{(\text{H}_2\text{O})} + F_{2005}^{(\text{H}_2\text{O})} + F_{2007}^{(\text{H}_2\text{O})}\right)$$

$$-\left(\text{STHYCELCONV} \times \left(F_{2001}^{(\text{Cellulose})} - F_{2004}^{(\text{Cellulose})}\right) \times (1) \times \frac{\text{mw(H}_2\text{O)}}{\text{mw(Cellulose)}}\right)$$

$$-\left(\text{STHYHECELCONV} \times \left(F_{2001}^{(\text{Hemicellulose})} - F_{2004}^{(\text{Hemicellulose})}\right) \times (1) \times \frac{\text{mw(H}_2\text{O)}}{\text{mw(Hemicellulose)}}\right)$$

$$-\left(\text{ENHYCELCONV} \times (1-\text{STHYCELCONV}) \times \left(F_{2001}^{(\text{Cellulose})} - F_{2004}^{(\text{Cellulose})}\right) \times (1) \times \frac{\text{mw(H}_2\text{O)}}{\text{mw(Cellulose)}}\right)$$

$$-\left(\text{ENHYHECELCONV} \times (1-\text{STHYHECELCONV}) \times \left(F_{2001}^{(\text{Hemicellulose})} - F_{2004}^{(\text{Hemicellulose})}\right) \times \frac{\text{mw(H}_2\text{O)}}{\text{mw(Hemicellulose)}}\right) = 0$$

Glucose

$$-F_{2007}^{(\text{Glucose})} + \left(\text{STHYCELCONV} \times \left(F_{2001}^{(\text{Cellulose})} - F_{2004}^{(\text{Cellulose})}\right) \times \frac{\text{mw(Glucose)}}{\text{mw(Cellulose)}}\right)$$

$$+\left(\text{ENHYCELCONV} \times (1-\text{STHYCELCONV}) \times \left(F_{2001}^{(\text{Cellulose})} - F_{2004}^{(\text{Cellulose})}\right) \times (1) \times \frac{\text{mw(Glucose)}}{\text{mw(Cellulose)}}\right) = 0$$

(continued)

TABLE 5.7 (continued)

Mass Balance Equations for Ethanol Production from Corn Stover Fermentation

Material Balance	IN−OUT + GENERATION−CONSUMPTION = 0
Xylose	$-F_{2007}^{(Xylose)} + \left(STHYHECELCONV \times \left(F_{2001}^{(Hemicellulose)} - F_{2004}^{(Hemicellulose)} \right) \times \dfrac{mw(Xylose)}{mw(Hemicellulose)} \right)$ $+ \left(ENHYHECELCONV \times (1 - STHYHECELCONV) \times \left(F_{2001}^{(Hemicellulose)} - F_{2004}^{(Hemicellulose)} \right) \times \dfrac{mw(Xylose)}{mw(Hemicellulose)} \right) = 0$
Ash	$F_{2001}^{(Ash)} - F_{2004}^{(Ash)} - F_{2005}^{(Ash)} - F_{2007}^{(Ash)} = 0$
Cellulase	$F_{2003}^{(Cellulase)} - F_{2007}^{(Cellulase)} = 0$
Other solids	$F_{2001}^{(Other\ Solids)} - F_{2004}^{(Other\ Solids)} - F_{2007}^{(Other\ Solids)} = 0$
Lignin	$F_{2001}^{(Lignin)} - F_{2004}^{(Lignin)} - F_{2007}^{(Lignin)} = 0$

Fermentation (Corn Stover)

Overall

$$(F_{2007} + F_{2008} + F_{2009} + F_{2010} + F_{2011}) - (F_{2012} + F_{2013} + F_{2014}) = 0$$

where

$$F_{2007} = F_{2007}^{(H_2O)} + F_{2007}^{(Cellulose)} + F_{2007}^{(Hemicellulose)} + F_{2007}^{(Lignin)} + F_{2007}^{(Ash)} + F_{2007}^{(Other\ Solids)} + F_{2007}^{(Glucose)} + F_{2007}^{(Xylose)} + F_{2007}^{(Cellulase)}$$

$$F_{2008} = F_{2008}^{(Nitrogen)} + F_{2008}^{(Oxygen)}$$

$$F_{2012} = F_{2012}^{(Nitrogen)} + F_{2012}^{(Oxygen)} + F_{2012}^{(H_2O)} + F_{2012}^{(CO_2)}$$

$$F_{2013} = F_{2013}^{(H_2O)} + F_{2013}^{(Ethanol)} + F_{2013}^{(CO_2)}$$

$$F_{2014} = F_{2014}^{(H_2O)} + F_{2014}^{(Cellulose)} + F_{2014}^{(Hemicellulose)} + F_{2014}^{(Lignin)} + F_{2014}^{(Ash)} + F_{2014}^{(Other\ Solids)} + F_{2014}^{(Cellulase)} + F_{2014}^{(Bacteria)} + F_{2014}^{(CSL)} + F_{2014}^{(DAP)} + F_{2014}^{(Ethanol)} + F_{2014}^{(CO_2)}$$

Species:

Glucose

$$
F_{2007}^{(\text{Glucose})} - \left(\left(\left(\text{SFEGLCONV} \times (\text{SSTFRAC} \times \text{SFSTFRAC}) \right) \times \frac{F_{2007}^{(\text{Glucose})}}{\text{mw}(\text{Glucose})} \times \text{mw}(\text{Glucose}) \right) \right.
$$

$$
- \left(\text{SFEGLCONV} \times \left(\left(\text{SSTFRAC} \times ((1-\text{SFSTFRAC}) + (1-\text{SFEGLCONV}) \times \text{SFSTFRAC}) \right) \right) \right.
$$
$$
\left. \times \frac{F_{2007}^{(\text{Glucose})}}{\text{mw}(\text{Glucose})} \times \text{mw}(\text{Glucose}) \right)
$$

$$
- \left((\text{FEGLECONV} + \text{FEGLBCONV}) \times \right.
$$
$$
\left. \left((1-\text{SSTFRAC}) + (1-\text{SFEGLCONV}) \times (\text{SSTFRAC} \times ((1-\text{SFSTFRAC}) + (1-\text{SFEGLCONV}) \times \text{SFSTFRAC})) \right) \right.
$$
$$
\left. \left. \times \frac{F_{2007}^{(\text{Glucose})}}{\text{mw}(\text{Glucose})} \times \text{mw}(\text{Glucose}) \right) \right) = 0
$$

Xylose

$$
F_{2007}^{(\text{Xylose})} - \left(\left(\left(\text{SFEXYCONV} \times (\text{SSTFRAC} \times \text{SFSTFRAC}) \right) \times \frac{F_{2007}^{(\text{Xylose})}}{\text{mw}(\text{Xylose})} \times \text{mw}(\text{Xylose}) \right) \right.
$$

$$
- \left(\text{SFEXYCONV} \times \left(\left(\text{SSTFRAC} \times ((1-\text{SFSTFRAC}) + (1-\text{SFEXYCONV}) \times \text{SFSTFRAC}) \right) \right) \right.
$$
$$
\left. \times \frac{F_{2007}^{(\text{Xylose})}}{\text{mw}(\text{Xylose})} \times \text{mw}(\text{Xylose}) \right)
$$

$$
- \left((\text{FEXYECONV} + \text{FEXYBCONV}) \times \right.
$$
$$
\left. \left((1-\text{SSTFRAC}) + (1-\text{SFEXYCONV}) \times (\text{SSTFRAC} \times ((1-\text{SFSTFRAC}) + (1-\text{SFEXYCONV}) \times \text{SFSTFRAC})) \right) \right.
$$
$$
\left. \left. \times \frac{F_{2007}^{(\text{Xylose})}}{\text{mw}(\text{Xylose})} \times \text{mw}(\text{Xylose}) \right) \right) = 0
$$

(continued)

TABLE 5.7 (continued)

Mass Balance Equations for Ethanol Production from Corn Stover Fermentation

Material Balance	IN−OUT + GENERATION−CONSUMPTION = 0
H_2O	$F_{2007}^{(H_2O)} - \left(F_{2012}^{(H_2O)} + F_{2013}^{(H_2O)} + F_{2014}^{(H_2O)} \right)$

$$+ \left((\mathrm{SFEGLCONV} \times (\mathrm{SSTFRAC} \times \mathrm{SFSTFRAC})) \times \left(\frac{3.33}{0.56}\right) \times \frac{F_{2007}^{(\mathrm{Glucose})}}{mw(\mathrm{Glucose})} \times mw(H_2O) \right)$$

$$+ \left(\mathrm{SFEGLCONV} \times \left(\mathrm{(SSTFRAC} \times \left((1-\mathrm{SFSTFRAC}) + (1-\mathrm{SFEGLCONV}) \times \mathrm{SFSTFRAC})\right)\right) \right.$$
$$\left. \times \left(\frac{3.33}{0.56}\right) \times \frac{F_{2007}^{(\mathrm{Glucose})}}{mw(\mathrm{Glucose})} \times mw(H_2O) \right)$$

$$+ \left((\mathrm{SFEXYCONV} \times (\mathrm{SSTFRAC} \times \mathrm{SFSTFRAC})) \times \left(\frac{3.33}{0.67}\right) \times \frac{F_{2007}^{(\mathrm{Xylose})}}{mw(\mathrm{Xylose})} \times mw(H_2O) \right)$$

$$+ \left(\mathrm{SFEXYCONV} \times \left(\mathrm{(SSTFRAC} \times \left((1-\mathrm{SFSTFRAC}) + (1-\mathrm{SFEXYCONV}) \times \mathrm{SFSTFRAC})\right)\right) \right.$$
$$\left. \times \left(\frac{3.33}{0.67}\right) \times \frac{F_{2007}^{(\mathrm{Xylose})}}{mw(\mathrm{Xylose})} \times mw(H_2O) \right) = 0$$

Bacteria

$$
\begin{aligned}
F_{2011} - F_{2014}^{(\text{Bacteria})} \\
+ \Bigg((\text{SFEGLCONV} \times (\text{SSTFRAC} \times \text{SFSTFRAC})) \times \left(\frac{0.229}{0.56}\right) \times \frac{F_{2007}^{(\text{Glucose})}}{\text{mw(Glucose)}} \times \text{mw(Bacteria)} \Bigg) \\
+ \Bigg(\text{SFEGLCONV} \times \Big((\text{SSTFRAC} \times ((1-\text{SFSTFRAC}) + (1-\text{SFEGLCONV}) \times \text{SFSTFRAC})) \\
\times \left(\frac{0.229}{0.56}\right) \times \frac{F_{2007}^{(\text{Glucose})}}{\text{mw(Glucose)}} \times \text{mw(Bacteria)} \Big) \Bigg) \\
+ \Bigg((\text{FEGLBCONV}) \times \\
\Big(((1-\text{SSTFRAC}) + (1-\text{SFEGLCONV}) \times (\text{SSTFRAC} \times ((1-\text{SFSTFRAC}) + (1-\text{SFEGLCONV}) \times \text{SFSTFRAC}))) \\
\times \left(\frac{3}{5}\right) \times \frac{F_{2007}^{(\text{Glucose})}}{\text{mw(Glucose)}} \times \text{mw(Bacteria)} \Big) \Bigg) \\
+ \Bigg((\text{SFEXYCONV} \times (\text{SSTFRAC} \times \text{SFSTFRAC})) \times \left(\frac{0.20}{0.67}\right) \times \frac{F_{2007}^{(\text{Xylose})}}{\text{mw(Xylose)}} \times \text{mw(Bacteria)} \Bigg) \\
+ \Bigg(\text{SFEXYCONV} \times \Big((\text{SSTFRAC} \times ((1-\text{SFSTFRAC}) + (1-\text{SFEXYCONV}) \times \text{SFSTFRAC})) \\
\times \left(\frac{0.20}{0.67}\right) \times \frac{F_{2007}^{(\text{Xylose})}}{\text{mw(Xylose)}} \times \text{mw(Bacteria)} \Big) \Bigg) \\
+ \Bigg((\text{FEXYBCONV}) \times \\
\Big(((1-\text{SSTFRAC}) + (1-\text{SFEXYCONV}) \times (\text{SSTFRAC} \times ((1-\text{SFSTFRAC}) + (1-\text{SFEXYCONV}) \times \text{SFSTFRAC}))) \\
\times \left(\frac{1}{2}\right) \times \frac{F_{2007}^{(\text{Xylose})}}{\text{mw(Xylose)}} \times \text{mw(Bacteria)} \Big) \Bigg) = 0
\end{aligned}
$$

(continued)

TABLE 5.7 (continued)

Mass Balance Equations for Ethanol Production from Corn Stover Fermentation

Material Balance	IN−OUT + GENERATION−CONSUMPTION = 0
Ethanol	$-F_{2013}^{(\text{Ethanol})} - F_{2014}^{(\text{Ethanol})}$

$$-F_{2013}^{(\text{Ethanol})} - F_{2014}^{(\text{Ethanol})}$$

$$+ \left(\left(\begin{array}{l} (\text{FEGLECONV}) \times \\ \left((1-\text{SSTFRAC}) + (1-\text{SFEGLCONV}) \times (\text{SSTFRAC} \times ((1-\text{SFSTFRAC}) + (1-\text{SFEGLCONV}) \times \text{SFSTFRAC})) \right) \\ \times \left(\dfrac{2}{1} \right) \times \dfrac{F_{2007}^{(\text{Glucose})}}{mw(\text{Glucose})} \times mw(\text{Ethanol}) \end{array} \right) \right.$$

$$\left. + \left(\begin{array}{l} (\text{FEXYECONV}) \times \\ \left((1-\text{SSTFRAC}) + (1-\text{SFEXYCONV}) \times (\text{SSTFRAC} \times ((1-\text{SFSTFRAC}) + (1-\text{SFEXYCONV}) \times \text{SFSTFRAC})) \right) \\ \times \left(\dfrac{1.654}{1} \right) \times \dfrac{F_{2007}^{(\text{Xylose})}}{mw(\text{Xylose})} \times mw(\text{Ethanol}) \end{array} \right) \right) = 0$$

CO_2

There are two separate equations for CO_2, the first one for the seed reactor section and the second one for the fermentation section.

$$-F_{2012}^{CO_2} + \left(\left(\text{SFEGLCONV} \times (\text{SSTFRAC} \times \text{SFSTFRAC}) \right) \times \left(\frac{3.4}{0.56} \right) \times \frac{F_{2007}^{(\text{Glucose})}}{mw(\text{Glucose})} \times mw(CO_2) \right)$$

$$+ \left(\begin{array}{l} \left(\text{SFEGLCONV} \times \left(\text{SSTFRAC} \times ((1-\text{SFSTFRAC}) + (1-\text{SFEGLCONV}) \times \text{SFSTFRAC}) \right) \right) \\ \times \left(\dfrac{3.4}{0.56} \right) \times \dfrac{F_{2007}^{(\text{Glucose})}}{mw(\text{Glucose})} \times mw(CO_2) \end{array} \right)$$

$$+\left(\left(\text{SFEXYCONV} \times (\text{SSTFRAC} \times \text{SFSTFRAC})\right) \times \left(\frac{3.52}{0.67}\right) \times \frac{F_{2007}^{(\text{Xylose})}}{\text{mw}(\text{Xylose})} \times \text{mw}(\text{CO}_2)\right)$$

$$+\left(\left(\text{SFEXYCONV} \times \left((\text{SSTFRAC} \times ((1-\text{SFSTFRAC}) + (1-\text{SFEXYCONV}) \times \text{SFSTFRAC}))\right)\right)\right.$$
$$\left. \times \left(\frac{3.52}{0.67}\right) \times \frac{F_{2007}^{(\text{Xylose})}}{\text{mw}(\text{Xylose})} \times \text{mw}(\text{CO}_2)\right) = 0$$

$$-F_{2013}^{(\text{CO}_2)} - F_{2014}^{\text{CO}_2}$$

$$+\left(\left(\left(\frac{2}{1}\right) \times \text{FEGLECONV} + \left(\frac{8.187}{5}\right) \times \text{FEGLBCONV}\right) \times \right.$$
$$\left. ((1-\text{SSTFRAC}) + (1-\text{SFEGLCONV}) \times (\text{SSTFRAC} \times ((1-\text{SFSTFRAC}) + (1-\text{SFEGLCONV}) \times \text{SFSTFRAC}))) \right.$$
$$\left. \times \frac{F_{2007}^{(\text{Glucose})}}{\text{mw}(\text{Glucose})} \times \text{mw}(\text{CO}_2)\right)$$

$$+\left(\left(\left(\frac{1.68}{1}\right) \times \text{FEXYECONV} + \left(\frac{2.729}{2}\right) \times \text{FEXYBCONV}\right) \times \right.$$
$$\left. ((1-\text{SSTFRAC}) + (1-\text{SFEXYCONV}) \times (\text{SSTFRAC} \times ((1-\text{SFSTFRAC}) + (1-\text{SFEXYCONV}) \times \text{SFSTFRAC}))) \right.$$
$$\left. \times \frac{F_{2007}^{(\text{Xylose})}}{\text{mw}(\text{Xylose})} \times \text{mw}(\text{CO}_2)\right) = 0$$

(continued)

TABLE 5.7 (continued)

Mass Balance Equations for Ethanol Production from Corn Stover Fermentation

Material Balance	IN–OUT + GENERATION–CONSUMPTION = 0
Cellulase	$F_{2007}^{\text{(Cellulase)}} - F_{2014}^{\text{(Cellulase)}} = 0$
CSL	$F_{2009} - F_{2014}^{\text{(CSL)}} = 0$
DAP	$F_{2010} - F_{2014}^{\text{(DAP)}} = 0$
Nitrogen	$F_{2008}^{\text{(Nitrogen)}} - F_{2012}^{\text{(Nitrogen)}} = 0$
Oxygen	$F_{2008}^{\text{(Oxygen)}} - F_{2012}^{\text{(Oxygen)}} \left(\left(\left(\text{SFEGLCONV} \times \left((\text{SSTFRAC} \times \text{SFSTFRAC}) \times \left(\dfrac{4.69}{0.56} \right) \times \dfrac{F_{2007}^{\text{(Glucose)}}}{\text{mw(Glucose)}} \times \text{mw(Oxygen)} \right) \right. \right. \right.$
	$\left. \left(\left(\text{SFEGLCONV} \times \left((\text{SSTFRAC} \times ((1-\text{SFSTFRAC}) + (1-\text{SFEGLCONV}) \times \text{SFSTFRAC}))) \right) \times \left(\dfrac{4.69}{0.56} \right) \right. \right. \right.$
	$\left. \times \dfrac{F_{2007}^{\text{(Glucose)}}}{\text{mw(Glucose)}} \times \text{mw(Oxygen)} \right)$
	$- \left(\left(\text{SFEXYCONV} \times (\text{SSTFRAC} \times \text{SFSTFRAC}) \right) \times \left(\dfrac{4.69}{0.67} \right) \times \dfrac{F_{2007}^{\text{(Xylose)}}}{\text{mw(Xylose)}} \times \text{mw(Oxygen)} \right)$
	$\left. \left(\left(\text{SFEXYCONV} \times \left((\text{SSTFRAC} \times ((1-\text{SFSTFRAC}) + (1-\text{SFEXYCONV}) \times \text{SFSTFRAC}))) \right) \times \left(\dfrac{4.69}{0.67} \right) \right. \right. \right.$
	$\left. \left. \times \dfrac{F_{2007}^{\text{(Xylose)}}}{\text{mw(Xylose)}} \times \text{mw(Oxygen)} \right) \right) = 0$
Cellulose	$F_{2007}^{\text{(Cellulose)}} - F_{2014}^{\text{(Cellulose)}} = 0$

Hemicellulose $\quad F_{2007}^{(\text{Hemicellulose})} - F_{2014}^{(\text{Hemicellulose})} = 0$

Ash $\quad F_{2007}^{(\text{Ash})} - F_{2014}^{(\text{Ash})} = 0$

Lignin $\quad F_{2007}^{(\text{Lignin})} - F_{2014}^{(\text{Lignin})} = 0$

Other solids $\quad F_{2007}^{(\text{Other Solids})} - F_{2014}^{(\text{Other Solids})} = 0$

Purification (Corn Stover EtOH)

Overall $\quad (F_{2013} + F_{2014}) - (F_{2015} + F_{2016} + F_{2017} + F_{2018}) = 0$

where

$$F_{2013} = F_{2013}^{(\text{H}_2\text{O})} + F_{2013}^{(\text{Ethanol})} + F_{2013}^{(\text{CO}_2)}$$

$$F_{2014} = F_{2014}^{(\text{H}_2\text{O})} + F_{2014}^{(\text{Cellulose})} + F_{2014}^{(\text{Hemicellulose})} + F_{2014}^{(\text{Lignin})} + F_{2014}^{(\text{Ash})} + F_{2014}^{(\text{Other Solids})} + F_{2014}^{(\text{Cellulase})} + F_{2014}^{(\text{Bacteria})} + F_{2014}^{(\text{CSL})} + F_{2014}^{(\text{DAP})} + F_{2014}^{(\text{Ethanol})} + F_{2014}^{(\text{CO}_2)}$$

$$F_{2015} = F_{2015}^{(\text{H}_2\text{O})} + F_{2015}^{(\text{Ethanol})}$$

$$F_{2016} = F_{2016}^{(\text{H}_2\text{O})} + F_{2016}^{(\text{CO}_2)} + F_{2016}^{(\text{Ethanol})}$$

$$F_{2017} = F_{2017}^{(\text{H}_2\text{O})} + F_{2017}^{(\text{Cellulose})} + F_{2017}^{(\text{Hemicellulose})} + F_{2017}^{(\text{Lignin})} + F_{2017}^{(\text{Ash})} + F_{2017}^{(\text{Other Solids})} + F_{2017}^{(\text{Cellulase})} + F_{2017}^{(\text{Bacteria})} + F_{2017}^{(\text{CSL})} + F_{2017}^{(\text{DAP})} + F_{2017}^{(\text{CO}_2)}$$

Species:

Ethanol $\quad \left(F_{2013}^{(\text{Ethanol})} + F_{2014}^{(\text{Ethanol})}\right) - \left(F_{2015}^{(\text{Ethanol})} + F_{2016}^{(\text{Ethanol})}\right) = 0$

$\text{H}_2\text{O} \quad \left(F_{2013}^{(\text{H}_2\text{O})} + F_{2014}^{(\text{H}_2\text{O})}\right) - \left(F_{2015}^{(\text{H}_2\text{O})} + F_{2016}^{(\text{H}_2\text{O})} + F_{2017}^{(\text{H}_2\text{O})} + F_{2018}^{(\text{H}_2\text{O})}\right) = 0$

$\text{CO}_2 \quad \left(F_{2013}^{(\text{CO}_2)} + F_{2014}^{(\text{CO}_2)}\right) - \left(F_{2016}^{(\text{CO}_2)} + F_{2017}^{(\text{CO}_2)}\right) = 0$

(continued)

TABLE 5.7 (continued)

Mass Balance Equations for Ethanol Production from Corn Stover Fermentation

Material Balance	IN–OUT + GENERATION–CONSUMPTION = 0
Cellulose	$F_{2014}^{(Cellulose)} - F_{2017}^{(Cellulose)} = 0$
Hemicellulose	$F_{2014}^{(Hemicellulose)} - F_{2017}^{(Hemicellulose)} = 0$
Ash	$F_{2014}^{(Ash)} - F_{2017}^{(Ash)} = 0$
Lignin	$F_{2014}^{(Lignin)} - F_{2017}^{(Lignin)} = 0$
Other solids	$F_{2014}^{(Other\ Solids)} - F_{2017}^{(Other\ Solids)} = 0$
Bacteria	$F_{2014}^{(Bacteria)} - F_{2017}^{(Bacteria)} = 0$
Cellulase	$F_{2014}^{(Cellulase)} - F_{2017}^{(Cellulase)} = 0$
CSL	$F_{2014}^{(CSL)} - F_{2017}^{(CSL)} = 0$
DAP	$F_{2014}^{(DAP)} - F_{2017}^{(DAP)} = 0$

TABLE 5.8

Energy Balance Equations for Ethanol Production from Corn Stover Fermentation

Energy Balance	IN$-$OUT$+$GENERATION$-$CONSUMPTION$=0$ $$Q_{in} - Q_{out} \times F_p + \sum_j H_j F_{j_{in}} - \sum_j H_j F_{j_{out}} = 0$$

Overall energy required from steam (Q_{FE}): $Q_{FE} = Q_{FEPRI} + Q_{FEI} + Q_{FEPUI}$

Pretreatment (Corn stover) (Q_{FEPRI})
$Q_{in} = Q_{FEPRI}$ (kJ/h)
$Q_{out} \times F_p = Q_{FEPRO} = 1515.75$ kJ/kg
$\phantom{Q_{out} \times F_p = Q_{FEPRO} =} \times F_{2007} \sqrt{b^2 - 4ac}$
$F_j = F_{2000}, F_{2001}, F_{2002}, F_{2003}, F_{2004}, F_{2005}, F_{2006},$
$ F_{2007}$ (kg/h)

$$Q_{FEPRI} - Q_{FEPRO} + \sum_j H_j F_{j_{in}} - \sum_j H_j F_{j_{out}} = 0$$

H_j	Mass Enthalpy (kJ/kg)
H_{2000}	$-1.58E+04$
H_{2001}	$-2.30E+03$
H_{2002}	$-1.31E+04$
H_{2003}	$-1.50E+04$
H_{2004}	$-9.06E+03$
H_{2005}	$-1.32E+04$
H_{2006}	$-1.33E+04$
H_{2007}	$-9.69E+03$

Fermentation (Corn Stover) (Q_{FEI})
$Q_{in} = Q_{FEI}$ (kJ/h)
$Q_{out} \times F_p = Q_{FEO} = 1725.04$ kJ/kg $\times (F_{2013}$
$\phantom{Q_{out} \times F_p = Q_{FEO} =} + F_{2014})$
$F_j = F_{2007}, F_{2008}, F_{2009}, F_{2010}, F_{2011}, F_{2012}, F_{2013},$
$ F_{2014}$ (kg/h)

$$Q_{FEI} - Q_{FEO} + \sum_j H_j F_{j_{in}} - \sum_j H_j F_{j_{out}} = 0$$

H_j	Mass Enthalpy (kJ/kg)
H_{2007}	$-9.69E+03$
H_{2008}	$0.00E+00$
H_{2009}	$-2.14E+03$
H_{2010}	$-2.14E+03$
H_{2011}	$-2.29E+03$
H_{2012}	$-1.52E+03$
H_{2013}	$-8.89E+03$
H_{2014}	$-1.12E+04$

Purification (Corn Stover EtOH) (Q_{FEPUI})
$Q_{in} = Q_{FEPUI}$ (kJ/h)
$Q_{out} \times F_p = Q_{FEPUO} = 12,948.28$ kJ/kg $\times F_{2015}$
$F_j = F_{2013}, F_{2014}, F_{2015}, F_{2016}, F_{2017}, F_{2018}$ (kg/h)

$$Q_{FEPUI} - Q_{FEPUO} + \sum_j H_j F_{j_{in}} - \sum_j H_j F_{j_{out}} = 0$$

H_j	Mass Enthalpy (kJ/kg)
H_{2013}	$-8.89E+03$
H_{2014}	$-1.12E+04$
H_{2015}	$-6.07E+03$
H_{2016}	$-9.01E+03$
H_{2017}	$-1.07E+04$
H_{2018}	$-1.54E+04$

5.3 Ethanol Production from Corn Dry-Grind Fermentation

The fermentation process for ethanol production from corn was designed in SuperPro Designer (Kwiatkowski et al., 2006; Intelligen, 2009). A description of the process was given in Chapter 4. The design was converted to the block flow diagram as shown in Figure 5.3. The block diagram for ethanol production from corn has three units: pretreatment (corn), fermentation (corn), and purification (corn EtOH). These denote the pretreatment section, fermentation section, and purification section from the SuperPro Designer model. These three processes are separately represented in three blocks in the optimization model. The reactions occurring in the process are given in Table 5.9. The NFDS denote the non-fermentable dissolved solids (other solids) from Table 4.33. NSP denotes non-starch polysaccharides, ProteinI denotes the insoluble proteins, and ProteinS denotes the soluble proteins from the same table. This nomenclature is used in this chapter. The molecular weights used in the design for protein, NFDS, and starch are given in Chapter 4. The streams are shown in Figure 5.3 and the stream descriptions

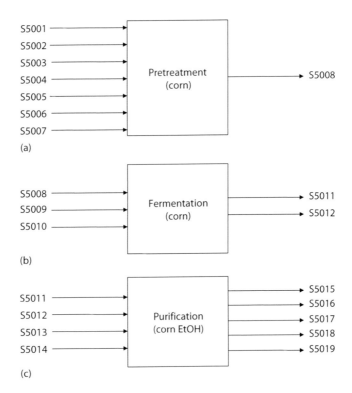

FIGURE 5.3
Block flow diagram of ethanol production from corn fermentation: (a) pretreatment (corn), (b) fermentation (corn), and (c) purification (corn EtOH).

TABLE 5.9

Reactions for Ethanol Production from Corn Fermentation

Step	Reaction	Conversion (%)
Starch pretreatment	8.9 Starch + $H_2O \rightarrow C_6H_{12}O_6$	99
Glucose fermentation	$C_6H_{12}O_6 \rightarrow 1.9C_2H_5OH + 1.9CO_2 + 0.05NFDS$	100
	NFDS $\rightarrow$ Protein	6.8

are given in Table 5.10. The parameters for the process are given in Table 5.11. The overall balance for each section and the individual species mass balance equations are given in Table 5.15. The inlet and outlet stream flow rates for the blocks from the SuperPro Designer corresponding to the streams in Table 5.10 are given in Appendix F.

The mass balance for the streams is according to the equation

$$F_{in}^{(i)} - F_{out}^{(i)} + F_{gen}^{(i)} - F_{cons}^{(i)} = 0$$

where
F^i is the flow rate of ith component in the system
the subscripts in, out, gen, and cons denote the flow rates into, out of, and generation of and consumption of component i in the system

The total energy requirement for all of the processes was not available from the design. The calculation for total energy for all the equipment could include error in estimating the total energy. However, the total cost for utility was available. The cost for energy per ton of ethanol produced from process was added to the utility costs equation in the superstructure to account for the cost of energy for the process.

5.3.1 Pretreatment (Corn)

The block for pretreatment was created in the chemical complex analysis system, with S5001–S5007 as inlet streams and S5008 as the outlet stream. The biomass composition for corn is given in Table 5.11. The flow rate of S5001 was fixed with the capacity constraint for 45,228 kg/h corn. The flow rate of the remaining inlet streams were obtained as fractions of the corn biomass stream. The saccharification (pretreatment step for obtaining fermentable glucose from biomass) of corn was carried out in one reactor, and the starch was hydrolyzed to glucose. The reaction for the conversion of starch to glucose was given by Intelligen (2009) and given in Table 5.9; 99% of the starch was converted to glucose in this step. The conversion parameter for this process is PRSTARCONV, given in Table 5.9. The outlet stream, S5008, contained fermentable glucose, unreacted biomass, and water. The stream was sent to the fermentation section for fermentation to ethanol.

TABLE 5.10

Description of Process Streams in Ethanol Production from Corn Fermentation

Name of Streams	Description
Pretreatment (Corn)	
Input streams	
S5001	Corn
S5002	α-Amylase
S5003	Liquid ammonia
S5004	Lime
S5005	Caustic
S5006	Gluco-amylase
S5007	Sulfuric acid
Output streams	
S5008	Pretreated corn biomass
Fermentation (Corn)	
Input streams	
S5008	Pretreated corn biomass
S5009	Yeast
S5010	Water to fermentor
Output streams	
S5011	Vapor from fermentor containing ethanol, CO_2, and water
S5012	Crude ethanol stream from fermentor
Purification (Corn EtOH)	
Input streams	
S5011	Vapor from fermentor containing ethanol, CO_2, and water
S5012	Crude ethanol stream from fermentor
S5013	Water to CO_2 scrubber
S5014	Hot air
Output streams	
S5015	Ethanol from the process
S5016	Process condensate water from the process
S5017	Crude distiller's dry grain solids (DDGS) from the process
S5018	Exhaust
S5019	CO_2 from scrubber

TABLE 5.11

Parameters in Ethanol Production from Corn Fermentation

Name	Meaning	Value
Biomass composition		
MFNFCOR	Mass fraction of non-fermentable dissolved solids (NFDS) in S5001 corn	0.067
MFNSP	Mass fraction of non-starch polyhydrate in S5001 corn	0.07
MFOIL	Mass fraction of oil in S5001 corn	0.034
MFPRI	Mass fraction of insoluble protein in S5001 corn	0.06
MFPRS	Mass fraction soluble protein in S5001 corn	0.024
MFSTARC	Mass fraction starch in S5001 corn	0.595
MFWATC	Mass fraction of water in S5001 corn	0.15
Caustic composition		
MFNFCIP	Mass fraction of NFDS in S5005 caustic	0.05
MFWACIP	Mass fraction of water in S5005 caustic	0.95
Conversion in reactors		
PRSTARCONV	Pretreatment conversion of starch to glucose	0.99
SGLECONV	Conversion of starch glucose to ethanol	1.00
SNFDCONV	Conversion of NFDS to protein	0.068
Stream fractions		
AAMYFRAC	Alpha-amylase fraction with respect to inlet corn stream S5001	0.0007
AMMFRAC	Ammonia fraction with respect to inlet corn stream S5001	0.0020
LIMFRAC	Lime fraction with respect to inlet corn stream S5001	0.0012
CAUFRAC	Caustic fraction with respect to inlet corn stream S5001	0.0499
GAMYFRAC	Glucoamylase fraction with respect to inlet corn stream S5001	0.0010
SACIDFRAC	Sulfuric acid fraction with respect to inlet corn stream S5001	0.0020
YEASFRAC	Yeast fraction to fermentor with respect to starch in stream S5008	0.0408
FERWFRAC	Water fraction to fermentor with respect to stream S5008	0.0004
CLNSPLT	Split fraction of components in 101U Cleaning	0.997
SPLTFRAC	Split fraction of components in Split1 going to stream S-174	0.9999
CENTSPLT1	Split fraction of CO_2, ethanol, NFDS, protein (sol.), and water in centrifuge 601U	0.1651
CENTSPLT2	Split fraction of NSP, oil, protein (insol.), and starch in centrifuge 601U	0.92
SPLIT1	Fraction of solids in recycle stream to the pretreatment section in stream S-154	0.2625
W5011FR	Fraction of water vapor in stream S5011 from fermentor	0.0023
C5011FR	Fraction of CO_2 in stream S5011 from fermentor	0.985
E5011FR	Fraction of ethanol in stream S5011 from fermentor	0.0255
RCY2WATFRAC	Fraction of water recycled in S-154 with respect to water into the purification section from the fermentation section	0.1690

(continued)

TABLE 5.11 (continued)

Parameters in Ethanol Production from Corn Fermentation

Name	Meaning	Value
SCRWAFRAC	Scrubber water fraction with respect to carbon dioxide in S5011 from fermentor	0.9867
RCY2ETHFRAC	Fraction of ethanol recycled in S-154 with respect to ethanol into the purification section from the fermentation section	0.0006
RCY1WATFRAC	Fraction of water recycled in S-127 with respect to water into the purification section from the fermentation section	0.6601
RCY1ETHFRAC	Fraction of ethanol recycled in S-127 with respect to ethanol into the purification section from the fermentation section	0.0278
RCY1CO2FRAC	Fraction of CO_2 recycled in S-127 with respect to CO_2 into the purification section from the fermentation section	0.0015
FRET5015	Fraction of ethanol in stream S5015 with respect to ethanol into the purification section from the fermentation section	0.9708
FRET5016	Fraction of ethanol in stream S5016 with respect to ethanol into the purification section from the fermentation section	0.0002
FRET5017	Fraction of ethanol in stream S5017 with respect to ethanol into the purification section from the fermentation section	0.0000
FRET5018	Fraction of ethanol in stream S5018 with respect to ethanol into the purification section from the fermentation section	0.0005
FRET5019	Fraction of ethanol in stream S5019 with respect to ethanol into the purification section from the fermentation section	0.0001
FRCO25016	Fraction of CO_2 in stream S5016 with respect to CO_2 into the purification section from the fermentation section	0.0000
FRCO25019	Fraction of CO_2 in stream S5019 with respect to CO_2 into the purification section from the fermentation section	0.9985
FRWA5015	Fraction of water in stream S5015 with respect to water into the purification section from the fermentation section	0.0005
FRWA5016	Fraction of water in stream S5016 with respect to water into the purification section from the fermentation section	0.0049
FRWA5017	Fraction of water in stream S5017 with respect to water into the purification section from the fermentation section	0.0116
FRWA5018	Fraction of water in stream S5018 with respect to water into the purification section from the fermentation section	0.1532
FRWA5019	Fraction of water in stream S5019 with respect to water into the purification section from the fermentation section	0.0009

There were 28 variables in the pretreatment section and 29 equations, including two overall equations. Therefore, the degree of freedom for the pretreatment section was $28 - (29 - 2) = 1$. The constraint for capacity of processing 45,228 kg/h corn in the plant in stream S5001 specified the degree of freedom. The stream flow rates for S5008 are compared with SuperPro Designer results to check the validity of the model and are given in Table 5.12.

TABLE 5.12

Pretreatment (Corn) Section Optimization Model Results
Validated with Data from SuperPro Designer (kg/h)

Stream Name	Data from SuperPro Designer	Data from the System	Percent Difference (%)
S5008	1.44E+05	1.44E+05	0
S5008CO_2	2.11E+01	2.11E+01	0
S5008ETOH	4.19E+02	4.20E+02	0
S5008G	2.95E+04	2.95E+04	0
S5008H_2O	1.00E+05	1.00E+05	0
S5008H_2SO_4	8.97E+01	8.97E+01	0
S5008NFDS	4.50E+03	4.50E+03	0
S5008NSP	3.22E+03	3.22E+03	0
S5008OIL	1.57E+03	1.57E+03	0
S5008PI	2.76E+03	2.76E+03	0
S5008PS	1.50E+03	1.50E+03	0
S5008S	2.68E+02	2.68E+02	0

5.3.2 Fermentation (Corn)

The input stream for this section was the S5008 stream containing glucose saccharified from corn starch. The component flow rates for this stream were calculated in the pretreatment section. The variables in this section were S5008–S5012. The rest of the input streams to this section were yeast and water to the fermentor. The glucose conversion in the fermentor was 100% and 6.8% of the NFDS (solids) was converted to soluble proteins. These conversions were SGLECONV and SNFDCONV, respectively.

The outlet streams in this section were S5011 and S5012. The S5011 stream was the vapor from the fermentor, containing carbon dioxide, ethanol, and water vapor. The ethanol was obtained in the S5012 stream along with water, unreacted solids, and proteins. Both of the streams were sent to the purification section.

There were 29 variables in the fermentation section and 18 equations, with 1 dependent equation. Therefore, the degree of freedom for the fermentation section was 29 − (18 − 1) = 12. The flow rate variables for individual components in stream S5008 was computed from the previous section, and these 12 mass flow rate variables were used to reduce the degree of freedom. The stream flow rates for S5011 and S5012 are compared with SuperPro Designer results to check the validity of the model and are given in Table 5.13.

5.3.3 Purification (Corn EtOH)

The input stream for the purification section was S5011 and S5012, containing ethanol, water, solids, and carbon dioxide. The variables in this section were S5011–S5019. The processes involved in this section included centrifugation

TABLE 5.13

Fermentation (Corn) Section Optimization Model Results
Validated with Data from SuperPro Designer (kg/h)

Stream Name	Data from SuperPro Designer	Data from the System	Percent Difference (%)
S5011	1.41E+04	1.41E+04	0
S5011CO_2	1.35E+04	1.35E+04	0
S5011ETOH	3.76E+02	3.77E+02	0
S5011H_2O	2.31E+02	2.31E+02	0
S5012	1.30E+05	1.30E+05	0
S5012CO_2	2.06E+02	2.06E+02	0
S5012ETOH	1.44E+04	1.44E+04	0
S5012H_2O	1.00E+05	1.00E+05	0
S5012H_2SO_4	8.97E+01	8.97E+01	0
S5012NFDS	5.57E+03	5.57E+03	0
S5012NSP	3.22E+03	3.22E+03	0
S5012OIL	1.57E+03	1.57E+03	0
S5012PI	2.76E+03	2.76E+03	0
S5012PS	1.91E+03	1.91E+03	0
S5012S	2.68E+02	2.68E+02	0

to remove the solids from the stream, followed by distillation and molecular sieve adsorption separation of ethanol from water. S5013 was the water required in the removal of carbon dioxide from the process in a CO_2 scrubber. The ethanol from the process was obtained by distillation and molecular sieve adsorption in stream S5015. The process condensate was obtained in stream S5016. The solids from the process, distiller's dry grain solids or DDGS, were obtained in stream S5017. The exhaust from the process was obtained in two streams, S5018, which was from the drying of DDGS, and S5019, which contained carbon dioxide from the fermentation process after water scrubbing. The fraction of water, carbon dioxide, and ethanol in each of the exit streams based on inlet to the purification section was computed from Petrides (2008) and is given in Table 5.11. These relations were used as constraints in the model.

There were 45 variables in the purification section and 35 equations, with 1 dependent equation. Therefore, the degree of freedom for the purification section was 45 − (35 − 1) = 11. The flow rate variables for individual components in stream S5012 were computed from the previous section, and the 11 mass flow rate variables in those streams solved the degree of freedom. The stream flow rates for S5015–S5019 were compared

TABLE 5.14

Purification (Corn EtOH) Section Optimization Model
Results Validated with Data from SuperPro Designer (kg/h)

Stream Name	Data from SuperPro Designer	Data from the System	Percent Difference (%)
S5015	1.44E+04	1.44E+04	0
S5015ETOH	1.43E+04	1.43E+04	0
S5015H$_2$O	5.44E+01	5.44E+01	0
S5016	5.55E+02	5.55E+02	0
S5016CO$_2$	1.55E–01	1.55E–01	0
S5016ETOH	3.02E+00	3.02E+00	0
S5016H$_2$O	5.52E+02	5.52E+02	0
S5017	1.50E+04	1.50E+04	0
S5017ETOH	1.18E–01	1.18E–01	0
S5017H2O	1.32E+03	1.32E+03	0
S5017H$_2$SO$_4$	8.97E+01	8.97E+01	0
S5017NFDS	4.36E+03	4.36E+03	0
S5017NSP	3.17E+03	3.17E+03	0
S5017OIL	1.54E+03	1.54E+03	0
S5017PI	2.71E+03	2.71E+03	0
S5017PS	1.49E+03	1.49E+03	0
S5017S	3.43E+02	3.43E+02	0
S5018	4.19E+04	4.19E+04	0
S5018ETOH	7.27E+00	7.27E+00	0
S5018H$_2$O	1.74E+04	1.74E+04	0
S5018N$_2$	1.87E+04	1.87E+04	0
S5018O$_2$	5.69E+03	5.69E+03	0
S5019	1.38E+04	1.38E+04	0
S5019CO$_2$	1.37E+04	1.37E+04	0
S5019ETOH	7.73E–01	7.73E–01	0
S5019H$_2$O	9.62E+01	9.62E+01	0

with SuperPro Designer results to check the validity of the model and are given in Table 5.14.

The mass balance equations for overall and species balances for the corn fermentation pretreatment section, fermentation section, and purification section are given in Table 5.15. The SuperPro design had three recycle streams that were included in the mass balances for the model. The energy cost associated with the production of 40 million gal/year of ethanol was $0.08/kg ethanol. This relation was used in the utilities cost to compute the cost of energy in the process.

TABLE 5.15

Mass Balance Equations for Ethanol Production from Corn Fermentation

Material Balance	IN–OUT + GENERATION–CONSUMPTION = 0

Pretreatment (Corn)

Overall

$$(F_{5001} + F_{5002} + F_{5003} + F_{5004} + F_{5005} + F_{5006} + F_{5007}) - (F_{5008}) = 0$$

where

$$F_{5001} = F_{5001}^{(NFDS)} + F_{5001}^{(NSP)} + F_{5001}^{(Oil)} + F_{5001}^{(ProteinI)} + F_{5001}^{(ProteinS)} + F_{5001}^{(Starch)} + F_{5001}^{(Water)}$$

$$F_{5005} = F_{5005}^{(NFDS)} + F_{5005}^{(Water)}$$

$$F_{5008} = F_{5008}^{(NFDS)} + F_{5008}^{(NSP)} + F_{5008}^{(Oil)} + F_{5008}^{(ProteinI)} + F_{5008}^{(ProteinS)} + F_{5008}^{(Starch)} + F_{5008}^{(Water)} + F_{5008}^{(Sulfuric\ acid)} + F_{5008}^{(CO_2)} + F_{5008}^{(Ethanol)} + F_{5008}^{(Glucose)}$$

Species:

NFDS

$$CLNSPLT \times F_{5001}^{(NFDS)} + F_{5003} + F_{5004} + F_{5005}^{(NFDS)} + \left((1 - CENTSPLT1) \times SPLIT1 \times F_{5012}^{(NFDS)}\right) - F_{5008}^{(NFDS)} = 0$$

NSP

$$CLNSPLT \times F_{5001}^{(NSP)} + \left((1 - CENTSPLT2) \times SPLIT1 \times F_{5012}^{(NSP)}\right) - F_{5008}^{(NSP)} = 0$$

Oil

$$CLNSPLT \times F_{5001}^{(Oil)} + \left((1 - CENTSPLT2) \times SPLIT1 \times F_{5012}^{(Oil)}\right) - F_{5008}^{(Oil)} = 0$$

ProteinI

$$CLNSPLT \times F_{5001}^{(ProteinI)} + \left((1 - CENTSPLT2) \times SPLIT1 \times F_{5012}^{(ProteinI)}\right) - F_{5008}^{(ProteinI)} = 0$$

ProteinS

$$CLNSPLT \times F_{5001}^{(ProteinI)} + \left((1 - CENTSPLT2) \times SPLIT1 \times F_{5012}^{(ProteinS)}\right) - F_{5008}^{(ProteinI)} = 0$$

Starch

$$CLNSPLT \times F_{5001}^{(Starch)} + \left((1 - CENTSPLT2) \times SPLIT1 \times F_{5012}^{(Starch)}\right) - F_{5012}^{()} - F_{5008}^{(Starch)}$$

$$- PRSTARCONV \times \left(CLNSPLT \times F_{5001}^{(Starch)} + \left((1 - CENTSPLT2) \times SPLIT1 \times F_{5012}^{(Starch)}\right)\right) = 0$$

Water

$$CLNSPLT \times F_{5001}^{(Water)} + F_{5002} + F_{5005}^{(Water)} + F_{5006}$$

$$+ \left(RCY2WATFRAC \times \left(F_{5011}^{(Water)} + F_{5012}^{(Water)} + F_{5013} + (1-CLNSPLT) \times F_{5001}^{(Water)}\right)\right)$$

$$+ \left(RCY1WATFRAC \times \left(F_{5011}^{(Water)} + F_{5012}^{(Water)} + F_{5013} + (1-CLNSPLT) \times F_{5001}^{(Water)}\right)\right) - F_{5008}^{(Water)}$$

$$- \left(PRSTARCONV \times \left(CLNSPLT \times F_{5001}^{(Starch)} + \left((1-CENTSPLT2) \times SPLIT1 \times F_{5012}^{(Starch)}\right) \times \left(\frac{1}{9}\right)\right)\right) = 0$$

Sulfuric acid

$$F_{5007} - F_{5008}^{(Sulfuric\ acid)} = 0$$

CO$_2$

$$\left(RCY1CO2FRAC \times \left(F_{5011}^{(CO_2)} + F_{5012}^{(CO_2)}\right)\right) - F_{5008}^{(CO_2)} = 0$$

Ethanol

$$\left(RCY1ETHFRAC \times \left(F_{5011}^{(Ethanol)} + F_{5012}^{(Ethanol)}\right)\right) + \left(RCY2ETHFRAC \times \left(F_{5011}^{(Ethanol)} + F_{5012}^{(Ethanol)}\right)\right) - F_{5008}^{(EtOH)} = 0$$

Glucose

$$-F_{5008}^{(Glucose)} + \left(PRSTARCONV \times \left(CLNSPLT \times F_{5001}^{(Starch)} + \left((1-CENTSPLT2) \times SPLIT1 \times F_{5012}^{(Starch)}\right) \times \left(\frac{10}{9}\right)\right)\right) = 0$$

Fermentation (Corn)

Overall

$$(F_{5008} + F_{5009} + F_{5010}) - (F_{5011} + F_{5012}) = 0$$

where

$$F_{5008} = F_{5008}^{(NFDS)} + F_{5008}^{(NSP)} + F_{5008}^{(Oil)} + F_{5008}^{(ProteinI)} + F_{5008}^{(ProteinS)} + F_{5008}^{(Starch)} + F_{5008}^{(Water)} + F_{5008}^{(Sulfuric\ acid)} + F_{5008}^{CO_2} + F_{5008}^{(Ethanol)} + F_{5008}^{(Glucose)}$$

$$F_{5011} = F_{5011}^{(CO_2)} + F_{5011}^{(Water)} + F_{5011}^{(Ethanol)}$$

$$F_{5012} = F_{5012}^{(CO_2)} + F_{5012}^{(Ethanol)} + F_{5012}^{(NFDS)} + F_{5012}^{(NSP)} + F_{5012}^{(Oil)} + F_{5012}^{(ProteinI)} + F_{5012}^{(ProteinS)} + F_{5012}^{(Starch)} + F_{5012}^{(Water)} + F_{5012}^{(Sulfuric\ acid)}$$

(continued)

TABLE 5.15 (continued)

Mass Balance Equations for Ethanol Production from Corn Fermentation

Material Balance	IN−OUT + GENERATION−CONSUMPTION = 0
Species:	
NFDS	$F_{5008}^{(NFDS)} + F_{5012}^{(NFDS)} \times \dfrac{1}{SPLTFRAC} - 1 \dfrac{F_{5012}^{(NFDS)}}{SPLTFRAC} + \left(SGLECONV \times F_{5008}^{(Glucose)} + \left(SGLECONV \times F_{5008}^{(Glucose)} \times \left(\dfrac{0.05}{1}\right) \times \dfrac{mwbio(NFDS)}{mwbio(Glucose)} \right) \right)$ $- \left(SNFDCONV \times \left(F_{5008}^{(NFDS)} + \left(SGLECONV \times F_{5008}^{(Glucose)} \times \left(\dfrac{0.05}{1}\right) \times \dfrac{mwbio(NFDS)}{mwbio(Glucose)} \right) \right) \right) = 0$
NSP	$F_{5008}^{(NSP)} F_{5012}^{(NSP)} \times \left(\dfrac{1}{SPLTFRAC} - 1\right) - \dfrac{F_{5012}^{(NSP)}}{SPLTFRAC} = 0$
Oil	$F_{5008}^{(Oil)} + F_{5012}^{(Oil)} \times \left(\dfrac{1}{SPLTFRAC} - 1\right) - \dfrac{F_{5012}^{(Oil)}}{SPLTFRAC} = 0$
ProteinI	$F_{5008}^{(ProteinI)} + F_{5012}^{(ProteinI)} \times \left(\dfrac{1}{SPLTFRAC} - 1\right) - \dfrac{F_{5012}^{(ProteinI)}}{SPLTFRAC} = 0$
ProteinS	$F_{5008}^{(ProteinS)} + F_{5012}^{(ProteinS)} \times \left(\dfrac{1}{SPLTFRAC} - 1\right) - \dfrac{F_{5012}^{(ProteinS)}}{SPLTFRAC}$ $+ \left(SNFDCONV \times F_{5008}^{(NFDS)} + \left(SGLECONV \times F_{5008}^{(Glucose)} \times \left(\dfrac{0.05}{1}\right) \times \dfrac{mwbio(NFDS)}{mwbio(Glucose)} \right) \right) \times \dfrac{mwbio(ProteinS)}{mwbio(NFDS)} = 0$
Starch	$F_{5008}^{(Starch)} + F_{5012}^{(Starch)} \times \left(\dfrac{1}{SPLTFRAC} - 1\right) - \dfrac{F_{5012}^{(Starch)}}{SPLTFRAC} = 0$
Water	$F_{5008}^{(Water)} + F_{5009} + F_{5010} - F_{5011}^{(Water)} + F_{5012}^{(Water)} \times \left(\dfrac{1}{SPLTFRAC} - 1\right) - \dfrac{F_{5012}^{(Water)}}{SPLTFRAC} = 0$

Sulfuric acid

$$F_{5008}^{(\text{Sulfuric acid})} + F_{5012}^{(\text{Sulfuric acid})} \times \left(\frac{1}{\text{SPLTFRAC}} - 1\right) - \frac{F_{5012}^{(\text{Sulfuric acid})}}{\text{SPLTFRAC}} = 0$$

CO₂

$$F_{5008}^{(\text{CO}_2)} + F_{5012}^{(\text{CO}_2)} \times \left(\frac{1}{\text{SPLTFRAC}} - 1\right) - F_{5011}^{(\text{CO}_2)} - \frac{F_{5012}^{(\text{CO}_2)}}{\text{SPLTFRAC}} + \left(\text{SGLECONV} \times F_{5008}^{(\text{Glucose})} \times \left(\frac{1.9}{1}\right) \times \frac{\text{mw}(\text{CO}_2)}{\text{mwbio}(\text{Glucose})}\right) = 0$$

Ethanol

$$F_{5008}^{(\text{Ethanol})} + F_{5012}^{(\text{Ethanol})} \times \left(\frac{1}{\text{SPLTFRAC}} - 1\right) - F_{5011}^{(\text{Ethanol})} - \frac{F_{5012}^{(\text{Ethanol})}}{\text{SPLTFRAC}}$$
$$+ \left(\text{SGLECONV} \times F_{5008}^{(\text{Glucose})} \times \left(\frac{1.9}{1}\right) \times \frac{\text{mwbio}(\text{Ethanol})}{\text{mwbio}(\text{Glucose})}\right) = 0$$

Glucose

$$F_{5008}^{(\text{Glucose})} - \left(\text{SGLECONV} \times F_{5008}^{(\text{Glucose})} \times \left(\frac{1.9}{1}\right) \times \frac{\text{mwbio}(\text{Glucose})}{\text{mwbio}(\text{Glucose})}\right) = 0$$

Purification (Corn EtOH)

Overall

$$(F_{5011} + F_{5012} + F_{5013} + F_{5014}) - (F_{5015} + F_{5016} + F_{5017} + F_{5018} + F_{5019}) = 0$$

where

$$F_{5011} = F_{5011}^{(\text{CO}_2)} + F_{5011}^{(\text{Water})} + F_{5011}^{(\text{Ethanol})}$$

$$F_{5012} = F_{5012}^{(\text{CO}_2)} + F_{5012}^{(\text{Ethanol})} + F_{5012}^{(\text{NFDS})} + F_{5012}^{(\text{NSP})} + F_{5012}^{(\text{Oil})} + F_{5012}^{(\text{ProteinI})} + F_{5012}^{(\text{ProteinS})} + F_{5012}^{(\text{Starch})} + F_{5012}^{(\text{Water})} + F_{5012}^{(\text{Sulfuric acid})}$$

$$F_{5014} = F_{5014}^{(\text{N}_2)} + F_{5014}^{(\text{O}_2)}$$

$$F_{5015} = F_{5015}^{(\text{Ethanol})} + F_{5015}^{(\text{Water})}$$

$$F_{5016} = F_{5016}^{(\text{Ethanol})} + F_{5016}^{(\text{Water})} + F_{5016}^{(\text{CO}_2)}$$

$$F_{5017} = F_{5017}^{(\text{Ethanol})} + F_{5017}^{(\text{Water})} + F_{5017}^{(\text{NFDS})} + F_{5017}^{(\text{NSP})} + F_{5017}^{(\text{Oil})} + F_{5017}^{(\text{ProteinI})} + F_{5017}^{(\text{ProteinS})} + F_{5017}^{(\text{Starch})} + F_{5017}^{(\text{Sulfuric acid})}$$

$$F_{5018} = F_{5018}^{(\text{N}_2)} + F_{5018}^{(\text{O}_2)} + F_{5018}^{(\text{Ethanol})} + F_{5018}^{(\text{Water})}$$

$$F_{5019} = F_{5019}^{(\text{CO}_2)} + F_{5019}^{(\text{Ethanol})} + F_{5019}^{(\text{Water})}$$

(*continued*)

TABLE 5.15 (continued)

Mass Balance Equations for Ethanol Production from Corn Fermentation

Material Balance	IN–OUT + GENERATION–CONSUMPTION = 0
Species:	
NFDS	$F_{5012}^{(NFDS)} + (1-CLNSPLT) \times F_{5001}^{(NFDS)} - \left((1-CENTSPLT1) \times SPLIT1 \times F_{5012}^{(NFDS)}\right) - F_{5017}^{(NFDS)} = 0$
NSP	$F_{5012}^{(NSP)} + (1-CLNSPLT) \times F_{5001}^{(NSP)} - \left((1-CENTSPLT2) \times SPLIT1 \times F_{5012}^{(NSP)}\right) - F_{5017}^{(NSP)} = 0$
Oil	$F_{5012}^{(Oil)} + (1-CLNSPLT) \times F_{5001}^{(Oil)} \left((1-CENTSPLT2) \times SPLIT1 \times F_{5012}^{(Oil)}\right) - F_{5017}^{(Oil)} = 0$
ProteinI	$F_{5012}^{(ProteinI)} + (1-CLNSPLT) \times F_{5001}^{(ProteinI)} - \left((1-CENTSPLT2) \times SPLIT1 \times F_{5012}^{(ProteinI)}\right) - F_{5017}^{(ProteinI)} = 0$
ProteinS	$F_{5012}^{(ProteinS)} + (1-CLNSPLT) \times F_{5001}^{(ProteinS)} - \left((1-CENTSPLT1) \times SPLIT1 \times F_{5012}^{(ProteinS)}\right) - F_{5017}^{(ProteinS)} = 0$
Starch	$F_{5012}^{(Starch)} + (1-CLNSPLT) \times F_{5001}^{(Starch)} - \left((1-CENTSPLT2) \times SPLIT1 \times F_{5012}^{(Starch)}\right) - F_{5017}^{(Starch)} = 0$
Water	$F_{5011}^{(Water)} + F_{5012}^{(Water)} + (1-CLNSPLT) \times F_{5001}^{(Water)} + F_{5013}$ $- \left(RCY1WATFRAC \times \left(F_{5011}^{(Water)} + F_{5012}^{(Water)} + F_{5013} + (1-CLNSPLT) \times F_{5001}^{(Water)}\right)\right)$ $- \left(RCY2WATFRAC \times \left(F_{5011}^{(Water)} + F_{5012}^{(Water)} + F_{5013} + (1-CLNSPLT) \times F_{5001}^{(Water)}\right) - F_{5015}^{(Water)} - F_{5016}^{(Water)} - F_{5017}^{(Water)} - F_{5018}^{(Water)} - F_{5019}^{(Water)} = 0$
Sulfuric acid	$F_{5012}^{(Sulfuric\ acid)} - F_{5017}^{(Sulfuric\ acid)} = 0$
CO_2	$F_{5011}^{(CO_2)} + F_{5012}^{(CO_2)} - F_{5016}^{(CO_2)} - F_{5019}^{(CO_2)} - \left(RCY1CO2FRAC \times \left(F_{5011}^{(CO_2)} + F_{5012}^{(CO_2)}\right)\right) = 0$
Ethanol	$F_{5011}^{(Ethanol)} + F_{5012}^{(Ethanol)} - F_{5015}^{(Ethanol)} - F_{5016}^{(Ethanol)} - F_{5017}^{(Ethanol)} - F_{5018}^{(Ethanol)} - F_{5019}^{(Ethanol)} - \left(RCY1ETHFRAC \times \left(F_{5011}^{(Ethanol)} + F_{5012}^{(Ethanol)}\right)\right)$ $- \left(RCY2ETHFRAC \times \left(F_{5011}^{(Ethanol)} + F_{5012}^{(Ethanol)}\right)\right) = 0$
Nitrogen	$F_{5014}^{(N_2)} - F_{5018}^{(N_2)} = 0$
Oxygen	$F_{5014}^{(N_2)} - F_{5018}^{(N_2)} = 0$

5.4 Ethylene Production from Dehydration of Ethanol

The process design for ethylene production was converted to the block flow diagram as shown in Figure 5.4. The block diagram for ethylene from ethanol had one unit to describe the reaction and purification section from the HYSYS model. The reaction occurring in the process is given in Table 5.16. The streams are shown in Figure 5.4 and the stream descriptions are given in Table 5.17. The parameters for the process are given in Table 5.18. The overall balance and the individual species mass balance and energy balance equations are given in Table 5.20 and Table 5.21, respectively. The inlet and outlet

FIGURE 5.4
Block flow diagram of ethylene production from dehydration of ethanol.

TABLE 5.16

Reaction for Ethylene Production
from Dehydration of Ethanol

Reaction	Conversion (%)
$C_2H_5OH \rightarrow C_2H_4 + H_2O$	99

TABLE 5.17

Description of Process Streams in Ethylene Production
from Dehydration of Ethanol

Name of Streams	Description
Input streams	
S2030	Ethanol from fermentation process to new ethylene process
Output streams	
S2031	Ethylene from new ethylene process
S2032	Purge from new ethylene process
Energy streams	
QEEO	Heat removed by cooling water in ethylene section
QEEI	Heat required from steam in ethylene section

TABLE 5.18

Parameters in Ethylene Production
from Dehydration of Ethanol

Name	Meaning	Value
EECONV	Ethanol-to-ethylene conversion	0.99
EEFRAC	Percent removal of ethylene from purification section in ethylene process	0.999

stream flow rates for the blocks from the HYSYS design corresponding to the streams in Table 5.17 are given in Appendix F.

The model formulation for optimization was done using the chemical complex analysis system. An iterative process was followed for the optimization model development. This is explained in the following sections. The species for the ethanol-to-ethylene process already existed in the model for the base case.

The mass balance for the streams is according to the equation

$$F_{in}^{(i)} - F_{out}^{(i)} + F_{gen}^{(i)} - F_{cons}^{(i)} = 0$$

where
F^i is the flow rate of ith component in the system
the subscripts in, out, gen, and cons denote the flow rates into, out of, generation of, and consumption of component i in the system

The energy balance for the process is according to the equation

$$Q_{in} - Q_{out} \times F_p + \sum_j H_j F_{j_{in}} - \sum_j H_j F_{j_{out}} = 0$$

H_j is the enthalpy per unit mass of jth stream computed from HYSYS. The enthalpy per unit mass of the stream from HYSYS was used. Q_{out} is the energy removed from the system per unit mass of product p having a flow rate of F_p. Q_{in} is the energy required by the system and calculated using the aforementioned equation with the values for the other terms specified from HYSYS.

The block for the ethylene process was created in the chemical complex analysis system, with S2030–S2032 as the streams. The parameter for the process was added to the scalar set (Scalar4, Description: Constant parameters for bioprocesses). The inlet ethylene stream, S2030, was dehydrated at 300°C. The resulting ethylene vapor was separated from water vapor and obtained in stream S2031. The purge stream, S2032, contained waste water and traces of ethylene. The parameters for ethylene conversion, EECONV and ethylene separation, EEFRAC were added to the constant parameters. The species mass and balance equations are given in Table 5.19.

There were six variables in the ethylene section and five equations. Therefore, the degree of freedom for the ethylene section was $6 - 5 = 1$. The constraint for capacity of ethylene produced from the plant, 200,000 MT/year in stream S2031, specified the degree of freedom. The stream flow rates for S2030 and S2032 are compared with HYSYS results to check the validity of the model and are given in Table 5.20.

The variables for the energy balance equation are H2030–H2032. A set with elements 2000 × 2050 was created in set "setbio." The enthalpy of biomass streams for ethylene process extension in the complex was included in the list H2 with the description "Enthalpy of biomass streams in complex." The enthalpy per unit mass was entered in the list for the corresponding streams.

TABLE 5.19

Mass Balance Equations for Ethylene Production from Dehydration of Ethanol

Material Balance	IN–OUT + GENERATION–CONSUMPTION = 0
Overall	$(F_{2030}) - (F_{2031} + F_{2032}) = 0$
	where
	$F_{2032} = F_{2032}^{(Ethylene)} + F_{2032}^{(H_2O)} + F_{2032}^{(Ethanol)}$
Species:	
Ethanol	$F_{2030} - F_{2032}^{(Ethanol)} - \left(EECONV \times F_{2030} \times \dfrac{mw(Ethanol)}{mw(Ethanol)} \right) = 0$
Ethylene	$-F_{2031} - F_{2032}^{(Ethylene)} + \left(EECONV \times F_{2030} \times \dfrac{mw(Ethylene)}{mw(Ethanol)} \right) = 0$
H_2O	$-F_{2032}^{(H_2O)} + \left(EECONV \times F_{2030} \times \dfrac{mw(H_2O)}{mw(Ethanol)} \right) = 0$

TABLE 5.20

Ethylene Section Optimization Model Results Validated with Data from HYSYS (kg/h)

Stream Name	Data from HYSYS	Data from the System	Percent Difference (%)
S2030	4.15E+04	4.15E+04	0
S2031	2.50E+04	2.50E+04	0
S2032	1.65E+04	1.65E+04	0
S2032E	2.53E+01	2.50E+01	−1
S2032ETOH	4.15E+02	4.15E+02	0
S2032H2O	1.61E+04	1.61E+04	0
QEEI (kJ/h)	1.03E+08	1.03E+08	0

TABLE 5.21

Energy Balance Equations for Ethylene Production from Dehydration of Ethanol

	IN – OUT + GENERATION – CONSUMPTION = 0
Energy Balance	$Q_{in} - Q_{out} \times F_p + \sum_j H_j F_{j_{in}} - \sum_j H_j F_{j_{out}} = 0$

Overall energy required from Steam (Q_{EEI}):	$Q_{EEI} - Q_{EEO} + \sum_j H_j F_{j_{in}} - \sum_j H_j F_{j_{out}} = 0$	
	H_j	**Mass Enthalpy (kJ/kg)**
$Q_{in} = Q_{EEI}$ (kJ/h)	H_{2030}	−1.58E+04
$Q_{out} \times F_p = Q_{EEO} = 2337.3$ (kJ/kg)	H_{2031}	−2.30E+03
$\times F_{2031}$	H_{2032}	−1.31E+04
$F_j = F_{2030}, F_{2031}, F_{2032}$ (kg/h)		

The external energy variable for this process was QEEI, where "I" denotes the input coefficient for heat supplied by steam. The heat removed from this process was QEEO, where "O" denotes the output coefficient for heat removed by cooling water. The value for QEEO and the enthalpy of the individual streams were specified from HYSYS. QFEFEI was calculated from the overall energy balance equation as given in Table 5.21. There were five unknown variables in the ethylene section and five equations. So the degree of freedom was 5 − 5 = 0. The value for QFEEI was also validated in this section, given by QFEEI in Table 5.19.

5.5 Acetic Acid Production from Corn Stover Anaerobic Digestion

The process design for acetic acid production from corn stover anaerobic digestion was converted to the block flow diagram as shown in Figure 5.5. The block diagram for acetic acid production from corn stover had three units: pretreatment (corn stover) anaerobic dig., anaerobic digestion, and purification (acetic acid). These denote the pretreatment section, anaerobic digestion section, and purification and recovery section from the HYSYS model. These three processes are separately represented in three blocks in the optimization model. The reactions occurring in the process are given in Table 5.22. The streams are shown in Figure 5.5 and the stream descriptions are given in Table 5.23. The parameters for the process are given in Table 5.24. The overall balance for each section and the individual species mass balance and energy balance equations are given in Tables 5.28 and 5.29, respectively. The inlet and outlet stream flow rates for the blocks from the HYSYS design corresponding to the streams in Table 5.23 are given in Appendix F.

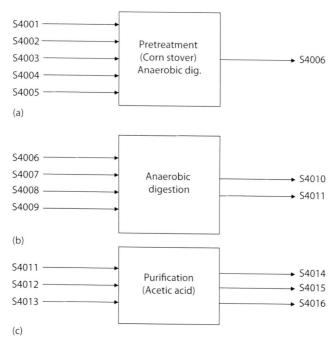

FIGURE 5.5

Block flow diagram of acetic acid production from corn stover anaerobic digestion: (a) pretreatment (corn stover) anaerobic dig., (b) anaerobic digestion, and (c) purification (acetic acid).

TABLE 5.22

Reactions for Acetic Acid Production from Corn Stover Anaerobic Digestion

Step	Reaction	Conversion (%)
Pretreatment	$(\text{Glucan})_n + n\text{H}_2\text{O} \rightarrow n\text{C}_6\text{H}_{12}\text{O}_6$	92
	$(\text{Xylan})_n + n\text{H}_2\text{O} \rightarrow n\text{C}_5\text{H}_{10}\text{O}_5$	92
Anaerobic digestion	$\text{C}_6\text{H}_{12}\text{O}_6 + 2\text{H}_2\text{O} \rightarrow 2\text{CH}_3\text{COOH} + 4\text{H}_2 + 2\text{CO}_2$	77
	$\text{C}_5\text{H}_{10}\text{O}_5 + 1.67\text{H}_2\text{O} \rightarrow 1.67\,\text{CH}_3\text{COOH} + 3.33\text{H}_2 + 1.664\text{CO}_2$	77

The mass balance for the streams is according to the equation

$$F_{\text{in}}^{(i)} - F_{\text{out}}^{(i)} + F_{\text{gen}}^{(i)} - F_{\text{cons}}^{(i)} = 0$$

where

F^i is the flow rate of ith component in the system

the subscripts in, out, gen, and cons denote the flow rates into, out of, generation of, and consumption of component i in the system

TABLE 5.23

Description of Process Streams in Acetic Acid Production
from Corn Stover Anaerobic Digestion

Name of Streams	Description
Pretreatment (Corn Stover) Anaerobic Dig.	
Input streams	
S4001	Dry biomass (corn stover)
S4002	Pig manure
S4003	Water added to dry biomass and pig manure
S4004	Steam added to pretreatment section
S4005	Lime added to pretreatment section
Output streams	
S4006	V-100 Out from pretreatment reactor
Energy streams	
QAAPRI	Heat required from steam in pretreatment section
Anaerobic Digestion	
Input streams	
S4006	V-100 Out from pretreatment reactor
S4007	Iodoform added to inhibit methane formation
S4008	Nutrients added for growth of mixed bacteria culture
S4009	Terrestrial inoculum added for anaerobic digestion
Output streams	
S4010	Gas mixture of CO_2 and H_2 from anaerobic digestion
S4011	MIX-103 Out from the anaerobic digestion section
Energy streams	
QAAO	Heat removed by cooling water in anaerobic digestion section
QAAI	Heat required from steam in anaerobic digestion section
Purification (Acetic Acid)	
Input streams	
S4011	MIX-103 Out from the anaerobic digestion section
S4012	Solvent used for extraction of acetic acid
S4013	Steam used for separation of solvent from water
Output streams	
S4014	Waste solids from the process
S4015	Acetic acid obtained from the process
S4016	Waste water from the process
Energy streams	
QAAPUO	Heat removed by cooling water in purification section
QAAPUI	Heat required from steam in purification section
Overall energy stream: QAAAD = QAAPRI + QAAI + QAAPUI	

TABLE 5.24

Parameters in Acetic Acid Production from Corn Stover Anaerobic Digestion

Name	Meaning	Value
Biomass composition		
MFCELP	Mass fraction of cellulose in S4001 corn stover	0.374
MFHEMP	Mass fraction of hemicellulose in S4001 corn stover	0.211
MFLIGP	Mass fraction of lignin in S4001 corn stover	0.180
MFASHP	Mass fraction of ash in S4001 corn stover	0.052
MFOTHP	Mass fraction of other solids in S4001 corn stover	0.183
Pig manure composition		
MFCELM	Mass fraction of cellulose in S4002 pig manure	0.525
MFASHM	Mass fraction of ash in S4002 pig manure	0.30
MFOTHM	Mass fraction of other solids in S4002 pig manure	0.175
Nutrient composition		
MFNUTAA	Mass fraction of nutrient in S4008	0.474
MFWATAA	Mass fraction of water in S4008	0.526
Conversion in reactors		
PRCELCONV	Pretreatment conversion of cellulose to glucose	0.92
PRHECELCONV	Pretreatment conversion of hemicellulose to xylose	0.92
AAGLCONV	Conversion of glucose to acetic acid	0.77
AAXYCONV	Conversion of xylose to acetic acid	0.77
Stream fractions		
MANFRAC	Manure fraction in biomass/manure mixture	0.20
AASTFRAC	Steam fraction with respect to biomass flow rate in S4001	0.025695
LIMEFRAC	Lime fraction with respect to biomass in stream S4001	0.10
IODFRAC	Iodoform fraction with respect to liquid in S4006	0.000023
NUTFRAC	Nutrient fraction with respect to liquid in S4006	0.001913
BACFRAC	Terrestrial inoculum fraction with respect to liquid in S4006	0.001913
SOLVFRAC	Fraction of solvent with respect to acetic acid in stream S4011	0.426312
AASSTFRAC	Fraction of steam required with respect to solvent in stream S4012	0.134
AAFRAC	Ratio of top and bottom acetic acid streams from extraction process	12.54

The energy balance for the process is according to the equation

$$Q_{in} - Q_{out} \times F_p + \sum_j H_j F_{j_{in}} - \sum_j H_j F_{j_{out}} = 0$$

H_j is the enthalpy per unit mass of jth stream computed from HYSYS. Q_{out} is the energy removed from the system per unit mass of product p having a flow rate of F_p. Q_{in} is the energy required by the system and calculated in the model with the other terms from the aforementioned equation specified from HYSYS.

5.5.1 Pretreatment (Corn Stover) Anaerobic Digestion

The block for pretreatment was created in the chemical complex analysis system, with S4001–S4005 as inlet streams and S4006 as outlet stream. The biomass composition for corn stover is the same as for the fermentation process, and the pig manure composition is added through the edit feature in the system. The flow rate of S4001 was fixed with the capacity constraint of 2000 MT/day of dry corn stover. The flow rate of stream S4002 was obtained from the relation of an 80% biomass–20% pig manure mixture. The water added to the stream was equal mass flow rate to the biomass and pig manure stream. This relation was used for the S4003 stream. The steam required for the pretreatment process was a fraction of the total biomass in the process. This relation was used in stream S4004. The lime used for the process was 0.1 g/g dry biomass, and this relation was used for the stream S4005.

The steam pretreatment reaction converted 20% of the biomass to monomeric form and the lime treatment converted 90% of the remaining biomass. So, an overall 92% conversion was attained in a single reactor and this was used as PRCELCONV and PRHECELCONV. The reactions are given in Table 5.22 for the cellulose and hemicellulose conversion to glucose and xylose, respectively.

There were 23 variables in the pretreatment section and 24 equations, with 2 dependent equations. Therefore, the degree of freedom for the pretreatment section was $23 - (24 - 2) = 1$. The constraint for the capacity of processing 2000 MT/day of corn stover in the plant in stream S4001 specified the degree of freedom. The stream flow rates for S4006 are compared with HYSYS results to check the validity of the model and are given in Table 5.25.

The energy balance equations are given in Table 5.29. The variables for the energy balance equation were H4001–H4006. A set with elements 4000 × 4020 was created in set "setbio." The mass enthalpy per unit mass of corresponding streams for anaerobic digestion process extension in the complex was included in the list H2, defined on "setbio," with the description "Enthalpy of biomass streams in complex."

The external energy variable for this section was QAAPRI, where "I" denotes the input coefficient for heat supplied by steam. There was no heat removed from the pretreatment process. The values for the enthalpy of the

TABLE 5.25

Pretreatment (Corn Stover) Anaerobic Digestion
Section Optimization Model Results Validated
with Data from HYSYS (kg/h)

Stream Name	Data from HYSYS	Data from the System	Percent Difference (%)
S4006	2.19E+05	2.19E+05	0
S4006A	1.06E+04	1.06E+04	0
S4006C	3.37E+03	3.37E+03	0
S4006CAOH$_2$	8.33E+03	8.33E+03	0
S4006G	4.30E+04	4.30E+04	0
S4006H	1.41E+03	1.41E+03	0
S4006H$_2$O	9.98E+04	9.98E+04	0
S4006L	1.50E+04	1.50E+04	0
S4006OS	1.89E+04	1.89E+04	0
S4006X	1.84E+04	1.84E+04	0

individual streams were specified from HYSYS. QAAPRI was calculated from the overall energy balance equation as given in Table 5.29. There were seven unknown variables in the energy balance for pretreatment section and seven equations. So the degree of freedom was $7 - 7 = 0$.

The total degree of freedom for the pretreatment section was $(23 + 7) - (24 - 2 + 7) = 1$.

5.5.2 Anaerobic Digestion

The input stream for this section was the S4006 pretreated biomass stream. The component flow rates for this stream were calculated in the pretreatment section. The variables in this section were S4006–S4011. The input iodoform addition rate was a fraction of the liquid medium in the anaerobic digestion process, so a fraction denoting the ratio between the iodoform stream and water content in the pretreated biomass stream was used as a parameter for calculating the flow rate of S4007. The nutrient addition and terrestrial inoculum addition rates to the process were 1 g/L of liquid medium. This was converted to mass ratio of the nutrients and inoculum with respect to water in the S4006 stream, and used as parameters to calculate the flow rates of S4008 and S4009.

The anaerobic digestion of biomass in this process produces acetic acid, carbon dioxide, and hydrogen according to the reactions given in Table 5.22. The conversion of volatile solids was 77% for both glucose and xylose, given by AAGLCONV and AAXYCONV. The carbon dioxide and hydrogen gases were vented from the process in stream S4010. The acetic acid and the waste biomass stream were sent to the purification section in stream S4011.

There are 32 variables in the anaerobic digestion section and 24 equations, with 2 dependent equations. Therefore, the degree of freedom for

TABLE 5.26

Anaerobic Digestion Section Optimization Model
Results Validated with Data from HYSYS (kg/h)

Stream Name	Data from HYSYS	Data from the System	Percent Difference (%)
S4010	2.52E+04	2.52E+04	0
S4010CO$_2$	2.31E+04	2.31E+04	0
S4010H$_2$	2.12E+03	2.12E+03	0
S4011	1.94E+05	1.94E+05	0
S4011A	1.06E+04	1.06E+04	0
S4011AA	3.15E+04	3.15E+04	0
S4011B	1.91E+02	1.91E+02	0
S4011C	3.37E+03	3.37E+03	0
S4011CAOH$_2$	8.33E+03	8.33E+03	0
S4011CHI$_3$	2.29E+00	2.29E+00	0
S4011G	9.90E+03	9.90E+03	0
S4011H	1.41E+03	1.41E+03	0
S4011H$_2$O	9.04E+04	9.04E+04	0
S4011L	1.50E+04	1.50E+04	0
S4011N	9.05E+01	9.05E+01	0
S4011OS	1.89E+04	1.89E+04	0
S4011X	4.23E+03	4.23E+03	0

the anaerobic digestion section is $32 - (24 - 2) = 10$. The flow rate variables for individual components in stream S4006 were computed from the previous section, and these ten mass flow rate variables were used to reduce the degree of freedom. The stream flow rates for S4010 and S4011 are compared with HYSYS results to check the validity of the model and are given in Table 5.26.

The variables for the energy balance equation are H4006–H4011. The enthalpy of biomass streams for acetic acid process extension in the complex was included in the list H2 with the description "Enthalpy of biomass streams in complex." The enthalpy per unit mass was entered in the list for the corresponding streams.

The external energy variable for this process was QAAI, where "I" denotes the input coefficient for heat supplied by steam. The heat removed from this process was QAAO, where "O" denotes the output coefficient for heat removed. The value for QAAO and the enthalpy of the individual streams were specified from HYSYS. QAAI was calculated from the overall energy balance equation, as given in Table 5.29. There were eight unknown variables in the anaerobic digestion section and eight equations. So the degree of freedom was $8 - 8 = 0$.

The total degree of freedom for the anaerobic digestion section was $(32 + 8) - (24 - 2 + 8) = 10$.

5.5.3 Purification (Acetic Acid)

The input stream for the purification section was the S4011 stream containing impure acetic acid. The component flow rates for this stream were calculated in the anaerobic digestion section. The variables in this section were S4011–S4016. The processes involved in this section included centrifugation to remove the solids from the stream followed by solvent extraction of acetic acid from water. The solvent was recycled in the process. The solvent addition rate was a fraction of acetic acid produced in the anaerobic digestion process. The steam required to strip the solvent from the water was a fraction of the solvent flow rate to the process. These relations were used as constraints in the model. The ratio of acetic acid removed in the top and the bottom streams of the solvent extraction column was used as a parameter to determine the split of acetic acid in streams S4015 and S4016. The product stream was S4015 containing acetic acid, with S4014 as the waste solids stream and S4016 as the waste water stream.

There are 33 variables in the purification section and 20 equations, with 1 dependent equation. Therefore, the degree of freedom for the purification section is $33 - (20 - 1) = 14$. The flow rate variables for individual components in stream S4011 were computed from the previous section, and the 14 mass flow rate variables in that stream solved the degree of freedom. The stream flow rates for S4014, S4015, and S4016 are compared with HYSYS results to check the validity of the model and are given in Table 5.27.

TABLE 5.27

Purification (Acetic Acid) Section Optimization Model Results Validated with Data from HYSYS (kg/h)

Stream Name	Data from HYSYS	Data from the System	Percent Difference (%)
S4014	7.20E+04	7.20E+04	0
S4014A	1.06E+04	1.06E+04	0
S4014B	1.91E+02	1.91E+02	0
S4014C	3.37E+03	3.37E+03	0
S4014CAOH$_2$	8.33E+03	8.33E+03	0
S4014CHI$_3$	2.29E+00	2.29E+00	0
S4014G	9.90E+03	9.90E+03	0
S4014H	1.41E+03	1.41E+03	0
S4014L	1.50E+04	1.50E+04	0
S4014N	9.05E+01	9.05E+01	0
S4014OS	1.89E+04	1.89E+04	0
S4014X	4.23E+03	4.23E+03	0
S4015	2.92E+04	2.92E+04	0
S4016	1.08E+05	1.08E+05	0
S4016AA	2.33E+03	2.33E+03	0
S4016H$_2$O	9.22E+04	9.22E+04	0
S4016MIBK	1.34E+04	1.34E+04	0
QAAAD (kJ/h)	4.59E+08	4.59E+08	0

TABLE 5.28

Mass Balance Equations for Acetic Acid Production from Corn Stover Anaerobic Digestion

Material Balance	IN−OUT + GENERATION−CONSUMPTION = 0

Pretreatment (Corn Stover) Anaerobic Dig.

Overall $(F_{4001} + F_{4002} + F_{4003} + F_{4004} + F_{4005}) - (F_{4006}) = 0$

where

$$F_{4001} = F_{4001}^{(Cellulose)} + F_{4001}^{(Hemicellulose)} + F_{4001}^{(Lignin)} + F_{4001}^{(Ash)} + F_{4001}^{(Other\ Solids)}$$

$$F_{4002} = F_{4002}^{(Cellulose)} + F_{4002}^{(Ash)} + F_{4002}^{(Other\ Solids)}$$

$$F_{4006} = F_{4006}^{(Ca(OH)_2)} + F_{4006}^{(H_2O)} + F_{4006}^{(Xylose)} + F_{4006}^{(Cellulose)} + F_{4006}^{(Hemicellulose)} + F_{4006}^{(Glucose)} + F_{4006}^{(Ash)} + F_{4006}^{(Other\ Solids)} + F_{4006}^{(Lignin)}$$

Species:

Cellulose
$$\left(F_{4001}^{(Cellulose)} + F_{4002}^{(Cellulose)} - F_{4006}^{(Cellulose)} - PRCELCONV \times \left(F_{4001}^{(Cellulose)} + F_{4002}^{(Cellulose)} \right) \times \frac{mw(Cellulose)}{mw(Cellulose)} = 0 \right)$$

Hemicellulose
$$F_{4001}^{(Hemicellulose)} - F_{4006}^{(Hemicellulose)} - \left(PRHECELCONV \times \left(F_{4001}^{(Hemicellulose)} \right) \times \frac{mw(Hemicellulose)}{mw(Hemicellulose)} \right) = 0$$

H_2O
$$F_{4003} + F_{4004} - F_{4006}^{H_2O} - \left(PRCELCONV \times \left(F_{4001}^{(Cellulose)} + F_{4002}^{(Cellulose)} \right) \times \frac{mw(H_2O)}{mw(Cellulose)} \right)$$
$$- \left(PRHECELCONV \times \left(F_{4001}^{(Hemicellulose)} \right) \times \frac{mw(H_2O)}{mw(Hemicellulose)} \right) = 0$$

Glucose
$$-F_{4006}^{(Glucose)} + \left(PRCELCONV \times \left(F_{4001}^{(Cellulose)} + F_{4002}^{(Cellulose)} \right) \times \frac{mw(Glucose)}{mw(Cellulose)} \right) = 0$$

Xylose
$$-F_{4006}^{(Xylose)} + \left(PRHECELCONV \times \left(F_{4001}^{(Hemicellulose)} \right) \times \frac{mw(Xylose)}{mw(Hemicellulose)} \right) = 0$$

Ca(OH)$_2$

$$F_{4005} - F_{4006}^{(Ca(OH)_2)} = 0$$

Ash

$$F_{4001}^{(Ash)} + F_{4002}^{(Ash)} - F_{4006}^{(Ash)} = 0$$

Other solids

$$F_{4001}^{(Other\,Solids)} + F_{4002}^{(Other\,Solids)} - F_{4006}^{(Other\,Solids)} = 0$$

Lignin

$$F_{4001}^{(Lignin)} - F_{4006}^{(Lignin)} = 0$$

Anaerobic Digestion

Overall

$$(F_{4006} + F_{4007} + F_{4008} + F_{4009}) - (F_{4010} + F_{4011}) = 0$$

where

$$F_{4006} = F_{4006}^{(Ca(OH)_2)} + F_{4006}^{(H_2O)} + F_{4006}^{(Xylose)} + F_{4006}^{(Cellulose)} + F_{4006}^{(Hemicellulose)} + F_{4006}^{(Glucose)} + F_{4006}^{(Ash)} + F_{4006}^{(Other\,Solids)} + F_{4006}^{(Lignin)}$$

$$F_{4008} = F_{4008}^{(H_2O)} + F_{4008}^{(Nutrients)}$$

$$F_{4010} = F_{4010}^{(H_2)} + F_{4010}^{(CO_2)}$$

$$F_{4011} = F_{4011}^{(Ca(OH)_2)} + F_{4011}^{(H_2O)} + F_{4011}^{(Xylose)} + F_{4011}^{(Cellulose)} + F_{4011}^{(Hemicellulose)} + F_{4011}^{(Glucose)} + F_{4011}^{(Ash)} + F_{4011}^{(Other\,Solids)} + F_{4011}^{(lignin)} + F_{4011}^{(CH_3)}$$
$$+ F_{4011}^{(Bacteria)} + F_{4011}^{(Nutrients)} + F_{4011}^{(Acetic\,Acid)}$$

Species:

Glucose

$$F_{4006}^{(Glucose)} - F_{4011}^{(Glucose)} - \left(AAGLCONV \times \frac{F_{4006}^{(Glucose)}}{mw(Glucose)} \times mw(Glucose) \right) = 0$$

Xylose

$$F_{4006}^{(Xylose)} - F_{4011}^{(Xylose)} - \left(AAXYCONV \times \frac{F_{4006}^{(Xylose)}}{mw(Xylose)} \times mw(Xylose) \right) = 0$$

H$_2$O

$$F_{4006}^{(H_2O)} + F_{4008}^{(H_2O)} - F_{4011}^{(H_2O)} - \left(AAGLCONV \times \left(\frac{2}{1}\right) \times \frac{F_{4006}^{(Glucose)}}{mw(Glucose)} \times mw(H_2O) \right)$$
$$- \left(AAXYCONV \times \left(\frac{1.67}{1}\right) \times \frac{F_{4006}^{(Xylose)}}{mw(Xylose)} \times mw(H_2O) \right) = 0$$

(continued)

TABLE 5.28 (continued)

Mass Balance Equations for Acetic Acid Production from Corn Stover Anaerobic Digestion

Material Balance	IN−OUT + GENERATION−CONSUMPTION = 0
Acetic acid	$-F_{4011}^{(Acetic\ Acid)} + \left(AAGLCONV \times \left(\dfrac{2}{1}\right) \times \dfrac{F_{4006}^{(Glucose)}}{mw(Glucose)} \times mw(Acetic\ Acid) \right)$ $+ \left(AAXYCONV \times \left(\dfrac{1.67}{1}\right) \times \dfrac{F_{4006}^{(Xylose)}}{mw(Xylose)} \times mw(Acetic\ Acid) \right) = 0$
CO_2	$-F_{4010}^{(CO_2)} \left(AAGLCONV \times \left(\dfrac{2}{1}\right) \times \dfrac{F_{4006}^{(Glucose)}}{mw(Glucose)} \times mw(CO_2) \right)$ $+ \left(AAXYCONV \times \left(\dfrac{1.664}{1}\right) \times \dfrac{F_{4006}^{(Xylose)}}{mw(Xylose)} \times mw(CO_2) \right) = 0$
H_2	$-F_{4010}^{(H_2)} \left(AAGLCONV \times \left(\dfrac{4}{1}\right) \times \dfrac{F_{4006}^{(Glucose)}}{mw(Glucose)} \times mw(H_2) \right)$ $+ \left(AAXYCONV \times \left(\dfrac{3.33}{1}\right) \times \dfrac{F_{4006}^{(Xylose)}}{mw(Xylose)} \times mw(H_2) \right) = 0$
$Ca(OH)_2$	$F_{4006}^{(Ca(OH)_2)} - F_{4011}^{(Ca(OH)_2)} = 0$
Cellulose	$F_{4006}^{(Cellulose)} - F_{4011}^{(Cellulose)} = 0$
Hemicellulose	$F_{4006}^{(Hemicellulose)} - F_{4011}^{(Hemicellulose)} = 0$
Lignin	$F_{4006}^{(Lignin)} - F_{4011}^{(Lignin)} = 0$

Ash	$F_{4006}^{(Ash)} - F_{4011}^{(Ash)} = 0$
Other solids	$F_{4006}^{(Other\ Solids)} - F_{4011}^{(Other\ Solids)} = 0$
Iodoform	$F_{4007} = F_{4011}^{(CHI_3)} = 0$
Nutrients	$F_{4008}^{(Nutrients)} - F_{4011}^{(Nutrients)} = 0$
Terrestrial inoculum	$F_{4009} - F_{4011}^{(Bacteria)} = 0$

Purification (Acetic Acid)

Overall

$$(F_{4011} + F_{4012} + F_{4013}) - (F_{4014} + F_{4015} + F_{4016}) = 0$$

where

$$F_{4011} = F_{4011}^{(Ca(OH)_2)} + F_{4011}^{(H_2O)} + F_{4011}^{(Xylose)} + F_{4011}^{(Cellulose)} + F_{4011}^{(Hemicellulose)} = F_{4011}^{(Glucose)} + F_{4011}^{(Ash)} + F_{4011}^{(Nutrients)} + F_{4011}^{(Other\ Solids)} + F_{4011}^{(Lignin)}$$
$$+ F_{4011}^{(CHI_3)} + F_{4011}^{(Bacteria)} + F_{4011}^{(Acetic\ Acid)}$$

$$F_{4014} = F_{4014}^{(Ca(OH)_2)} + F_{4014}^{(Xylose)} + F_{4014}^{(Cellulose)} + F_{4014}^{(Hemicellulose)} + F_{4014}^{(Glucose)} + F_{4014}^{(Ash)} + F_{4014}^{(Nutrients)} + F_{4014}^{(Other\ Solids)} + F_{4014}^{(Lignin)} + F_{4014}^{(CHI_3)} + F_{4014}^{(Bacteria)}$$

$$F_{4016} = F_{4016}^{(H_2O)} + F_{4016}^{(Acetic\ Acid)} + F_{4016}^{(MIBK)}$$

Species:

Ca(OH)$_2$	$F_{4011}^{(Ca(OH)_2)} - F_{4014}^{(Ca(OH)_2)} = 0$
Cellulose	$F_{4011}^{(Cellulose)} - F_{4014}^{(Cellulose)} = 0$
Hemicellulose	$F_{4011}^{(Hemicellulose)} - F_{4014}^{(Hemicellulose)} = 0$
Lignin	$F_{4011}^{(Lignin)} - F_{4014}^{(Lignin)} = 0$
Ash	$F_{4011}^{(Ash)} - F_{4014}^{(Ash)} = 0$

(continued)

TABLE 5.28 (continued)

Mass Balance Equations for Acetic Acid Production from Corn Stover Anaerobic Digestion

Material Balance	IN–OUT + GENERATION–CONSUMPTION = 0
Other solids	$F_{4011}^{(\text{Other Solids})} - F_{4014}^{(\text{Other Solids})} = 0$
Iodoform	$F_{4011}^{(\text{CHI}_3)} - F_{4014}^{(\text{CHI}_3)} = 0$
Nutrients	$F_{4011}^{(\text{Nutrients})} - F_{4014}^{(\text{Nutrients})} = 0$
Terrestrial inoculum	$F_{4011}^{(\text{Bacteria})} - F_{4014}^{(\text{Bacteria})} = 0$
Xylose	$F_{4011}^{(\text{Xylose})} - F_{4014}^{(\text{Xylose})} = 0$
Glucose	$F_{4011}^{(\text{Glucose})} - F_{4014}^{(\text{Glucose})} = 0$
Solvent	$F_{4012} - F_{4016}^{(\text{MIBK})} = 0$
H_2O	$F_{4011}^{(\text{H}_2\text{O})} + F_{4013} - F_{4016}^{(\text{H}_2\text{O})} = 0$
Acetic acid	$F_{4011}^{(\text{Acetic Acid})} - F_{4015} - F_{4016}^{(\text{Acetic Acid})} = 0$

The variables for the energy balance equation are H4011–H4016. The enthalpy of biomass streams for acetic acid process extension in the complex was included in the list H2 with the description "Enthalpy of biomass streams in complex." The mass enthalpy per unit mass was entered in the list for the corresponding streams.

The external energy variable for this process was QAAPUI, where "I" denotes the input coefficient for heat supplied by steam. The heat removed from this process was QAAPUO, where "O" denotes the output coefficient for heat removed. The value for QAAPUO and the enthalpy of the individual streams were specified from HYSYS. QAAPUI was calculated from the overall energy balance equation as given in Table 5.29. There were nine unknown variables in the anaerobic digestion section and nine equations. So the degree of freedom was 9 – 9 = 0.

The overall energy required from steam for the anaerobic digestion process was QAAAD, which was equal to the sum of QAAPRI, QAAI, and QAAPUI. The value for QAAAD was also validated in this section, given by QAAAD in Table 5.27.

The total degree of freedom for the purification section was (33 + 9) – (20 – 1 + 9) = 14.

5.6 Fatty Acid Methyl Ester and Glycerol from Transesterification of Natural Oil

The process design for FAME and glycerol production from natural oils (soybean oil) was converted to the block flow diagram as shown in Figure 5.6. The block diagram for FAME production had one block, which included the three sections from HYSYS (transesterification reaction section, FAME purification section, and glycerol recovery and purification section) combined into one. The reactions occurring in the process are given in Table 5.30. The streams are shown in Figure 5.6 and the stream descriptions are given in Table 5.31. The parameters for the process are given in Table 5.32. The overall balance for each section and the individual species mass balance and energy balance equations are given in Table 5.34 and Table 5.35, respectively. The inlet and outlet stream flow rates for the blocks from the HYSYS design corresponding to the streams in Table 5.31 are given in Appendix F.

The mass balance for the streams is according to the equation

$$F_{in}^{(i)} - F_{out}^{(i)} + F_{gen}^{(i)} - F_{cons}^{(i)} = 0$$

where
F^i is the flow rate of ith component in the system
the subscripts in, out, gen, and cons denote the flow rates into, out of, generation of, and consumption of component i in the system

TABLE 5.29

Energy Balance Equations for Acetic Acid Production from Corn
Stover Anaerobic Digestion

	IN – OUT + GENERATION – CONSUMPTION = 0
Energy Balance	$Q_{in} - Q_{out} \times F_p + \sum_j H_j F_{jin} - \sum_j H_j F_{jout} = 0$

Overall energy required from steam: $Q_{AAAD} = Q_{AAPRI} + Q_{AAI} + Q_{AAPUI}$

Pretreatment (Corn Stover) Anaerobic Dig. (Q_{AAPRI})	$Q_{AAPRI} - Q_{AAPRO} + \sum_j H_j F_{jin} - \sum_j H_j F_{jout} = 0$	
$Q_{in} = Q_{AAPRI}$ (kJ/h) $Q_{out} \times F_p = Q_{AAPRO} = 0 \times F_{4006}$ $F_j = F_{4001}, F_{4002}, F_{4003}, F_{4004}, F_{4005}, F_{4006}$ (kg/h)	H_j	Mass Enthalpy (kJ/kg)
	H_{4001}	−2.30E+03
	H_{4002}	−2.30E+03
	H_{4003}	−1.58E+04
	H_{4004}	−1.32E+04
	H_{4005}	−7.36E+03
	H_{4006}	−8.08E+03
Anaerobic Digestion (Q_{AAI})	$Q_{AAI} - Q_{AAO} + \sum_j H_j F_{jin} - \sum_j H_j F_{jout} = 0$	
$Q_{in} = Q_{AAI}$(kJ/h) $Q_{out} \times F_p = Q_{AAO} = 1477.99$ (kJ/kg) × F_{4011} $F_j = F_{4006}, F_{4007}, F_{4008}, F_{4009}, F_{4010}, F_{4011}$ (kg/h)	H_j	Mass Enthalpy (kJ/kg)
	H_{4006}	−8.08E+03
	H_{4007}	−5.88E+03
	H_{4008}	−9.40E+03
	H_{4009}	−2.29E+03
	H_{4010}	−8.17E+03
	H_{4011}	−9.54E+03
Purification (Acetic Acid) (Q_{AAPUI})	$Q_{AAPUI} - Q_{AAPUO} + \sum_j H_j F_{jin} - \sum_j H_j F_{jout} = 0$	
$Q_{in} = Q_{AAPUI}$ (kJ/h) $Q_{out} \times F_p = Q_{AAPU} = 3416.07$ (kJ/kg) × F_{4015} $F_j = F_{4011}, F_{4012}, F_{4013}, F_{4014}, F_{4015}, F_{4016}$ (kg/h)	H_j	Mass Enthalpy (kJ/kg)
	H_{4011}	−9.54E+03
	H_{4012}	−3.25E+03
	H_{4013}	−1.32E+04
	H_{4014}	−2.54E+03
	H_{4015}	−7.52E+03
	H_{4016}	−1.26E+04

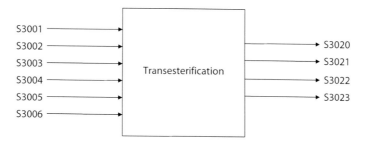

FIGURE 5.6
Block flow diagram of FAME and glycerol production from transesterification.

TABLE 5.30

Reactions for FAME and Glycerol Production from Transesterification

Step	Reaction	Conversion (%)
Transesterification	$C_{57}H_{98}O_6 + 3CH_3OH \rightarrow C_3H_8O_3$ $+ 3C_{19}H_{34}O_2$	90
Catalyst neutralization	$HCl + NaOCH_3 \rightarrow NaCl + CH_3OH$	100
Acid neutralization	$NaOH + HCl \rightarrow NaCl + H_2O$	100

TABLE 5.31

Description of Process Streams in FAME and Glycerol
Production from Transesterification

Name of Streams	Description
Transesterification	
Input streams	
S3001	Oil supplied to the transesterification process
S3002	Catalyst ($NaOCH_3$) added to the transesterification process
S3003	Methanol added to the transesterification process
S3004	Water added to the wash process
S3005	Hydrochloric acid added to neutralize the catalyst
S3006	Sodium hydroxide added to neutralize excess HCl
Output streams	
S3020	Fatty acid methyl ester (FAME) produced from process
S3021	Glycerol by-product produced from process
S3022	Water removed from process
S3023	Free fatty acids removed from process
Energy streams	
QFAMEO	Heat removed by cooling water in transesterification section
QFAMEI	Heat required from steam in transesterification section

TABLE 5.32

Parameters in FAME and Glycerol Production from Transesterification

Name	Description	Value
Stream composition		
MFCAT	Mass fraction of $NaOCH_3$ catalyst stream	0.25
MFHCL	Mass fraction of hydrochloric acid in HCl stream	0.35
Conversion in reactors		
TROICONV	Transesterification oil conversion	0.90
CATCONV	Conversion of catalyst in neutralization with HCl	1.00
NAOHCONV	Conversion of NaOH in excess HCl neutralization	1.00
Stream fractions		
CATFRAC	Fraction of catalyst stream with respect to inlet oil	0.0125
METFRAC	Fraction of methanol with respect to inlet oil	0.0994
HCLFRAC	Fraction of HCl stream with respect to inlet catalyst	0.571
NAOHFRAC	Fraction of NaOH stream with respect to inlet HCl	0.0632
GLYFRAC	Fraction of glycerol removed as free fatty acids	0.1111
FFFRAC	Fraction of total oil out in fatty acid stream	0.11
OILFRAC	Ratio of Oil in waste water with respect to FAME stream	0.23

The energy balance for the process was according to the equation

$$Q_{in} - Q_{out} \times F_p + \sum_j H_j F_{j_{in}} - \sum_j H_j F_{j_{out}} = 0$$

H_j is the enthalpy per unit mass of *j*th stream computed from HYSYS. Q_{out} is the energy removed from the system per unit mass of product *p* having a flow rate of F_p. Q_{in} is the energy required by the system and calculated using the aforementioned equation with the values for the other terms specified from HYSYS.

The block for transesterification was created in the chemical complex analysis system, with S3001–S3006 as inlet streams and S3020–S3023 as outlet streams. The flow rate of S3020 was fixed with the capacity constraint for 10 million gal/year (or 4257 kg/h) of FAME production. The streams S3002 and S3003 were fractions of the inlet flow rate of oil for transesterification. These relations were obtained from the HYSYS design and entered in the parameters table as CATFRAC and METFRAC for S3002 and S3003, respectively. Acid was added to the process for catalyst neutralization, and this was a fraction of the catalyst stream, given by HCLFRAC. Water used to wash the acid was a fraction of the acid stream, S3005, and this relation gave the flow rate for stream S3004. The excess acid was neutralized with sodium hydroxide and was a fraction of stream S3005, given by NAOHFRAC.

The conversion of the oil was 90% each in two sequential reactors. The excess catalyst was neutralized using acid and the excess acid was neutralized with sodium hydroxide. The conversion and neutralization reactions are given in Table 5.30, and the respective parameters are given in Table 5.32.

The outlet stream, S3020, was the FAME stream containing trace amounts of oil. The stream, S3021, was the glycerol by-product stream, containing sodium chloride. The water removed from the process was S3022, also containing oil and methanol. The stream, S3023, contained glycerol and oil removed from the process to simulate the removal of fatty acids from the process. The fraction of glycerol and oil removed in this stream were computed from the parameters, GLYFRAC and FFFRAC.

There are 39 variables in the transesterification section and 38 equations. Therefore, the degree of freedom for the transesterification section was $39 - 38 = 1$. The constraint for capacity of the plant to produce FAME in stream S3020 specified the degree of freedom. The stream flow rates for S3001–S3006 are compared with HYSYS results to check the validity of the model and are given in Table 5.33.

The energy balance equations are given in Table 5.35. The variables for the energy balance equation were H3001–H3006 for the inlet streams and H3020–H3023 for the outlet streams. A set with elements 3000 × 3050 was created in set "setbio." The mass enthalpy per unit mass of corresponding streams for transesterification process extension in the complex was included in the list H2, defined on "setbio," with the description "Enthalpy of biomass streams in complex."

The external energy variable for this process was QFAMEI, where "I" denotes the input coefficient for heat supplied by steam. The heat removed from this process was QFAMEO, where "O" denotes the output coefficient for heat removed by cooling water. The value for QFAMEO and the enthalpy of the individual streams were specified from HYSYS. QFAMEI was calculated

TABLE 5.33

Transesterification Section Optimization Model Results
Validated with Data from HYSYS (kg/h)

Stream Name	Data from HYSYS	Data from the System	Percent Difference (%)
S3001	4.25E+03	4.25E+03	0
S3002	5.31E+01	5.31E+01	0
S3002CH$_3$OH	3.98E+01	3.98E+01	0
S3002NAOCH$_3$	1.33E+01	1.33E+01	0
S3003	4.23E+02	4.23E+02	0
S3004	8.55E+01	8.49E+01	−1
S3005	3.03E+01	3.03E+01	0
S3005H$_2$O	1.97E+01	1.97E+01	0
S3005HCL	1.06E+01	1.06E+01	0
S3006	1.91E+00	1.92E+00	0
QFAMEI (kJ/h)	1.14E+07	1.14E+07	0

TABLE 5.34

Mass Balance Equations for FAME and Glycerol Production from Transesterification

Material Balance	IN–OUT + GENERATION–CONSUMPTION = 0

Transesterification

Overall

$$(F_{3001} + F_{3002} + F_{3003} + F_{3004} + F_{3005} + F_{3006}) - (F_{3020} + F_{3021} + F_{3022} + F_{3023}) = 0$$

where

$$F_{3002} = F_{3002}^{(\mathrm{NaOCH_3})} + F_{3002}^{(\mathrm{CH_3OH})}$$

$$F_{3005} = F_{3005}^{(\mathrm{H_2O})} + F_{3005}^{(\mathrm{HCl})}$$

$$F_{3020} = F_{3020}^{(\mathrm{FAME})} + F_{3020}^{(\mathrm{Oil})}$$

$$F_{3021} = F_{3021}^{(\mathrm{Glycerol})} + F_{3021}^{(\mathrm{NaCl})}$$

$$F_{3022} = F_{3022}^{(\mathrm{CH_3OH})} + F_{3022}^{(\mathrm{H_2O})} + F_{3022}^{(\mathrm{Oil})}$$

$$F_{3023} = F_{3023}^{(\mathrm{Glycerol})} + F_{3023}^{(\mathrm{Oil})}$$

Species:

Trilinolein

$$F_{3001} - F_{3020}^{(\mathrm{Oil})} - F_{3022}^{(\mathrm{Oil})} - F_{3023}^{(\mathrm{Oil})}$$

$$- \left(\mathrm{TROICONV} \times \frac{F_{3001}}{\mathrm{mw(Oil)}} \times \mathrm{mw(Oil)}\right) - \left(\mathrm{TROICONV} \times (1 - \mathrm{TROICONV}) \times \frac{F_{3001}}{\mathrm{mw(Oil)}} \times \mathrm{mw(Oil)}\right) = 0$$

$\mathrm{CH_3OH}$

$$F_{3002}^{(\mathrm{CH_3OH})} + F_{3003} - F_{3022}^{(\mathrm{CH_3OH})}$$

$$- \left(\mathrm{TROICONV} \times \left(\frac{3}{1}\right) \times \frac{F_{3001}}{\mathrm{mw(Oil)}} \times \mathrm{mw(Methanol)}\right)$$

$$- \left(\mathrm{TROICONV} \times (1 - \mathrm{TROICONV}) \times \left(\frac{3}{1}\right) \times \frac{F_{3001}}{\mathrm{mw(Oil)}} \times \mathrm{mw(Methanol)}\right)$$

$$+ \left(\mathrm{CATCONV} \times \frac{F_{3002}^{(\mathrm{NaOCH_3})}}{\mathrm{mw(NaOCH_3)}} \times \mathrm{mw(Methanol)}\right) = 0$$

NaOCH₃

$$F_{3002}^{-(NaOCH_3)} - \left(CATCONV \times \frac{F_{3002}^{-(NaOCH_3)}}{mw(NaOCH_3)} \times mw(NaOCH_3) \right) = 0$$

H₂O

$$F_{3004} + F_{3005}^{-(H_2O)} - F_{3022}^{-(H_2O)} - \left(NAOHCONV \times \frac{F_{3006}}{mw(NaOH)} \times mw(H_2O) \right) = 0$$

HCl

$$F_{3005}^{-HCl} - \left(CATCONV \times \frac{F_{3002}^{-(NaOCH_3)}}{mw(NaOCH_3)} \times mw(HCl) \right) - \left(NAOHCONV \times \frac{F_{3006}}{mw(NaOH)} \times mw(HCl) \right) = 0$$

NaOH

$$F_{3006} - \left(NAOHCONV \times \frac{F_{3006}}{mw(NaOH)} \times mw(NaOH) \right) = 0$$

FAME

$$-F_{3020}^{-(FAME)} + \left(TROICONV \times \frac{F_{3001}}{mw(Oil)} \times mw(FAME) \right)$$
$$+ \left(TROICONV \times (1 - TROICONV) \times \left(\frac{3}{1}\right) \times \frac{F_{3001}}{mw(Oil)} \times mw(FAME) \right) = 0$$

Glycerol

$$-F_{3021}^{-(Glycerol)} - F_{3023}^{-(Glycerol)}$$
$$+ \left(TROICONV \times \frac{F_{3001}}{mw(Oil)} \times mw(Glycerol) \right) + \left(TROICONV \times (1 - TROICONV) \times \frac{F_{3001}}{mw(Oil)} \times mw(Glycerol) \right) = 0$$

NaCl

$$-F_{3021}^{-(NaCl)} + \left(CATCONV \times \frac{F_{3002}^{-(NaOCH_3)}}{mw(NaOCH_3)} \times mw(NaCl) \right) + \left(NAOHCONV \times \frac{F_{3006}}{mw(NaOH)} \times mw(NaCl) \right) = 0$$

TABLE 5.35

Energy Balance Equations for FAME and Glycerol Production
from Transesterification

	IN – OUT + GENERATION – CONSUMPTION = 0
Energy Balance	$Q_{in} - Q_{out} \times F_p + \sum_j H_j F_{j_{in}} - \sum_j H_j F_{j_{out}} = 0$

Overall energy requirement from steam (Q_{FAMEI}):	$Q_{FAMEI} - Q_{FAMEO} + \sum_j H_j F_{j_{in}} - \sum_j H_j F_{j_{out}} = 0$	
$Q_{in} = Q_{FAMEI}$ (kJ/h)	H_j	**Mass Enthalpy (kJ/kg)**
$Q_{out} \times F_p = 2856.97$ (kJ/kg) $\times F_{3020}$		
$F_j = F_{3001}, F_{3002}, F_{3003}, F_{3004}, F_{3005}, F_{3006},$	H_{3001}	−1.68E+03
$F_{3020}, F_{3021}, F_{3022}, F_{3023}$ (kg/h)	H_{3002}	−6.39E+03
	H_{3003}	−7.47E+03
	H_{3004}	−1.58E+04
	H_{3005}	−1.12E+04
	H_{3006}	−2.09E+03
	H_{3020}	−1.96E+03
	H_{3021}	−6.57E+03
	H_{3022}	−1.36E+04
	H_{3023}	−6.81E+03

from the overall energy balance equation as given in Table 5.35. There were 12 unknown variables in the transesterification section and 12 equations. So the degree of freedom was 12 − 12 = 0. The value for QFAMEI was also validated in this section, given by QFAMEI in Table 5.33.

The total degree of freedom for the transesterification section was (39 + 12) − (38 + 12) = 1.

5.7 Propylene Glycol Production from Hydrogenolysis of Glycerol

The process design for propylene glycol production from hydrogenolysis of glycerol was converted to the block flow diagram as shown in Figure 5.7. The block diagram for propylene glycol had one block that included the sections from HYSYS. The reaction occurring in the process is given in Table 5.36. The streams are shown in Figure 5.7 and the stream descriptions are given in Table 5.37. The parameters for the process are given in Table 5.38. The overall balance for each section and the individual species mass balance and energy balance equations are given in Table 5.40 and Table 5.41, respectively. The inlet and outlet stream flow rates for the blocks from the HYSYS design corresponding to the streams in Table 5.37 are given in Appendix F.

FIGURE 5.7
Block flow diagram of propylene glycol production from glycerol.

TABLE 5.36

Reaction for Propylene Glycol
Production from Glycerol

Reaction	Conversion (%)
$C_3H_8O_3 + H_2 \rightarrow C_3H_8O_2$	54.80

TABLE 5.37

Description of Process Streams in Propylene Glycol
Production from Glycerol

Name of Streams	Description
Input streams	
S3030	Glycerol from transesterification process to new propylene glycol process
S3031	Hydrogen to new propylene glycol process
Output streams	
S3032	Propylene glycol from new propylene glycol process
S3033	Waste water from new propylene glycol process
Energy streams	
QPGO	Heat removed by cooling water in propylene glycol section
QPGI	Heat required from steam in propylene glycol section

TABLE 5.38

Parameters in Propylene Glycol Production from Glycerol

Name	Description	Value
GPGCONV	Glycerol to propylene glycol conversion	0.548
MFGLYPG	Mass fraction of glycerol in input stream of propylene glycol process	0.8
MFWATERPG	Mass fraction of water in input stream of propylene glycol process	0.2

The mass balance for the streams is according to the equation

$$F_{in}^{(i)} - F_{out}^{(i)} + F_{gen}^{(i)} - F_{cons}^{(i)} = 0$$

where
F^i is the flow rate of ith component in the system
the subscripts in, out, gen, and cons denote the flow rates into, out of, generation of and consumption of component i in the system

The energy balance for the process was according to the equation

$$Q_{in} - Q_{out} \times F_p + \sum_j H_j F_{j_{in}} - \sum_j H_j F_{j_{out}} = 0$$

H_j is the enthalpy per unit mass of jth stream computed from HYSYS. Q_{out} is the energy removed from the system per unit mass of product p having a flow rate of F_p. Q_{in} is the energy required by the system and calculated using the aforementioned equation with the values for the other terms specified from HYSYS.

The block for propylene glycol was created in the chemical complex analysis system, with S3030–S3033 as the stream variables. The flow rate of S3032 was fixed with the capacity constraint for 65,000 MT/year production of propylene glycol. The S3030 and S3031 were the glycerol and hydrogen streams to the system, respectively. The species mass balance equations are given in Table 5.40, and the model included recycle of methanol stream. This was incorporated by using a constant, CON40, which was computed from the conversion in the sequential reactors. The propylene glycol was obtained in S3032 and the waste water was obtained in S3033. The conversion parameter for glycerol was GPGCONV and is given in Table 5.38.

There were seven variables in the propylene glycol section and seven equations with one overall equation. Therefore, the degree of freedom for the propylene glycol section was $7 - (7 - 1) = 1$. The constraint for capacity of the plant to produce propylene glycol in stream S3032 specified the degree of freedom. The stream flow rates for S3030, S3031, and S3033 are compared with HYSYS results to check the validity of the model and are given in Table 5.39.

The energy balance equations are given in Table 5.41. The variables for the energy balance equation were H3030–H3033. The values for enthalpy of the stream were added to the list, H2, with the description "Enthalpy of biomass streams in complex."

The external energy variable for this process was QPGI, where "I" denotes the input coefficient for heat supplied by steam. The heat removed from this process was QPGO, where "O" denotes the output coefficient for heat removed by cooling water. The value for QPGO and the enthalpy of the individual streams were specified from HYSYS. QPGI was calculated from

TABLE 5.39

Propylene Glycol Section Optimization Model
Results Validated with Data from HYSYS (kg/h)

Stream Name	Data from HYSYS	Data from the System	Percent Difference (%)
S3030	1.48E+04	1.48E+04	0
S3030GLY	1.12E+04	1.12E+04	0
S3030H2O	3.54E+03	3.53E+03	0
S3031	2.46E+02	2.46E+02	0
S3033	5.74E+03	5.73E+03	0
QPGI (kJ/h)	1.02E+08	1.02E+08	0

the overall energy balance equation as given in Table 5.41. There were five unknown variables in the propylene glycol section and five equations. So the degree of freedom was $5 - 5 = 0$. The value for QPGI was also validated in this section, given by QPGI in Table 5.39.

The total degree of freedom for the propylene glycol section was $(7 + 5) - (7 - 1 + 5) = 1$.

5.8 Algae Oil Production

Algae have the potential for being an important source of oil and carbohydrates for the production of fuels, chemicals, and energy. Carbon dioxide and sunlight can be used to cultivate algae and produce algae with 60% triglycerides and 40% carbohydrates and protein (Pienkos and Daezins, 2009). A model algal lipid production system with algae growth, harvesting, extraction, separation, and uses is shown in Figure 2.24. A process for production of algae is also outlined by Pokoo-Aikins et al. (2010). Methods to convert whole algae into biofuels exist through anaerobic digestion to biogas, supercritical fluid extraction and pyrolysis to liquid or vapor fuels, and gasification process for production of syngas-based fuels and chemicals. Algae oil can supplement refinery diesel in hydrotreating units, or be used as feedstock for the biodiesel process. The research on algae as a biomass feedstock is a very dynamic field currently, and the potential of algae seems promising as new results are presented continuously.

Algae can be produced in an open system (raceway ponds) or a closed system (photobioreactors). In the superstructure, the production of algae oil from carbon dioxide is considered for the transesterification process. The algae biomass uses 1.8 ton of carbon dioxide per ton of algae produced (Oilgae, 2010). This relation is used for calculating the amount of carbon dioxide that can be utilized in the process.

TABLE 5.40

Mass Balance Equations for Propylene Glycol Production from Glycerol

Material Balance	IN−OUT + GENERATION−CONSUMPTION = 0
Overall	$(F_{3030} + F_{3031}) - (F_{3032} + F_{3033}) = 0$ where $F_{3030} = F_{3030}^{(H_2O)} + F_{3030}^{(Glycerol)}$
Glycerol	$F_{3030}^{(Glycerol)} - \left(GPGCONV \times \left(\left[1 + \dfrac{(1 - GPGCONV) \times (1 - GPGCONV)}{1 - (1 - GPGCONV) \times (1 - GPGCONV)} \right] \times F_{3030}^{(Glycerol)} \right) \times \dfrac{mw(Glycerol)}{mw(Glycerol)} \right) = 0$ $- \left(GPGCONV \times (1 - GPGCONV) \times \left(\left[1 + \dfrac{(1 - GPGCONV) \times (1 - GPGCONV)}{1 - (1 - GPGCONV) \times (1 - GPGCONV)} \right] \times F_{3030}^{(Glycerol)} \right) \times \dfrac{mw(Glycerol)}{mw(Glycerol)} \right)$
Hydrogen	$F_{3031} - \left(GPGCONV \times \left(\left[1 + \dfrac{(1 - GPGCONV) \times (1 - GPGCONV)}{1 - (1 - GPGCONV) \times (1 - GPGCONV)} \right] \times F_{3030}^{(Glycerol)} \right) \times \dfrac{mw(Hydrogen)}{mw(Glycerol)} \right)$ $- \left(GPGCONV \times (1 - GPGCONV) \times \left(\left[1 + \dfrac{(1 - GPGCONV) \times (1 - GPGCONV)}{1 - (1 - GPGCONV) \times (1 - GPGCONV)} \right] \times F_{3030}^{(Glycerol)} \right) \times \dfrac{mw(Hydrogen)}{mw(Glycerol)} \right) = 0$
H_2O	$F_{3030}^{H_2O} - F_{3033} + \left(GPGCONV \times \left(\left[1 + \dfrac{(1 - GPGCONV) \times (1 - GPGCONV)}{1 - (1 - GPGCONV) \times (1 - GPGCONV)} \right] \times F_{3030}^{(Glycerol)} \right) \times \dfrac{mw(Water)}{mw(Glycerol)} \right)$ $+ \left(GPGCONV \times (1 - GPGCONV) \times \left(\left[1 + \dfrac{(1 - GPGCONV) \times (1 - GPGCONV)}{1 - (1 - GPGCONV) \times (1 - GPGCONV)} \right] \times F_{3030}^{(Glycerol)} \right) \times \dfrac{mw(Water)}{mw(Glycerol)} \right) = 0$
Propylene glycol	$-F_{3032} + \left(GPGCONV \times \left(\left[1 + \dfrac{(1 - GPGCONV) \times (1 - GPGCONV)}{1 - (1 - GPGCONV) \times (1 - GPGCONV)} \right] \times F_{3030}^{(Glycerol)} \right) \times \dfrac{mw(Propylene\ Glycol)}{mw(Glycerol)} \right)$ $+ \left(GPGCONV \times (1 - GPGCONV) \times \left(\left[1 + \dfrac{(1 - GPGCONV) \times (1 - GPGCONV)}{1 - (1 - GPGCONV) \times (1 - GPGCONV)} \right] \times F_{3030}^{(Glycerol)} \right) \times \dfrac{mw(Propylene\ Glycol)}{mw(Glycerol)} \right) = 0$

TABLE 5.41

Energy Balance Equations for Propylene Glycol Production from Glycerol

Energy Balance	IN – OUT + GENERATION – CONSUMPTION = 0 $Q_{\text{in}} - Q_{\text{out}} \times F_p + \sum_j H_j F_{j_{\text{in}}} - \sum_j H_j F_{j_{\text{out}}} = 0$
Overall Energy Required from Steam (Q_{PGI}):	$Q_{\text{PGI}} - Q_{\text{PGO}} + \sum_j H_j F_{j_{\text{in}}} - \sum_j H_j F_{j_{\text{out}}} = 0$

$Q_{\text{in}} = Q_{\text{PGI}}$ (kJ/h)	H_j	Mass Enthalpy (kJ/kg)
$Q_{\text{out}} \times F_p = Q_{\text{PGO}} = 10781.28$ (kJ/kg) $\times F_{3032}$	H_{3030}	−9.38E+03
	H_{3031}	0
$F_j = F_{3030}, F_{3031}, F_{3032}, F_{3033}$ (kg/h)	H_{3032}	−6.52E+03
	H_{3033}	−1.32E+04

The industrial-scale production of algae oil is under extensive research and the results are widely varying, so a black box model for the production of algae oil is developed and used for this research. The oil production can be with two yields: low yield of 30% oil from algae and high yield of 50% oil from algae (Pokoo-Aikins et al., 2010). The equation for conversion of carbon dioxide to algae oil is given in Table 5.42. The block flow diagram for algae oil production is shown in Figure 5.8.

The algae oil composition was considered same as that for soybean oil. The stream descriptions are given in Table 5.43. The mass balance equations for the process are given in Table 5.44.

TABLE 5.42

Reactions for Algae Oil Production

Step	Reaction	Yield (Pokoo-Aikins et al. 2010)
30% algae oil from CO_2	$1.8CO_2 \rightarrow$ Algae Algae $\rightarrow$ Algae Oil/0.3	30% algae to oil yield (mass conversion)
50% algae oil from CO_2	$1.8CO_2 \rightarrow$ Algae Algae $\rightarrow$ Algae Oil/0.5	50% algae to oil yield (mass conversion)

FIGURE 5.8
Block flow diagram of algae oil production.

TABLE 5.43

Description of Process Streams in Algae Oil Production

Name of Streams	Description
Input stream	
S3050	Carbon dioxide stream to algae oil production process
Output stream	
S3051	Algae oil from algae oil production process

TABLE 5.44

Mass Balance Equations for Algae Oil Production

Material Balance	IN–OUT + GENERATION–CONSUMPTION = 0
30% oil yield	$1.8F_{3050} - F_{3051}/0.3 = 0$
50% oil yield	$1.8F_{3050} - F_{3051}/0.5 = 0$

The energy balance relations for this process were not available. For algae production from carbon dioxide, the raw material is carbon dioxide and sunlight. Thus, the raw materials cost can be considered zero.

Currently, for large-scale production of algae oil, there is a significant cost for drying and separation. However, if the algae strain is selected such that it secretes oil, for example, *Botryococcus braunii*, the utility costs can be substantially reduced. *B. braunii* species of algae has been engineered to produce the terpenoid C30 botryococcene, a hydrocarbon similar to squalene in structure (Arnaud, 2008). The species has been engineered to secrete the oil, and the algae can be reused in the bioreactor.

Low-cost photobioreactors are being developed, which promises to bring down the cost of algae production. 32A vertical reactor system is being developed by Valcent Products, Inc. of El Paso, Texas, using the 340 annual days of sunshine and carbon dioxide available from power plant exhaust. The company uses vertical bag bioreactors made of polythene to grow algae.

Also, new methods for algae oil separation have been developed where algae cells are ruptured and oil is liberated without the need for dewatering or solvents (Ondrey, 2009). This process, developed by Origin Oil Inc., has reduced energy costs by 90% and substantial savings have been made to capital cost for oil extraction. In this process, algae ready for harvesting is pumped into an extraction tank through a static mixer, which induces cavitation in the water. Simultaneously, a low-power pulsed electromagnetic field is applied to the algae stream, and CO_2 is introduced to lower the pH. The combination of these measures ruptures the cell walls and releases the oil, which rises to the surface in the tank and the biomass sinks to the bottom of the tank. The final separation is achieved in a clarification tank, where gravity settling is used to separate the biomass (solids after oil extraction) and oil. Thus, a future plant with typically low costs (zero production cost) for production of algae oil is considered in the superstructure in Chapter 6. To account for changes in the optimal solution for inclusion of algae oil production costs, a case study is developed in Chapter 7, which uses the algae oil production costs from Pokoo-Aikins et al. (2010). Vertical bag reactors can reduce the capital and equipment cost substantially. If the major contribution for producing a process is the operating cost, then the production costs can be considered same as operating costs. Moreover, the major contributors to operating costs are raw material costs and utility costs. So, the cost for production can be approximated as the utility costs for the process. Thus, the production costs from Pokoo-Aikins et al. (2010) were included as utility costs in the economic model for optimization.

5.9 Gasification of Corn Stover

The commercial process for hydrogen production is steam reforming of natural gas, involving reforming and a shift conversion. The gasification process for synthesis gas production from biomass was described in Klass (1998). Gasification can be carried out in the absence of oxygen, known as pyrolysis; in the presence of oxygen, known as partial oxidation; or in the presence of steam, known as steam reforming. The steam reforming process of corn stover was included in the optimization model for the production of syngas. The equation for steam reforming of hydrocarbons (Equation 5.1) was given by Ciferno and Marano (2002) and for cellulose as representative biomass (Equation 5.2) was given by Klass (1998). The equation for hemicellulose is similar to cellulose and given in Equation 5.3. The equations for cellulose and hemicellulose as given in Table 5.45 were used in the model formulation for gasification in the chemical complex analysis system. The streams are shown in Figure 5.9 and the stream descriptions are given in Table 5.46. A 100% conversion of cellulose and hemicellulose to syngas was considered for the process. The overall balance and the individual species mass balance are given in Table 5.47. The energy balance equation used in the model is given in Table 5.48.

TABLE 5.45

Reactions for Gasification of Corn Stover

Step	Reaction	Conversion (%)
Cellulose steam reforming	$C_6H_{10}O_5 + H_2O \rightarrow 6CO + 6H_2$	100
Hemicellulose steam reforming	$C_5H_8O_4 + H_2O \rightarrow 5CO + 5H_2$	100

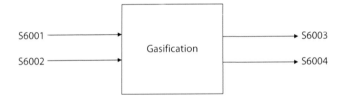

FIGURE 5.9
Block flow diagram of gasification of corn stover process.

TABLE 5.46

Description of Process Streams in Gasification of Corn Stover

Name of Streams	Description
Gasification	
Input streams	
S6001	Biomass (corn stover) to gasification process
S6002	Steam for gasification process
Output streams	
S6003	Syngas from gasification
S6004	Tar and other products
Energy streams	
QSYNGBIOO	Heat removed by cooling water in gasification section
QSYNGBIO	Heat required from steam in gasification section

$$C_nH_{2m} + nH_2O \rightarrow nCO + (m+n)H_2 \qquad (5.1)$$

$$C_6H_{10}O_5 + H_2O \rightarrow 6CO + 6H_2 \qquad (5.2)$$

$$C_5H_8O_4 + H_2O \rightarrow 5CO + 5H_2 \qquad (5.3)$$

The production capacity for the gasification plant was set at 13,400 MT of H_2 per year. This was based on a conventional hydrogen plant of Air Products and Chemicals Inc. located in Geismar, Louisiana, with the capacity of 15 million ft³/day (Louisiana Chemicals and Petroleum Products List, 1998). The mass balance for the streams is according to the equation

$$F_{in}^{(i)} - F_{out}^{(i)} + F_{gen}^{(i)} - F_{cons}^{(i)} = 0$$

where
  F^i is the flow rate of ith component in the system
  the subscripts in, out, gen, and cons denote the flow rates into, out of, generation of, and consumption of component i in the system

TABLE 5.47

Mass Balance Equations for Gasification of Corn Stover

Material Balance	IN–OUT + GENERATION–CONSUMPTION = 0

Gasification

Overall

$$(F_{6001} + F_{6002}) - (F_{6003} + F_{6004}) = 0$$

where

$$F_{6001} = F_{6001}^{(Cellulose)} + F_{6001}^{(Hemicellulose)} + F_{6001}^{(Lignin)} + F_{6001}^{(Ash)} + F_{6001}^{(Other\ Solids)}$$

$$F_{6003} = F_{6003}^{(CO)} + F_{6003}^{(H_2)}$$

$$F_{6004} = F_{6004}^{(Lignin)} + F_{6004}^{(Ash)} + F_{6004}^{(Other\ Solids)}$$

Species:

Cellulose

$$F_{6001}^{(Cellulose)} - \left(1 \times \frac{F_{6001}^{(Cellulose)}}{mw(Cellulose)} \times mw(Cellulose)\right) = 0$$

Hemicellulose

$$F_{6001}^{(Hemicellulose)} - \left(1 \times \frac{F_{6001}^{(Hemicellulose)}}{mw(Hemicellulose)} \times mw(Hemicellulose)\right) = 0$$

Carbon monoxide

$$-F_{6003}^{(CO)} + \left(\frac{6}{1} \times \frac{F_{6001}^{(Cellulose)}}{mw(Cellulose)} \times mw(CO)\right)$$

$$+ \left(\frac{5}{1} \times \frac{F_{6001}^{(Hemicellulose)}}{mw(Hemicellulose)} \times mw(CO)\right) = 0$$

Hydrogen

$$-F_{6003}^{(H_2)} + \left(\frac{6}{1} \times \frac{F_{6001}^{(Cellulose)}}{mw(Cellulose)} \times mw(H_2)\right)$$

$$+ \left(\frac{5}{1} \times \frac{F_{6001}^{(Hemicellulose)}}{mw(Hemicellulose)} \times mw(H_2)\right) = 0$$

Lignin

$$F_{6001}^{(Lignin)} - F_{6004}^{(Lignin)} = 0$$

Ash

$$F_{6001}^{(Ash)} - F_{6004}^{(Ash)} = 0$$

Other solids

$$F_{6001}^{(Other\ Solids)} - F_{6004}^{(Other\ Solids)} = 0$$

The energy requirement for the process was according to the equation

$$Q_{in} - Q_{out} \times F_p + \sum_{j} H_j F_{j_{in}} - \sum_{j} H_j F_{j_{out}} = 0$$

H_j is the enthalpy per unit mass of jth stream computed from HYSYS. Q_{out} is the energy removed from the system per unit mass of product p having a flow rate of F_p and this is determined from HYSYS. Q_{in} is the energy required by the system and calculated using the aforementioned equation.

The mass balance equations for overall and species balances for the corn stover gasification process are given in Table 5.47. The energy requirement, $Q_{SYNGBIO}$, for the gasification of corn stover process was calculated using

TABLE 5.48

Energy Balance Equations for Gasification of Corn Stover

	IN − OUT + GENERATION − CONSUMPTION = 0
Energy Balance	$Q_{\text{in}} - Q_{\text{out}} \times F_p + \sum_j H_j F_{j_{\text{in}}} - \sum_j H_j F_{j_{\text{out}}} = 0$

Overall Energy Required by the Process: Q_{SYNGBIO}

$Q_{\text{SYNGBIO}} - Q_{\text{SYNGBIOO}} + (F_{6003}^{(CO)}/mw^{(CO)}H_{6003}^{(CO)} + F_{6003}^{(CO)}/mw^{(CO)}H_{6003}^{(CO)} + F_{6004} \times H_{6004})$

$\quad - (F_{6001} \times H_{6001} + F_{6002} \times H_{6002}) = 0$

$Q_{\text{in}} = Q_{\text{SYNGBIO}} \text{ (kJ/h)}$

$Q_{\text{out}} \times F_p = Q_{\text{SYNGBIOO}} = 3099.3 \text{ (kJ/kg)} \times F_{6003}^{(H_2)}; F_j = F_{6001}, F_{6002}$

H_j	Mass Enthalpy (kJ/kg)
H_{6001}	−1.58E+04
H_{6002}	−1.31E+04

Enthalpy functions for CO and H_2 were used from the gasification process for natural gas, which were already incorporated in the chemical complex analysis system.

the enthalpy per unit mass of biomass stream, enthalpy per unit mass of steam, and the enthalpy per unit mass of the carbon monoxide and hydrogen streams. The energy removed by cooling water, Q_{SYNGBIO}, was assumed similar to the hydrogen process described by Indala (2004) and Xu (2004). Data for the enthalpy of the tars and other products were not available and were not considered in the model equation. This term, if considered, would reduce the value of Q_{SYNGBIO}, and a lower utility cost would be calculated for the gasification process. So, it can be assumed that if the gasification process is selected with a higher utility cost for Q_{SYNGBIO}, the process will also be selected with a lower utility cost.

5.10 Summary of Bioprocess Model Formulation

The process flow designs described in Chapter 4 were converted to block flow models in this chapter. The ethanol fermentation from corn and corn stover, acetic acid production from corn stover, FAME and glycerol from transesterification of oils, algae oil production, gasification of corn stover, ethylene from ethanol, and propylene glycol from glycerol processes were modeled to give the plants as shown in Figure 5.10. Comparing to Figure 5.1, there are streams that need to be connected within the biochemical complex. These streams and the formulation of the superstructure will be explained in Section 5.11.

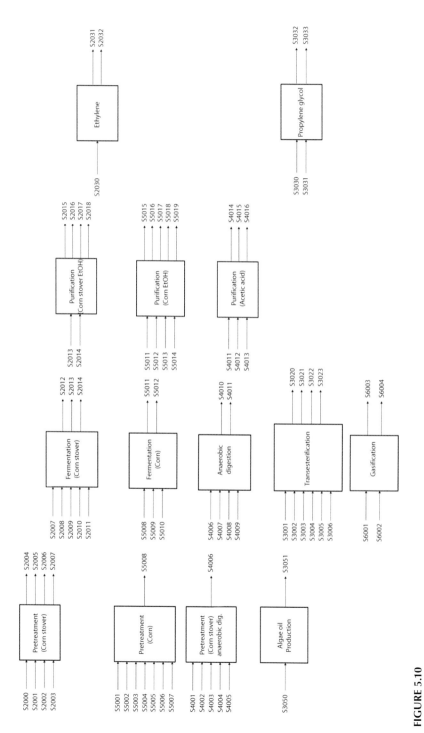

FIGURE 5.10

Arrangement of process block flow diagrams to form the overall biochemical processes block flow diagram.

5.11 Interconnections for Bioprocesses

The streams and number of plants not defined in the plant models described in the previous section are described in Table 5.49. The total corn stover required for the bioprocesses is given by S4000. The soybean oil required in the transesterification process is given by S3052. The water and steam required for bioprocesses is given by S7000 and S7001. The waste water streams obtained by pooling all the waste water from the bioprocesses is given by S7002. The pure carbon dioxide having mass fraction greater than 90% produced from the bioprocesses is given by SBIOCO2P. The impure carbon dioxide emission (less than 90% mass fraction) from the bioprocesses is given by SBIOCO2I.

The number of corn-stover-feedstock-based ethanol plants required to meet bioethylene demand is EP1, and the number of corn-feedstock-based ethanol plants required to meet bioethylene demand is EP2. The number of FAME plants producing 10 million gal/year FAME required to meet the glycerol demand for propylene glycol production is FA. Ethanol and glycerol were for the production of bioethylene and biopropylene glycol, respectively. The plants for ethanol and FAME are constrained by the capacity of the processing corn and producing FAME, respectively. The number of plants required to meet the demand for biochemicals are computed from the optimal structure. The plants were considered as package plants having a

TABLE 5.49

Description of Variables in Superstructure

Name of Streams	Description
S4000	Total corn stover to the biochemical production complex
S3052	Soybean oil required in the transesterification process
S7000	Water required for bioprocesses pretreatment, process reactions, and purification
S7001	Steam required for bioprocess pretreatment, process reactions, and purification
S7002	Waste water from bioprocesses
SBIOCO$_2$P	Pure carbon dioxide from bioprocesses (>90% (mass) in stream)
SBIOCO$_2$I	Impure carbon dioxide from bioprocesses (<90% (mass) in stream)
EP1	Number of corn stover ethanol plants, each of maximum capacity 667,000 MT/year of corn stover processing, required for ethylene production
EP2	Number of corn ethanol plants, each of maximum capacity 360,000 MT/year of corn processing, required to produce ethanol for ethylene production
FA	Number of FAME plants of maximum capacity 33,700 MT/year each required to meet the glycerol requirement in the propylene glycol production plant

specific capacity. The products and energy requirements were multiplied by the number of plants to obtain the optimal results from the process.

The stream relationships for the biochemical production complex are given in Table 5.50. The units in the chemical complex analysis system are given in brackets at the end of the description for reference to the unit where the equation is included.

TABLE 5.50

Stream Relationship for Biochemical Production Complex

Relationship	Description
1. $F_{3051} + F_{3052} = FA \times F_{3001}$	Oil obtained from algae and purchased soybean oil going into the transesterification process (U19).
2. $F_{4000} = EP1 \times F_{2001} + F_{4001} + F_{6001}$	Total corn stover to the superstructure distributed among ethanol fermentation, acetic acid from anaerobic digestion, and gasification process, respectively (U20).
3. $EP1 \times F_{2015}^{(Ethanol)} + EP2 \times F_{5015}^{(Ethanol)} = F_{2030}$	Ethanol from corn stover fermentation and corn fermentation to the ethylene process in the superstructure (U21).
4. $F_{3030}^{(Glycerol)} = FA \times F_{3021}^{(Glycerol)}$	Glycerol required to meet the propylene glycol plant requirement (U22).
5. $F_{BIOCO2P} = EP1 \times F_{2016}^{(CO_2)} + EP2 \times F_{5019}^{(CO_2)}$	Pure carbon dioxide from corn stover ethanol and corn ethanol fermentation bioprocesses (U24).
6. $F_{BIOCO2I} = EP1 \times F_{2012}^{(CO_2)} + F_{4010}^{(CO_2)}$	Impure carbon dioxide from seed generation in corn stover fermentation and acetic acid from anaerobic digestion bioprocesses (U25).
7. $F_{7000} = EP1 \times F_{2000} + FA \times F_{3004} + F_{4003} + EP2 \times F_{5010}$ $+ EP2 \times F_{5013}$	Water required for corn stover fermentation, FAME purification, acetic acid from anaerobic digestion, from corn stover ethanol, transesterification to FAME, anaerobic digestion to acetic acid, corn ethanol, ethanol to ethylene dehydration, and propylene glycol processes (U26).
8. $F_{7001} = EP1 \times F_{2002} + F_{4004} + F_{4013} + F_{6002}$	HP steam required for corn stover fermentation pretreatment, corn stover anaerobic digestion pretreatment, and corn stover steam reforming (gasification) (U27).
9. $F_{7002} = EP1 \times F_{2018} + FA \times F_{3022} + F_{4016}$ $+ EP2 \times F_{5016} + F_{2032} + F_{3033}$	Waste water from corn stover fermentation, transesterification, anaerobic digestion, new ethylene process, and new propylene glycol process (U23).

For the biomass integration process, the Equations 1 through 9 in Table 5.50 described the relation among the units and streams. Equation 1 calculates the total natural oil that can be obtained from purchasing soybean oil, and the production of algae oil using carbon dioxide from the chemical complex (unit U19). Equation 2 calculates the total corn stover used in fermentation, anaerobic digestion, and biomass gasification processes (unit U20). Equation 3 gives the total ethanol produced from the complex from the corn stover fermentation ethanol plants and the corn ethanol fermentation plants (unit U21). This equation determines the total number of corn stover ethanol plants (EP1) and the total number of corn ethanol plants (EP2) required in the complex to meet the demand for ethylene. Equation 4 gives the total amount of glycerol required from the transesterification process, and this equation also determines the total number of FAME plants (FA) required in the complex to meet the requirement for glycerol for the propylene glycol process (unit U22). Equation 5 determines the total pure carbon dioxide emissions from the bioprocesses, BIOCO2P, and this includes the carbon dioxide from corn stover and corn fermentation sections (unit U24). Equation 6 gives the total impure CO_2 emissions from the bioprocesses (unit U25). This includes the CO_2 produced from corn stover ethanol seed generation section, and the carbon dioxide produced from anaerobic digestion section. Equation 7 gives the total water required in the bioprocesses (unit U26). Equation 8 gives the total steam required in the bioprocesses (unit U27) and Equation 9 gives the total waste water from the processes (unit U23). Thus, the bioprocess block flow units as shown in Figure 5.1 were obtained from the aforementioned stream relations.

━━━━━━━━━━

5.12 Summary

The plant model formulation for superstructure was described in this chapter. The processes developed in HYSYS in Chapter 4 were converted to input–output block models, as shown in Figure 5.10. Stream relations were defined among the biochemical plants (Table 5.50) to obtain the biochemical complex shown in Figure 5.1.

Chapter 6 describes the formulation of the superstructure of chemical plants using the bioprocess models described in this chapter. The superstructure was constructed by integrating the bioprocess models into a base case of plants (Xu, 2004). Interconnections between the base case and the bioprocesses are defined. Carbon dioxide from the integrated chemical production complex was used for algae oil production and the carbon dioxide–consuming processes discussed by Indala (2004) and Xu (2004). The superstructure optimization and the optimal structure obtained from the superstructure will be discussed in Chapter 6.

6

Formulation and Optimization
of the Superstructure

6.1 Introduction

This chapter describes the results from the formulation of the superstructure by integrating bioprocesses into the base case of existing chemical plants in the lower Mississippi River corridor. Carbon dioxide produced from the integrated biochemical and chemical complex was used for the production of algae oil and for the production of chemicals from carbon dioxide. The base case of existing chemical plants is shown in Figure 6.1. The superstructure is shown in Figure 6.2. The superstructure is a mixed-integer nonlinear programming problem (MINLP), solved with global solvers in GAMS. Details on MINLP problem formulation, superstructure definition, and mathematical representation are given in Appendix B. Logical constraints were included for the selection of competing processes in the model. Lower and upper bounds on the flow rates of the production capacities of the plants in the complex were specified.

The chemical complex analysis system (Appendix E) was used and multi-criteria optimization for Pareto optimal sets of profit versus sustainable was obtained. Sensitivity analysis of the optimal structure using Monte Carlo simulation methods in the chemical complex analysis system gave a cumulative probability distribution of the triple bottom line.

6.2 Integrated Biochemical and Chemical Production Complex Optimization

Renewable raw materials and bioprocesses are needed, and industrial-scale chemical plants from biomass were designed, as explained in Chapter 4. Chapter 5 discussed the mathematical model formulation for these biochemical plants. The biochemical production units are shown in Figure 5.1.

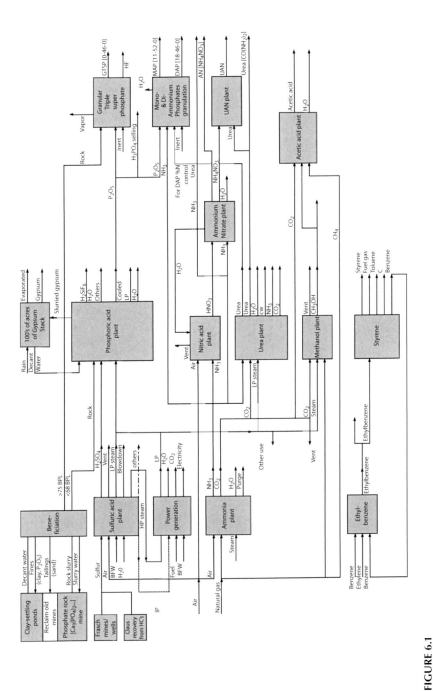

FIGURE 6.1

(See color insert.) Base case of chemical plants in the lower Mississippi River corridor. (From Xu, A., Chemical Production Complex Optimization, Pollution Reduction and Sustainable Development, PhD dissertation, Louisiana State University, Baton Rouge, LA, 2004.)

(continued)

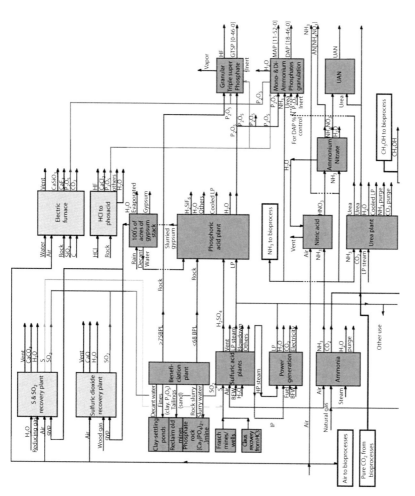

FIGURE 6.2
(See color insert.) Integrated chemical production complex, superstructure.

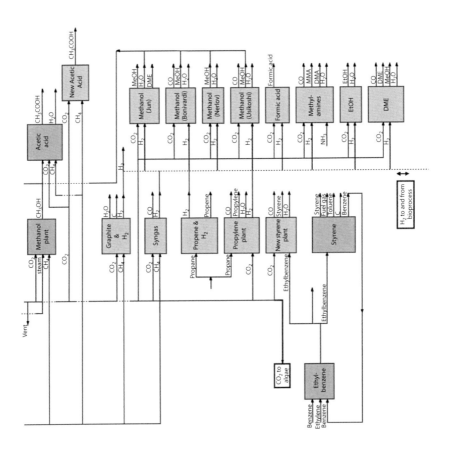

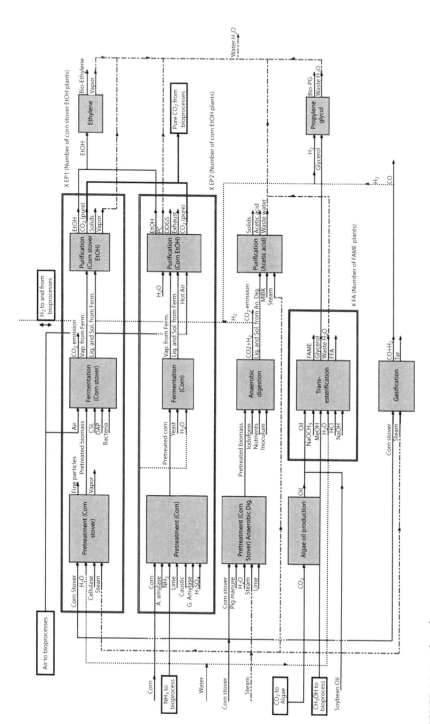

FIGURE 6.2 (continued)

In this chapter, the input–output block models of the biochemical plants are integrated into a base case of existing plants in the lower Mississippi River corridor to form a superstructure for optimization. Also included in the superstructure are 14 processes developed by Indala (2004), which can use high-purity carbon dioxide to produce chemicals. These were integrated into the complex for CO_2 utilization.

Figure 6.1 shows the base case of existing chemical plants in the lower Mississippi River corridor developed with information provided by the cooperating companies and other published sources (Xu, 2004). This complex is representative of the current operations and practices in the chemical industry. There are 13 production units plus associated utilities for power, steam, and cooling water, and facilities for waste treatment in the base case.

A superstructure of chemical plants and biochemical plants was constructed by integrating biochemical processes described in Chapter 5 into the base case of chemical plants and the processes for CO_2 described by Xu (2004). Figure 6.2 shows the superstructure of plants in the integrated chemical production complex. The biochemical processes are shown in green, the existing plants in the base case are shown in blue, and the processes consuming carbon dioxide to produce chemicals are shown in red. The plants in yellow are four additional units, two of which give alternative processes for producing phosphoric acid, and two units for sulfur and SO_2 recovery. Table 6.1 is a convenient way to show the plants added to form the superstructure given in Figure 6.2. The table gives the plants along with the color used to represent in Figure 6.2.

Additional equations were required to link the processes in the superstructure, and these are given in Table 6.2. These were used to connect the plants in the base case to the biochemical plants. Equations 1 through 7 in Table 6.2 give the relationships for biomass process integration in the base case of plants with carbon dioxide consumption in new processes. Xu (2004) provides complete stream definitions from the base case and carbon dioxide–consuming processes for chemical production. The unit number in brackets is the unit reference from the chemical complex analysis system.

In Table 6.2, Equation 1 gives the requirement of air to sulfuric acid, nitric acid, ammonia, electric furnace, SO_2 recovery, S and SO_2 recovery plant, corn stover ethanol, and corn ethanol fermentation plant. Equation 2 gives the ammonia required from ammonia plant to nitric acid, ammonium nitrate, ammonium phosphate, urea plant, for sale, methylamines plant, and corn ethanol plant. Equation 3 gives the relation for CO_2 from ammonia plant and pure ethanol from bioprocesses (BIOCO$_2$P) to urea, methanol, acetic acid, emission to atmosphere, new acetic acid, and new CO_2-consuming processes. This equation is important for calculating the total pure carbon dioxide emissions from the integrated biochemical production complex. Equation 4 gives the methanol from methanol plant and other methanol production plants to existing acetic acid plant, transesterification to FAME,

TABLE 6.1

Processes in Chemical Production Complex Base Case and Superstructure

Plants in Base Case (Xu, 2004)	Plants Added to Form the Superstructure
(Blue in Figure 6.2)	**Bioprocesses and CO_2 Consumption by Algae**
Ammonia	**(Green in Figure 6.2)**
Nitric acid	Fermentation ethanol (corn stover)
Ammonium nitrate	Fermentation ethanol (corn)
Urea	Anaerobic digestion to acetic acid (corn stover)
UAN	Algae oil production
Methanol	Transesterification to FAME and glycerol
Granular triple super	(Soybean oil and algae)
Phosphate (GTSP)	Gasification to syngas (corn stover)
MAP and DAP	Ethylene from dehydration of ethanol
Contact process for sulfuric acid	Propylene glycol from glycerol
Wet process for phosphoric acid	
Acetic acid—conventional method	**CO_2 Consumption for Chemicals (Indala, 2004)**
Ethyl benzene	**(Red in Figure 6.2)**
Styrene	Methanol (Bonivardi et al., 1998)
Power generation	Methanol (Jun et al., 1998)
	Methanol (Ushikoshi et al., 1998)
	Methanol (Nerlov and Chorkendorff, 1999)
	Ethanol
	Dimethyl ether
	Formic acid
	Acetic acid—new method
	Styrene—new method
	Methylamines
	Graphite
	Hydrogen/synthesis gas
	Propylene from CO_2
	Propylene from propane dehydrogenation
	Choice for Phosphoric Acid Production and
	SO_2 Recovery (Xu, 2004) (Yellow in Figure 6.2)
	Electric furnace process for phosphoric acid
	Haifa process for phosphoric acid
	SO_2 recovery from gypsum waste
	S and SO_2 recovery from gypsum waste

and for sale. Equation 5 gives the impure CO_2 emissions from power plant, urea, nitric acid, sulfuric acid, methanol, electric furnace, SO_2 recovery, S and SO_2 recovery, methylamines plants, and impure emissions from bioprocesses. Equation 6 gives the split for pure CO_2 to new CO_2-consuming processes in stream S922. This equation calculates the amount of carbon dioxide required for algae oil production process. Equation 7 gives the H_2 produced from conventional H_2 manufacture via natural gas process, new

TABLE 6.2

Stream Relations for Superstructure (Integrated Chemical Production Complex)

Equation	Relationship	Description
1.	$F_5 = F_7 + F_8 + F_9 + F_{200}$ $+ F_{402} + F_{410} + EP1 \times F_{2008}$ $+ EP2 \times F_{5014}$	Air to sulfuric acid, nitric acid, ammonia, electric furnace, SO_2 recovery, S and SO_2 recovery plant, corn stover ethanol, and corn ethanol fermentation plant (U2)
2.	$F_{19} = F_{29} + F_{30} + F_{31} + F_{42}$ $+ F_{948} + EP2 \times F_{5003}$	Ammonia from ammonia plant to nitric acid, ammonium nitrate, ammonium phosphate, urea plant, for sale, methylamines plant, and corn ethanol plant (U6)
3.	$F_{20} + F_{BIOCO2P} = F_{32} + F_{33}$ $+ F_{64} + F_{82} + F_{700} + F_{922}$	CO_2 from ammonia plant and pure ethanol from bioprocesses to urea, methanol, acetic acid, emission to atmosphere, new acetic acid, and new CO_2-consuming processes (U7)
4.	$F_{47} + F_{991} = F_{423} + F_{424}$ $+ FA \times F_{3003}$	Methanol from methanol plant and other methanol production plants to acetic acid plant, transesterification to FAME and for sale (U10)
5.	$F_{CDEM} = F_{301} + F_{801}$ $+ F_{81}^{(CO_2)} + F_{15}^{(CO_2)} + F_{802}^{(CO_2)}$ $+ F_{166} + F_{151}^{(CO_2)} + F_{403}^{(CO_2)}$ $+ F_{413}^{(CO_2)} + F_{949}^{(CO_2)} + F_{BIOCO2I}$	Impure CO_2 emissions from power plant, urea, nitric acid, sulfuric acid, methanol, electric furnace, SO_2 recovery, S and SO_2 recovery, methylamines plants, and impure emissions from bioprocesses (U7)
6.	$F_{922} = F_{912} + F_{935} + F_{942} + F_{946}$ $+ F_{953} + F_{958} + F_{963} + F_{967}$ $+ F_{972} + F_{980} + F_{984} + F_{993}$ $+ F_{3050}$	CO_2 from ammonia plant to new CO_2-consuming processes, such as propane dehydrogenation with CO_2, H_2, formic acid, methylamines, methanol (Jun), methanol (Bonivardi), methanol (Nerlov), methanol (Ushikoshi), new styrene, ethanol, DME, graphite, and algae oil production for transesterification processes (U15)
7.	$F_{936} + F_{6003}^{(Hydrogen)} + F_{6003}^{(Hydrogen)}$ $+ F_{4010}^{(Hydrogen)} + F_{918} + F_{916}$ $+ F_{994} = F_{943} + F_{947} + F_{981}$ $+ F_{985} + F_{954} + F_{959} + F_{964}$ $+ F_{968} + F_{3031} + F_{903}$	H_2 produced from conventional H_2 manufacture via natural gas process, new gasification of biomass process, anaerobic digestion to acetic acid process, propane dehydrogenation, propane dehydrogenation with CO_2 and graphite processes to formic acid, methylamines, ethanol, DME, methanol (Jun), methanol (Bonivardi), methanol (Nerlov), and methanol (Ushikoshi) processes, new propylene glycol process, and for sale (U17)
8.	$F_2 + F_3 = 0$	Sulfur from Frasch mines/wells and Claus recovery to sulfuric acid plant (U1)
9.	$F_6 = F_{10} + F_{11} + F_{300} + F_{83}$ $+ F_{701} + F_{924}$	Natural gas to ammonia, methanol, power plant, acetic acid, new acetic acid plant, and other CO_2-consuming plants (U5)
10.	$F_{16} + F_{18} = F_{24} + F_{27} + F_{918}$ $+ F_{apply}$	LP steam from sulfuric acid and power plant to phosphoric acid, urea, and other plants as heat input (U3)

TABLE 6.2 (continued)

Stream Relations for Superstructure (Integrated Chemical Production Complex)

Equation	Relationship	Description
11.	$F_{30} = F_{43} + F_{44}$	Ammonia to ammonium phosphate plant and for sale (U8)
12.	$F_{46} = F_{54} + F_{59}$	Urea from urea plant to UAN plant and for sale (U9)
14.	$F_{60} = F_{39} + F_{40} + F_{41}$	Phosphoric acid from phosphoric acid plant to GTSP, ammonium phosphate plant, and for sale (U4)
15.	$F_{1069} = F_{1070} + F_{1071} + F_{971}$	Ethylbenzene from ethylbenzene plant to styrene plant, for sale, and new styrene process (U11)
16.	$F_{22} = F_{408} + F_{400} + F_{416}$	Gypsum from wet process for phosphoric acid to electric furnace and Haifa processes, and to the gypsum stack
17.	$F_{112} = F_{114} + F_{115}$	Phosphoric acid from electric furnace to GTSP, MAP, and DAP plants (U13)
18.	$F_{87} = F_{117} + F_{118}$	Phosphoric acid from Haifa process to GTSP, MAP, and DAP plants (U14)
19.	$F_{924} = F_{934} + F_{992}$	Natural gas to new processes, such as graphite and H_2(U16)
20.	$F_{955} + F_{961} + F_{965} + F_{969} = F_{991}$	Methanol produced from methanol (Jun), methanol (Bonivardi), methanol (Nerlov), and methanol (Ushikoshi) processes (U18)

gasification of biomass process, anaerobic digestion, and other processes consuming CO_2. The hydrogen produced was used for new processes consuming CO_2, bio-propylene glycol process, and for sale. Equations 8 through 20 were the equations for CO_2 utilization in the integrated chemical complex from Xu (2004).

6.3 Binary Variables and Logical Constraints for MINLP

The model of the superstructure is a MINLP problem. For mixed-integer optimization, binary variables are associated with the production capacities of each plant. If the binary variable for a process is one, then the plant operates at least at its lower bound on the production capacity. If the binary variable of a process is zero, then the production capacity of that process is zero, and the plant is not in the optimal structure. The binary variables associated with the superstructure are given in Table 6.3.

Relations among the binary variables and the logic constraints used in the superstructure are given in Table 6.4. The binary variables are used to decide which plants among the competing plants for carbon dioxide

TABLE 6.3

Binary Variables Associated with the Superstructure

Binary Variable	Process Description
Y_{11}	Acetic acid
Y_{12}	Acetic acid—new process
Y_{13}	SO_2 recovery from gypsum
Y_{14}	S and SO_2 recovery from gypsum
Y_1	Phosphoric acid—electric furnace
Y_2	Phosphoric acid—Haifa process
Y_3	Phosphoric acid—wet process
Y_{16}	Methanol
Y_{31}	Methanol (Jun et al., 1998)
Y_{32}	Methanol (Bonivardi et al., 1998)
Y_{33}	Methanol (Nerlov and Chorkendorff, 1999)
Y_{34}	Methanol (Ushikoshi et al., 1998)
Y_{35}	Styrene—new process
Y_{40}	Styrene
Y_{41}	Ethylbenzene
Y_{29}	Formic acid
Y_{30}	Methylamines
Y_{37}	Ethanol
Y_{38}	Dimethyl ether
Y_{23}	Propylene from CO_2
Y_{24}	Propylene from propane dehydrogenation
Y_{27}	Synthesis gas
Y_{39}	Graphite
Y_{60}	Synthesis gas from corn stover gasification
Y_{61}	Acetic acid from corn stover anaerobic digestion

utilization will be chosen in the optimal structure based on mass balances and energy requirements. The superstructure is designed to always include the biomass-feedstock-based processes for fermentation and transesterification. The binary variable added for the choice of the biomass-feedstock-based gasification process was Y60. This was to determine whether the biomass gasification process would be chosen among the competing processes for hydrogen production.

The binary variable added for the biomass-feedstock-based anaerobic digestion process was Y61. This was to determine whether the acetic acid produced from the biomass-feedstock-based process would be selected over the conventional acetic acid plant or the new acetic acid plant from the carbon dioxide utilization process. The constraint was chosen such that one of the three acetic acid plants may be chosen, that is, any one should have the value of 1. It is also a possibility that none of the plants will be chosen, where all the binary variables are zero, satisfying the condition of ≤ 1.

TABLE 6.4

Logical Relations Used to Select the Optimal Structure from Superstructure

Logic Expression	Logic Meaning
$Y_{11} + Y_{12} + Y_{61} \leq 1$	At most, one of these three acetic acid plants is selected (or none may be selected)
$Y_{13} + Y_{14} \leq Y_3$	At most, one of these two S and SO_2 recovery plants is selected only if phosphoric acid (wet process) is selected (or none may be selected)
$Y_{16} + Y_{31} + Y_{32} + Y_{33} + Y_{34} \leq 1$	At most, one of the five methanol plants is selected, the existing one or one of the four proposed plants (or none may be selected)
$Y_{11} \leq Y_{16} + Y_{31} + Y_{32} + Y_{33} + Y_{34}$	Only if at least one of these five methanol plants is selected, the conventional acetic acid may be selected
$Y_{35} + Y_{40} \leq Y_{41}$	At most one of these two styrene plants is selected only if ethylbenzene plant is selected
$Y_{29} \leq Y_{23} + Y_{24} + Y_{27} + Y_{39} + Y_{60}$	Only if at least one of the five plants that produce H_2 is selected, the formic acid plant may be selected
$Y_{30} \leq Y_{23} + Y_{24} + Y_{27} + Y_{39} + Y_{60}$	Only if at least one of the five plants that produce H_2 is selected, the methylamines plant may be selected
$Y_{31} \leq Y_{23} + Y_{24} + Y_{27} + Y_{39} + Y_{60}$	Only if at least one of the five plants that produce H_2 is selected, the new methanol plant may be selected
$Y_{32} \leq Y_{23} + Y_{24} + Y_{27} + Y_{39} + Y_{60}$	Only if at least one of the five plants that produce H_2 is selected, the new methanol plant may be selected
$Y_{33} \leq Y_{23} + Y_{24} + Y_{27} + Y_{39} + Y_{60}$	Only if at least one of the five plants that produce H_2 is selected, the new methanol plant may be selected
$Y_{34} \leq Y_{23} + Y_{24} + Y_{27} + Y_{39} + Y_{60}$	Only if at least one of the five plants that produce H_2 is selected, the new methanol plant may be selected
$Y_{37} \leq Y_{23} + Y_{24} + Y_{27} + Y_{39} + Y_{60}$	Only if at least one of the five plants that produce H_2 is selected, the new ethanol plant may be selected
$Y_{38} \leq Y_{23} + Y_{24} + Y_{27} + Y_{39} + Y_{60}$	Only if at least one of the five plants that produce H_2 is selected, the dimethyl ether plant may be selected

Referring to Table 6.4, the conventional process from base case for acetic acid and corresponding new processes for acetic acid production was compared to each other. Processes for S and SO_2 recovery, methanol, and styrene were also compared and the best processes were selected. Also, hydrogen must be available for plants that require hydrogen for them to be included in the complex.

6.4 Constraints for Capacity and Demand

For optimization, upper and lower bounds of the production capacities of plants in the complex are required. The upper bounds for the potentially new processes were from the HYSYS simulations that were based on actual plants.

TABLE 6.5

Plant Demand and Capacities in the Superstructure

Plant Names	Capacity Constraints (Metric Tons per Year)
Biochemical processes	
Fermentation (corn stover)	$333,000 \leq F_{2001} \leq 667,000$
Fermentation (corn)	$180,000 \leq F_{5001} \leq 360,000\partial$
Transesterification (FAME)	$16,850 \leq F_{3020} \leq 33,700$
Anaerobic digestion (acetic acid)	$4,080 \leq F_{4015} \leq 8,160$
Bioethylene	$100,000 \leq F_{2031} \leq 200,000$
Biopropylene glycol	$37,000 \leq F_{3032} \leq 74,000$
Gasification (syngas from corn stover)	$6,700 \leq F_{6003}^{(H_2)} \leq 13,400$
Base-case plants	
Ammonia	$329,000 \leq F_{19} \leq 658,000$
Nitric acid	$89,000 \leq F_{45}^{(HNO_3)} \leq 178,000$
Ammonium nitrate	$113,000 \leq F_{56} + F_{62}^{(AN)} \leq 227,000$
Urea	$49,900 \leq F_{46} + F_{53}^{(UREA)} \leq 99,800$
Methanol	$91,000 \leq F_{47} \leq 181,000$
UAN	$30,000 \leq F_{58} \leq 60,000$
MAP	$146,000 \leq F_{52} \leq 293,000$
DAP	$939,000 \leq F_{57} \leq 1,880,000$
GTSP	$374,000 \leq F_{51} \leq 749,000 \in$
Contact process sulfuric acid	$1,810,000 \leq F_{14}^{(H_2SO_4)} \leq 3,620,000$
Wet process phosphoric acid	$635,000 \leq F_{60}^{(P_2O_5)} \leq 1,270,000$
Acetic acid (conventional)	$4,080 \leq F_{84} \leq 8,160$
Ethylbenzene	$431,000 \leq F_{1069} \leq 862,000$
Styrene	$386,000 \leq F_{1072} \leq 771,000$
CO$_2$-consuming processes and other processes	
Acetic acid (new)	$4,080 \leq F_{702} \leq 8,160$
Electric furnace phosphoric acid	$635,000 \leq F_{112}^{(P_2O_5)} \leq 1,270,000$
Haifa phosphoric acid	$635,000 \leq F_{87}^{(P_2O_5)} \leq 1,270,000$
SO$_2$ recovery from gypsum	$987,000 \leq F_{405} \leq 1,970,000$
S and SO$_2$ recovery from gypsum	$494,000 \leq \dfrac{32.06}{64.06} F_{405} + F_{412} \leq 988,000$
New styrene	$181,000 \leq F_{974} \leq 362,000$
New methanol (Bonivardi)	$240,000 \leq F_{961} \leq 480,000$
New methanol (Jun)	$240,000 \leq F_{955} \leq 480,000$
New methanol (Nerlov)	$240,000 \leq F_{965} \leq 480,000$
New methanol (Ushikoshi)	$240,000 \leq F_{969} \leq 480,000$
New formic acid	$39,000 \leq F_{944} \leq 78,000$
New methylamines	$13,200 \leq F_{950} \leq 26,400$

TABLE 6.5 (continued)

Plant Demand and Capacities in the Superstructure

Plant Names	Capacity Constraints (Metric Tons per Year)
New ethanol	$52{,}000 \le F_{982} \le 104{,}000$
New dimethyl ether (DME)	$22{,}900 \le F_{987} \le 45{,}800$
New graphite	$23{,}000 \le F_{995} \le 46{,}000$
New hydrogen	$6{,}700 \le F_{936} \le 13{,}400$
New propylene by CO_2	$21{,}000 \le F_{914} \le 41{,}900$
New propylene	$20{,}900 \le F_{919} \le 41{,}800$

For convenience, the lower bound for the production capacity was selected as half the value of the upper bound. The upper bound signifies the capacity of the plant, beyond which it cannot produce a product or process a raw material. The lower bound signifies the demand of the chemicals that a particular plant must meet.

If a process is selected, it has to operate at least at the lower bound of its production capacity. The upper bounds and lower bounds of the production capacities of all the plants in the chemical complex are shown in Table 6.5. The capacity of the acetic acid plant from anaerobic digestion of corn stover was based on corn stover processing in the HYSYS design, but it was chosen as the existing acetic acid plant in the optimization model.

6.5 Optimization Economic Model—Triple Bottom Line

The optimum configuration of plants from the superstructure was obtained by maximizing a triple bottom line model, explained as follows. The triple bottom line included a value-added economic model given by Equation 6.1. The triple bottom line also included environmental and sustainable costs. Environmental costs are costs required to comply with federal and state environmental regulations including permits, monitoring emissions, fines, etc., as described in the AIChE/TCA report (Constable et al., 1999). Sustainable costs are costs to society from damage to the environment by emissions discharged within permitted regulations. Sustainable credits are credits that may be given to a particular process that avoids damages to the environment. The Intergovernmental Panel on Climate Change identified that changes in atmospheric concentration of greenhouse gases (GHGs), aerosols, land cover, and solar radiation alter the energy balance of the climate system (IPCC, 2007). Thus, if the emission of GHGs can be avoided, then sustainable costs are avoided, and the processes that contribute to avoiding the cost, may be rewarded with credits.

$$\text{Triple bottom line} = \text{Profits} - \sum \text{Environmental costs}$$

$$+ \sum \text{Sustainable (Credits} - \text{Costs)} \qquad (6.1)$$

$$\text{Triple bottom line} = \sum \text{Product sales} - \sum \text{Raw material}$$

$$- \sum \text{Energy costs} - \sum \text{Environmental costs}$$

$$+ \sum \text{Sustainable (Credits} - \text{Costs)} \qquad (6.2)$$

$$\text{Profit} = \sum \text{Product sales} - \sum \text{Raw material costs} - \sum \text{Energy costs} \qquad (6.3)$$

The triple bottom line is the difference between sales and sustainable credits and economic costs (raw materials and utilities), environmental costs, and sustainable costs as given by Equation 6.2. The sales prices for products and the costs of raw materials are given in Table 6.6 along with sustainable costs and credits.

Environmental costs were estimated to be 67% of the raw material costs based on the data provided by Amoco, DuPont, and Novartis in the AIChE/TCA report (Constable et al., 1999). This report lists environmental costs and raw material costs as approximately 20% and 30% of the total manufacturing costs, respectively.

Sustainable costs were estimated from results given for power generation in the AIChE/TCA report where CO_2 emissions had a sustainable cost of $3.25 per MT of CO_2. As shown in Table 6.6, a cost of $3.25 was charged as a cost to plants that emitted CO_2, and a credit of twice this cost ($6.50) was given to plants that utilized CO_2. In this report, SO_2 and NOX emissions had sustainable costs of $192 per MT of SO_2 and $1030 per MT of NOX. In addition, for gypsum production and use, an arbitrary but conservative sustainable cost of $2.5 per MT for gypsum production was used, and a credit of $5.0 per MT for gypsum consumption was used (Xu, 2004).

The equations in the economic model are given in Table 6.7. Equation 1 in Table 6.7 gives the objective function, which requires maximizing the triple bottom line. This equation is the same as Equation 6.3. Equation 2 gives the value-added profit where the economic costs (Econcost) and environmental costs (Envcost) are subtracted from the income from sales (Groprofit). Equation 3 gives the sustainable costs and credits for the complex. The gross income from sales is given in Equation 4. Equation 5 gives the economic costs considered in this complex analysis. The economic costs considered are the raw material costs (Rawcost) and the utility costs (Utilcost). Equation 6 gives the environmental costs considered in this process. Equation 7 gives the raw material costs for the process. Equation 8 gives the utilities costs in

TABLE 6.6

Raw Material Costs and Product Prices (Appendix C)

Raw Materials	Cost ($/mt)	Std. Dev. ($/mt)	Products	Cost ($/mt)	Std. Dev. ($/mt)
Corn stover	60.83	9.4	Bioethylene	930	—
Corn	108.26	36	FAME	968	213
Soybean oil	616	240	Biopropylene glycol	1636	84
Cellulase	146	—	Acetic acid	515	35
Corn steep Liquor	177	—	DDGS	99	—
Bacteria	146	—	Ammonia	424	237
Sodium methylate	980	—	Methanol	435	211
HCl	215	—	Acetic acid	515	35
NaOH	617	—	GTSP	370	—
Lime	90	—	MAP	423	—
Iodoform	3300	—	DAP	457	7.89
MIBK	1290	—	Ammonium nitrate	373	—
α-Amylase	3300	—	Urea	354	17.4
Caustic	12	—	UAN	237	—
Gluco-amylase	3300	—	Phosphoric acid	772	—
Sulfuric acid	110	—	Hydrogen	1490	460
Yeast	5510	—	Ethylbenzene	1543	—
Steam	9.83	—	Styrene	1260	—
Water	0.02	—	Propylene	1207	442
Natural gas	382	105	Formic acid	735	—
Phosphate rock			MMA	1610	—
Wet process	27	—	DMA	1610	—
Electric furnace	34	—	DME	946	—
Haifa process	34	—	Ethanol	1224	108
GTSP process	32	—	Toluene	813	222
HCl	215	—	Graphite	2500	—
Sulfur			Fuel gas	1274	—
Frasch	53	9.5	CO	70	19
Claus	21	3.55	**Sustainable Cost and Credits**	**Cost/Credit ($/mt)**	
Coke electric furnace	124	—	Credit for CO_2 consumption	6.50	
Propane	180	—	Debit for CO_2 production	3.25	
Benzene	914	337	Debit for NOx production	1030	
Ethylene	1071	378	Debit for SO_2 production	192	
Reducing gas	75	—	Credit for gypsum consumption	5.0	
Wood gas	88	—	Debit for gypsum production	2.5	

TABLE 6.7

Economic Model for Superstructure

Optimization Model Equations

1. PR = Convprofit – Suscost

2. Convprofit = Groprofit – Econcost – Encost

3. Suscost =

$$-\left(\left(\begin{array}{l} -(F_{64} + F_{\text{SCDEM}}) \times P(\text{'CO2P'}) - (F_{81}^{NO} + F_{800}) \times P(\text{'NOP'}) + (F_{32} + F_{33} + F_{82} + F_{700} + F_{922}) \times P(\text{'CO2C'}) \\ + F_{17} \times P(\text{'HPP'}) + F_{77} \times P(\text{'TPP'}) + (F_{400} + F_{408}) \times P(\text{'GYPC'}) + F_{15}^{SO_2} \times P(\text{'SO2P'}) + F_{416} \times P(\text{'GYPP'}) \end{array} \right) \right)$$

4. Groprofit =

$$\left(\begin{array}{l} F_{41} \times P(\text{'PA'}) + F_{43} \times P(\text{'NH}_3\text{'}) + F_{423} \times P(\text{'MeOH'}) + F_{84} \times P(\text{'AA'}) + F_{702} \times P(\text{'AA'}) + F_{51} \times P(\text{'GTSP'}) + F_{52} \times P(\text{'MAP'}) \\ + F_{57} \times P(\text{'DAP'}) + F_{56} \times P(\text{'NN'}) + F_{58} \times P(\text{'UAN'}) + F_{59} \times P(\text{'UREA'}) + (F_{995} + F_{1076}) \times P(\text{'C'}) + (F_{974} + F_{1072}) \times P(\text{'STYRENE'}) \\ + (F_{914} + F_{919}) \times P(\text{'C3H6'}) + F_{903} \times P(\text{'H2'}) + (F_{913} + F_{937} + F_{949}^{CO} + F_{960} + F_{973} + F_{986} + F_{990} + F_{6003}^{CO}) \times P(\text{'CO'}) + F_{944} \times P(\text{'HCOOH'}) \\ + F_{950} \times P(\text{'MMA'}) + F_{951} \times P(\text{'DMA'}) + (F_{987} + F_{956}) \times P(\text{'DME'}) + F_{988} \times P(\text{'MeOH'}) + F_{982} \times P(\text{'EtOH'}) \\ + F_{1075} \times P(\text{'TOLUENE'}) + F_{1073} \times P(\text{'FG'}) + F_{1070} \times P(\text{'EB'}) + F_{2031} \times P(\text{'BIOETHY'}) + FA \times F_{3020} \times P(\text{'FAME'}) \\ + F_{3032} \times P(\text{'PG'}) + F_{4015} \times P(\text{'AA'}) + \text{EP2} \times F_{5017} \times P(\text{'DDGS'}) \end{array} \right)$$

5. Econcost = Rawcost + Utilcost

 Econcost = Rawcost + Utilcost

6. Encost =

$$
\begin{aligned}
&\left\{ \begin{aligned}
&\left(F_6 - F_{300}\right) \times P('CH_4') + F_{12} \times P('RGTSP') + F_{13} \times P('RWET') + F_2 \times P('SF') + F_3 \times P('SC') \\
&+ F_{109} \times P('RELE') + F_{165} \times P('COKE') + F_{86} \times P('HCL') + F_{85} \times P('RHCL') + F_{407} \times P('RGAS') + F_{401} \times P('WGAS') \\
&+ \left(F_{911} + F_{917}\right) \times P('C_3H_8') + F_{1067} \times P('BENZENE') + F_{1068} \times P('C_2H_4') + F_{4000} \times P('CSTOV') \\
&+ EP1 \times \left(F_{2003} \times P('CELLE') + F_{2009} \times P('CSL') + F_{2010} \times P('DAP') + F_{2011} \times P('BAC')\right)
\end{aligned} \right\} \times \frac{2}{3} \\[6pt]
&+ FA \times \left(F_{3002} \times P('NAOCH_3') + F_{3005} \times P('HCL') + F_{3006} \times P('NAOH')\right) + F_{3052} \times P('SOYOIL') + F_{4005} \times P('LIME') \\
&+ F_{4007} \times P('CHI_3') + F_{4012} \times P('MIBK') \\
&+ EP2 \times \left(\begin{aligned}
&F_{5001} \times P('CORN') + F_{5002} \times P('AAMY') + F_{5004} \times P('LIME') + F_{5005} \times P('CAUS') \\
&+ F_{5006} \times P('GAMY') + F_{5007} \times P('H_2SO_4') + F_{5009} \times P('YEAST')
\end{aligned} \right) \\
&+ F_{7001} \times P('STEAM') + F_{7000} \times P('WATER')
\end{aligned}
$$

(continued)

TABLE 6.7 (continued)

Economic Model for Superstructure

Optimization Model Equations

7. Rawcost $=$
$$\begin{aligned}
&\big(F_6 - F_{300}\big) \times P('CH_4') + F_{12} \times P('RGTSP') + F_{13} \times P('RWET') + F_2 \times P('SF') + F_3 \times P('SC') \\
&+ F_{109} \times P('RELE') + F_{165} \times P('COKE') + F_{86} \times P('HCL') + F_{85} \times P('RHCL') + F_{407} \times P('RGAS') + F_{401} \times P('WGAS') \\
&+ \big(F_{911} + F_{017}\big) \times P('C_3H_8') + F_{1067} \times P('BENZENE') + F_{1068} \times P('C_2H_4') + F_{4000} \times P('CSTOV') \\
&+ EP1 \times \big(F_{2003} \times P('CELLE') + F_{2009} \times P('CSL') + F_{2010} \times P('DAP') + F_{2011} \times P('BAC')\big) \\
&+ FA \times \big(F_{3002} \times P('NAOCH_3) + F_{3005} \times P('HCL') + F_{3006} \times P('NAOH)\big) + F_{3052} \times P('SOYOIL') + F_{4005} \times P('LIME') \\
&+ F_{4007} \times P('CHI_3') + F_{4012} \times P('MIBK') \\
&+ EP2 \times \left(\begin{aligned} &F_{5001} \times P('CORN') + F_{5002} \times P('AAMY') + F_{5004} \times P('LIME') + F_{5005} \times P('CAUS') \\ &+ F_{5006} \times P('GAMY') + F_{5007} \times P('H_2SO_4') + F_{5009} \times P('YEAST') \end{aligned}\right) \\
&+ F_{7001} \times P('STEAM') + F_{7000} \times P('WATER')
\end{aligned}$$

8. Utilcost $= \big(F_{300} \times P('CH_4') + 80 \times EP2 \times F_{5015}\big)$

9. $Q = \left(\begin{aligned} &Q_{AMM} + Q_{NIT} + Q_{an} + Q_{met} + Q_{apg} + Q_{gtsp} + Q_{ppa} + Q_{EF} + Q_{CH} + Q_{AA} \\ &+ Q_{SR} + Q_{SSR} + Q_{AA2} + Q_{AMM} + \left(\dfrac{F_{16}}{18.02} \times (-285830) - F16 \times HLP\right) \times 10^{-1} + Q_{GENextn} \end{aligned}\right)$

10. $Q_{GENextn} = \left(\begin{aligned} &Q_{SYNGBIO} + Q_{FE} + Q_{EEI} + FA \times Q_{FAME} + Q_{PGI} + Q_{AAAD} + Q_{PIPEN} + Q_{PPEND} + Q_{SYNGC} + Q_{FA} \\ &+ Q_{MA} + Q_{NMEB} + Q_{NSTYB} + Q_{ETB} + Q_{DME} + Q_{GH} + Q_{NMEA} + Q_{NMEC} + Q_{NMED} + Q_{STY} + Q_{EB} \end{aligned}\right)$

the process. The cost for utilities was calculated from the natural gas require-ment to supply steam for heat exchange in all the processes, except for the corn ethanol process. For the corn ethanol process, the utility cost per ton of ethanol produced ($80 per MT ethanol) from the process was considered. Equation 9 gives the total energy, Q, required by the existing processes in the base plus additional energy required by the proposed biochemical complex extension and carbon dioxide utilization processes. Equation 10 gives the energy required by the biochemical complex extension and carbon dioxide–consuming processes only, given by $Q_{GENextn}$.

6.6 Optimal Structure

The optimum configuration of plants was obtained from the superstructure by maximizing the triple bottom line, Equation 1 in Table 6.7, subject to the equality and inequality constraints. The characteristics of the superstruc-ture are given in Table 6.8. There were 978 equality constraints that describe material and energy balances for the plants (including dependent equations). Also, there are 91 inequality constraints that describe the product demand, raw material availability, and capacities of the plants in the chemical com-plex. There were 25 integer variables in the superstructure.

The optimal structure from the superstructure is shown in Figure 6.3, and a convenient way to show the plants in the optimal structure is given in Table 6.9. The corn fermentation process was selected in the optimal struc-ture and the corn stover fermentation process was not selected. The higher utility costs associated with the corn stover process made the corn ethanol process selection more profitable, even with corn prices being higher than corn stover. Six corn ethanol plants (5.84 in optimal structure) producing 57,000 tons/year bioethanol each (179,000 MT/year corn processing) were required to meet the demand for bioethylene. The bioethylene plant oper-ated at full capacity of 200,000 MT/year in the optimal structure.

TABLE 6.8

Characteristics of Superstructure

Superstructure Characteristics
978 equality constraints that describe material and energy balances for the plants (including dependent equations)
91 inequality constraints that describe availability of raw materials, demand for product, capacities of the plants, and logical relations in the chemical complex
969 continuous variables
25 integer variables
2 tables

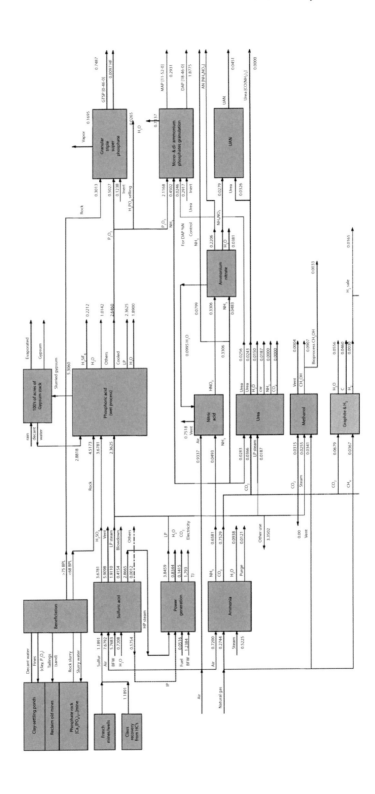

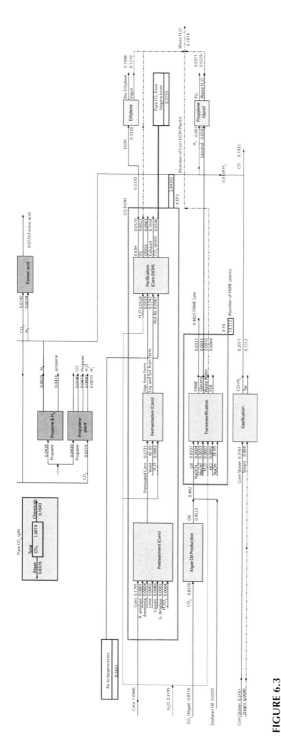

FIGURE 6.3
(See color insert.) Optimal configuration of integrated chemical production complex, flow rates million metric tons per year.

TABLE 6.9

Plants in the Optimal Structure from the Superstructure

Existing Plants in the Optimal Structure	New Plants in the Optimal Structure
Ammonia	Fermentation to ethanol (Corn)
Nitric acid	Bioethylene from dehydration of bioethanol
Ammonium nitrate	Transesterification to FAME and glycerol (Soy oil and algae)
Urea	
UAN	Algae oil production; biopropylene glycol from glycerol
Methanol	
Granular triple super phosphate (GTSP)	Gasification to syngas (corn stover)
MAP and DAP	Formic acid
Contact process for sulfuric acid	Graphite
Wet process for phosphoric acid	Propylene from CO_2
Power generation	Propylene from propane dehydrogenation
Existing Plants Not in the Optimal Structure	**New Plants Not in the Optimal Structure**
Acetic acid	Fermentation to ethanol (corn stover)
Ethylbenzene	Anaerobic digestion to acetic acid (corn stover)
Styrene	Methanol (Bonivardi et al., 1998)
	Methanol (Jun et al., 1998)
	Methanol (Ushikoshi et al., 1998)
	Methanol (Nerlov and Chorkendorff, 1999)
	Methylamines (MMA and DMA)
	Ethanol
	Dimethyl ether
	Hydrogen/synthesis gas
	Acetic acid—new process
	Styrene—new method
	Electric furnace process for phosphoric acid
	Haifa process for phosphoric acid
	SO_2 recovery from gypsum waste
	S and SO_2 recovery from gypsum waste

The transesterification process utilized multiple feedstocks; 450,000 MT/year of algae oil produced from the algae production unit and 29,000 MT/year of soybean oil purchased were used to meet the demand for glycerol. The demand for glycerol was determined by the use of glycerol in the propylene glycol process. Fifteen plants (14.312 in optimal structure), each producing 33,700 MT/year (10 million gallons per year) of fatty acid methyl ester (FAME) for a total of 483,000 MT/year of FAME, were required to produce the glycerol necessary to meet the capacity of 37,000 MT/year of propylene glycol. The propylene glycol plant operated at the lower bound for the capacity.

There were three competing processes for acetic acid, with the logic that at most one (or none) of the three plants will operate. The existing acetic acid plant, the acetic acid plant from biomass anaerobic digestion, and the acetic acid plant from carbon dioxide consumption process were all excluded in the optimal structure. There were other options for utilizing carbon dioxide (the raw material for acetic acid conventional process and new process). The low selling price for acetic acid is the reason for the exclusion of the acetic acid processes in the optimal structure.

The existing ethylbenzene and styrene plants were excluded from the optimal structure. The cost for benzene as raw material for these processes was significantly high, and the operation of these plants was unprofitable. This exclusion of the ethylbenzene plant reduced the cost for total raw material use in the optimal structure.

A comparison of the sales and costs associated with the triple bottom line is shown in Table 6.10 for the base case and the optimal structure. The triple bottom line increased from $854 to $1650 million per year or about 93% from the base case to the optimal structure. Sales increased from new products from biomass-feedstock-based processes like FAME, propylene glycol, bioethylene, hydrogen from gasification process, and by-product (DDGS) sales from corn ethanol processes. Additional products from carbon dioxide–consuming processes like formic acid, graphite, and propylene also increased the income from sales. The economic costs decreased from $697 million per year to $516 million per year, which was approximately a 26% reduction from the base case. A breakdown of the economic costs shows that the raw material costs decreased by 31% while the utility costs increased from $12 million per year to $46 million per year. The environmental costs decreased from $457 million per year to $313 million per year due to decrease in the use of raw materials from the base case. The cost to society improved since sustainable costs decreased from $18 million per year to $10 million per year from the credits given for using carbon dioxide in bioprocesses and chemical processes.

TABLE 6.10

Sales and Costs Associated with the Triple Bottom Line for the Base Case and Optimal Structure (Million Dollars per Year)

	Base Case	Optimal Structure
Income from sales	2026	2490
Economic costs (raw materials and utilities)	697	516
Raw material costs	685	470
Utility costs	12	46
Environmental cost (67% of raw material cost)	457	313
Sustainable credits (+)/costs (−)	−18	−10
Triple bottom line	**854**	**1650**

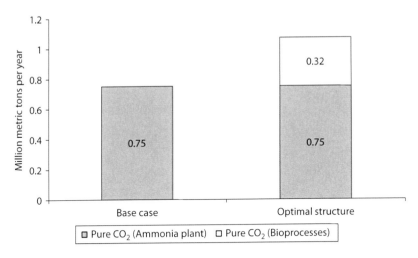

FIGURE 6.4
Pure carbon dioxide sources in base case and optimal structure.

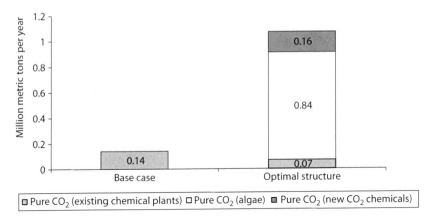

FIGURE 6.5
Pure carbon dioxide consumption in base case and optimal structure.

The various sources and consumption of CO_2 in the base case and superstructure are shown in Figures 6.4 and 6.5, respectively. The data are given in Table 6.11. The pure carbon dioxide produced from the bioprocesses was 0.32 million MT/year. This was in addition to the 0.75 million MT/year of pure carbon dioxide that is already being produced from the existing ammonia plant in the complex. The total 1.07 million MT/year of pure carbon dioxide was now available to the carbon dioxide–consuming processes. Existing plants in the base-case (urea, methanol, and acetic acid) process utilize 0.07 million MT/year of the pure carbon dioxide.

There were 15 new processes for carbon dioxide consumption in the superstructure, which included the algae oil production process. From these, four

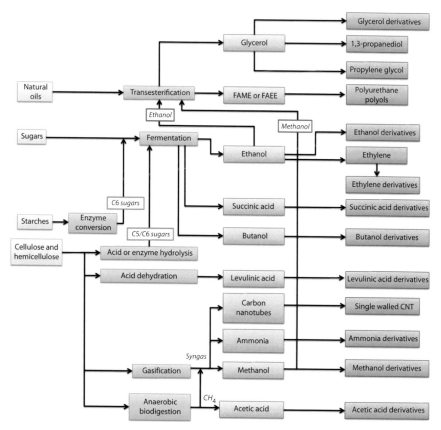

FIGURE 4.1
Conceptual design of biomass-feedstock-based chemical production.

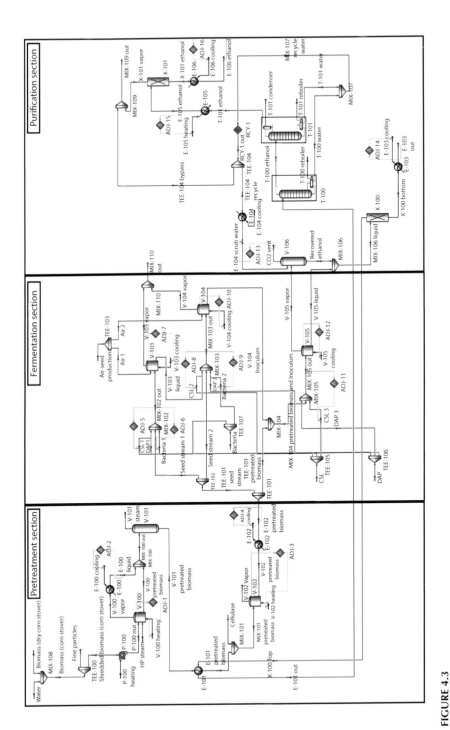

FIGURE 4.3
Overall process design diagram for ethanol production from corn stover fermentation.

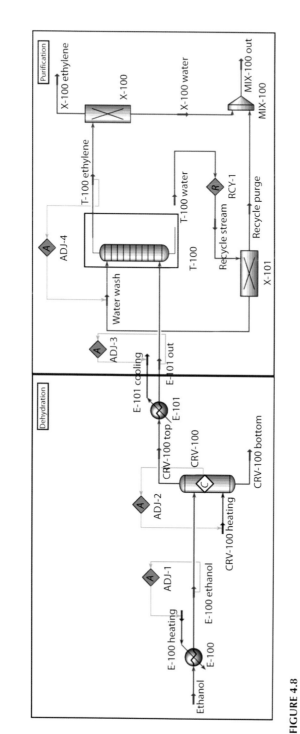

FIGURE 4.8

Overall process design diagram for ethylene production from dehydration of ethanol.

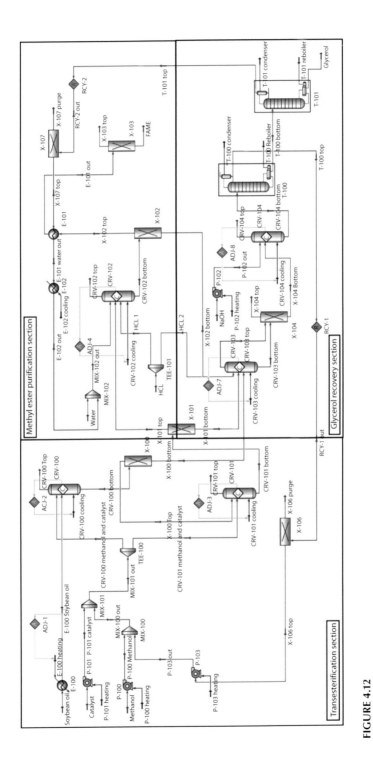

FIGURE 4.12
Overall process design diagram for fatty acid methyl ester and glycerol from transesterification of soybean oil.

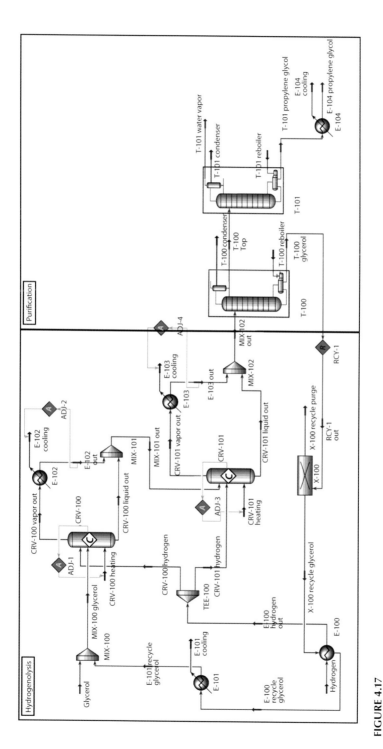

FIGURE 4.17

Overall process design diagram for propylene glycol production from hydrogenolysis of glycerol.

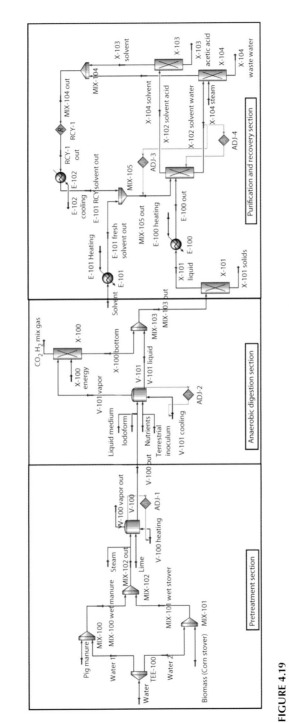

FIGURE 4.19

Overall process design diagram for acetic acid production from corn stover anaerobic digestion.

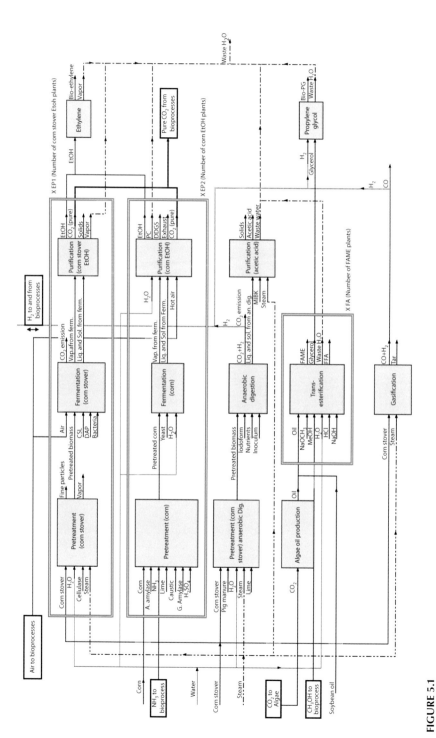

FIGURE 5.1
Overall biochemical processes block flow diagram.

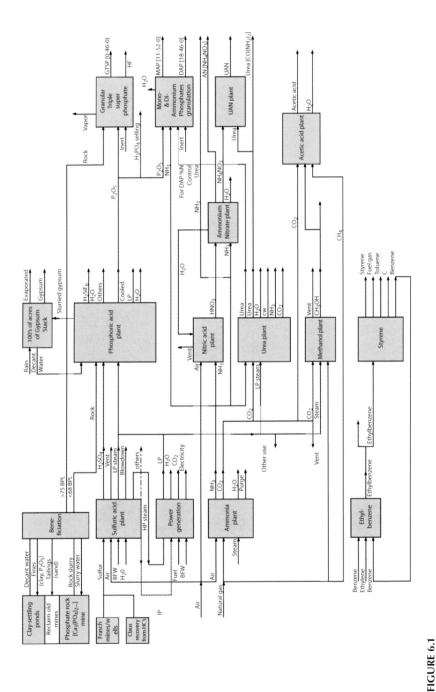

FIGURE 6.1
Base case of chemical plants in the lower Mississippi River corridor (From Xu, A., Chemical Production Complex Optimization, Pollution Reduction and Sustainable Development, PhD dissertation, Louisiana State University, Baton Rouge, LA, 2004.)

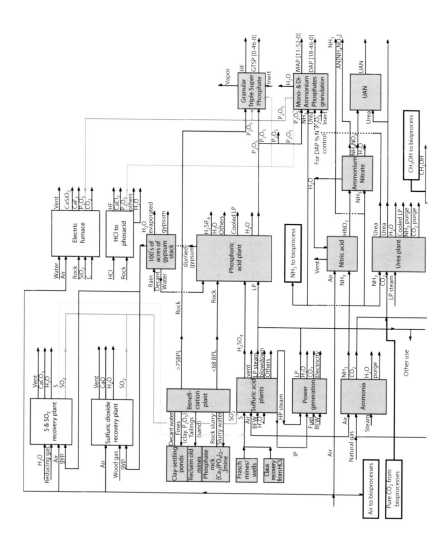

FIGURE 6.2
Integrated chemical production complex, superstructure.

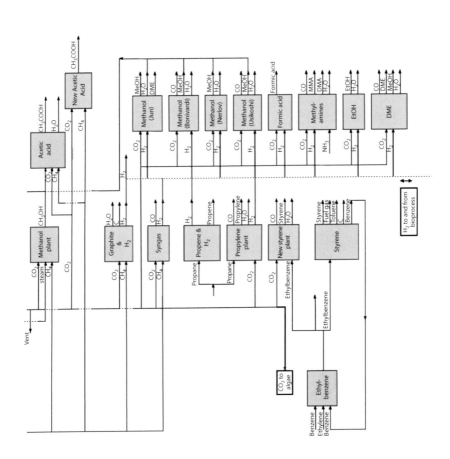

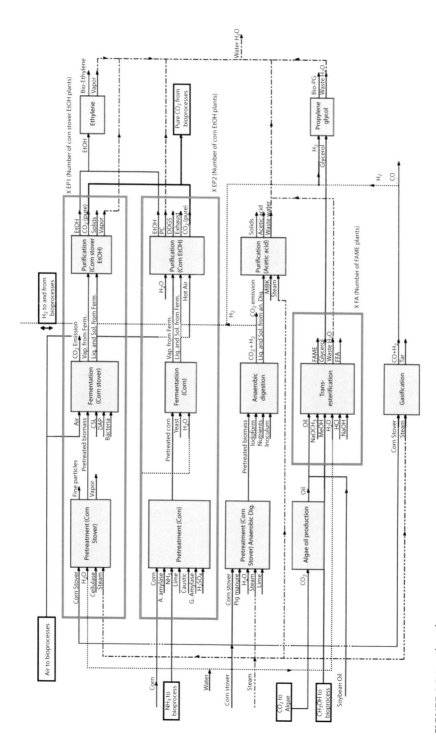

FIGURE 6.2 (continued)

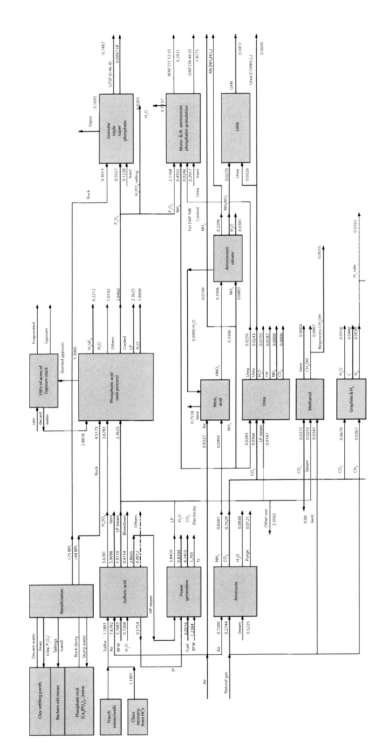

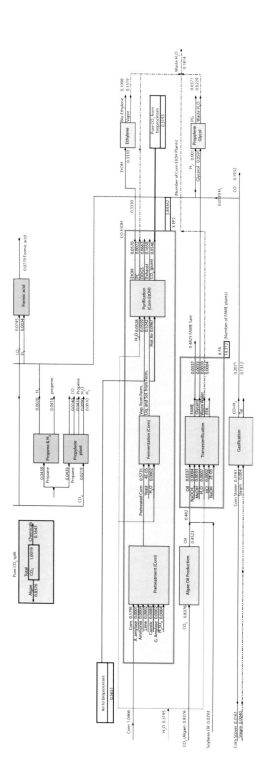

FIGURE 6.3
Optimal configuration of integrated chemical production complex, flow rates million metric tons per year.

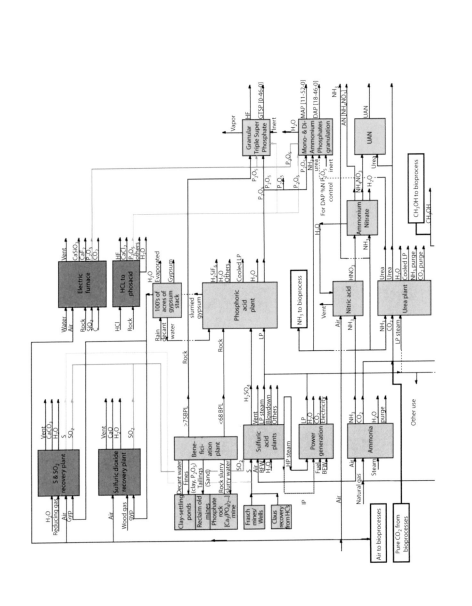

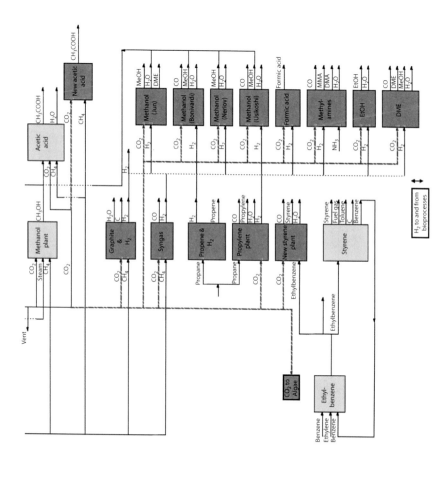

FIGURE 7.1
Integrated chemical production complex, superstructure w/o CO₂ use.

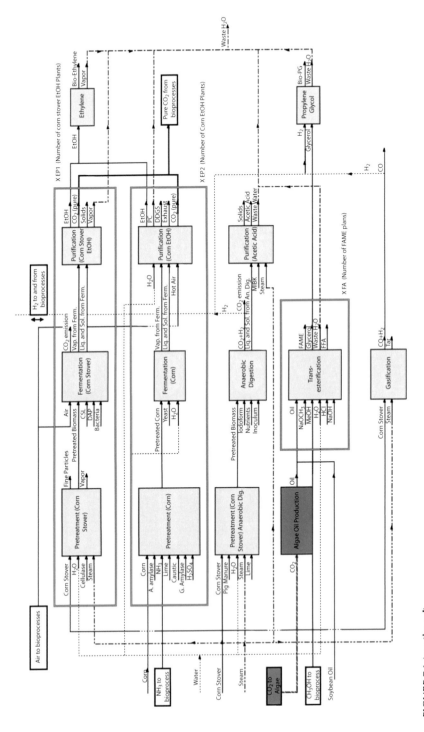

FIGURE 7.1 (continued)

TABLE 6.11

Carbon Dioxide Sources and Consumption in the Base Case and Optimal Structure (Million Metric Tons per Year)

	Base Case	Optimal Structure
Pure CO_2 produced by ammonia plant	0.75	0.75
Pure CO_2 produced by bioprocesses	na	0.32
Pure CO_2 consumed by existing chemical plants in base case	0.14	0.07
Pure CO_2 consumed by algae production process and new CO_2-consuming processes	na	1.00
Pure CO_2 consumed by algae production process	na	0.84
Pure CO_2 vented to atmosphere	0.61	0
Impure CO_2 emission from power plant	0.08	0.14

new processes were selected for the consumption of carbon dioxide, and 1.0 million MT/year of carbon dioxide was utilized in these new carbon processes; 0.84 million MT/year of pure carbon dioxide was used in the production of algae and 0.16 million MT/year was used in the production of other chemicals from carbon dioxide. The total pure carbon dioxide emission in the optimal structure was reduced to zero with all the carbon dioxide utilized in the algae oil production process and other carbon dioxide–consuming processes.

A relevant study was to see whether the impure CO_2 emission increased from the power plant for inclusion of the bioprocesses. From Table 6.11 it can be seen that the impure emission from the power plant increased from 0.08 million MT/year to 0.14 million MT/year (an increase of 75% from the base case).

Ten new processes were selected in the optimal structure, of which five were carbon dioxide–consuming processes as shown in Tables 6.9 and 6.12. Table 6.12 gives the plant capacities for the base case and the optimal structure, and the energy required in both these cases.

The energy used or produced for each process in the complex is given in Table 6.12. The energy required in the base case was 898 TJ/year, which increased to 2340 TJ/year excluding the energy required for the corn ethanol process. The energy was supplied from firing boilers with natural gas that had a sustainable cost of $3.25 per MT of carbon dioxide emitted from the power generation process. The corn ethanol process considered purchased utility (not from the power generation plant in the complex). An estimate of the energy required by the corn ethanol process was computed from the energy content of high pressure steam and cost of steam from ICARUS utility specifications, as given in Table 6.12. Thus, the total energy requirement by the plants in the optimal structure was 6405 TJ/year. Also, from Table 6.12, it can be seen that the sulfuric acid plant is an important source of energy, and operating this plant is as important as the production of sulfuric acid from the plant.

The flow rates of important chemicals from the optimal solution are given in the Table 6.13. This is not an exhaustive list of all the chemicals produced in

TABLE 6.12

Comparison of Capacities for the Base Case and Optimal Structure

Plant Name	Capacity (Upper–Lower Bounds) (MT/year)	Optimal Capacity (MT/year)	Energy Requirement (Base Case) (TJ/year)	Energy Requirement (Optimal Structure) (TJ/year)
Ammonia	329,000–658,000	658,000	3820	3820
Nitric acid	89,000–178,000	178,000	−775	−775
Ammonium nitrate	113,000–227,000	227,000	229	245
Urea	49,900–99,800	49,900	128	64
Methanol	91,000–181,000	91,000	2165	1083
UAN	30,000–60,000	45,100	0	0
MAP	146,000–293,000	293,000		
DAP	939,000–1,880,000	1,880,000	1901	1901
GTSP	374,000–749,000	749,000	1311	1311
Sulfuric acid	1,810,000–3,620,000	3,620,000	−14,642	−14,642
Wet process phosphoric acid	635,000–1,270,000	1,270,000	5181	5181
Ethylbenzene	431,000–862,000	0	−386	0
Styrene	386,000–771,000	0	1698	0
Acetic acid	4,080–8,160	0	268	0
Fermentation (corn stover)	333,000–667,000 (each)	0	na	0
Number of corn stover ethanol plants	—	0	na	0
Fermentation (corn)	180,000–360,000 (each)	180,000	na	693
Number of corn ethanol plants	—	6	na	4,158
Transesterification (FAME)	16,850–33,700	33,700	na	90
Number of FAME plants	—	15	na	1293
Anaerobic digestion (corn stover)	333,000–667,000	0	na	0
Gasification (hydrogen)	6,700–13,400	13,400	na	−594
Bioethylene	1000,000–200,000	200,000	na	820
Biopropylene glycol	37,000–74,000	37,000	na	409
Algae oil production	Constrained by availability of CO_2	452,300	na	0
Electric furnace phosphoric acid	635,000–1,270,000	0	na	0
Haifa phosphoric acid	635,000–1,270,000	0	na	0
New acetic acid	4090–8180	0	na	0
SO_2 recovery from gypsum	987,000–1,970,000	0	na	0

TABLE 6.12 (continued)

Comparison of Capacities for the Base Case and Optimal Structure

Plant Name	Capacity (Upper–Lower Bounds) (MT/Year)	Optimal Capacity (MT/Year)	Energy Requirement (Base Case) (TJ/Year)	Energy Requirement (Optimal Structure) (TJ/Year)
S and SO₂ recovery from gypsum	494,000–988,000	0	na	0
Graphite and H₂ from CO₂ and CH₄	23,000–46,000	46,000	na	1046
Syngas	6,700–13,400	0	na	0
Propene and H₂	20,900–41,800	41,800	na	658
Propene using CO₂	21,000–41,900	41,900	na	413
New styrene	181,000–362,000	0	na	0
New methanol—Bonivardi	240,000–480,000	0	na	0
New methanol—Jun	240,000–480,000	0	na	0
New methanol—Nerlov	240,000–480,000	0	na	0
New methanol—Ushikoshi	240,000–480,000	0	na	0
Formic acid	39,000–78,000	78,000	na	14
Methylamines	13,200–26,400	0	na	0
Ethanol	52,000–104,000	0	na	0
Dimethyl ether	22,900–45,800	0	na	0
Total energy requirement			898	6405

TABLE 6.13

Base Case and Optimal Flow Rates of Products (Metric Tons per Year)

Product	Base Case	Optimal Structure
Ammonia sale	53,600	80,000
Ammonium nitrate sale	218,000	221,000
Urea sale	41,600	0
Wet process phosphoric acid sale	26,500	26,500
Ethylbenzene sale	441,000	0
Bioethylene sale	—	200,000
Biopropylene glycol sale	—	37,000
Bioacetic acid sale	—	0
FAME sale	—	483,000
Hydrogen sale	—	16,500
Methanol sale	177,000	43,000
Pure CO₂ vented	61,200	0

the complex, but it represents some of the more important optimal flow rates in the complex. From Table 6.13, it can be seen that methanol sales decreased from the base case to the optimal structure. This can be attributed to the use of methanol in the transesterification process, thus decreasing the amount of methanol available for sale in the complex.

The optimal solution from the superstructure was used for further case studies, given in Chapter 7. In the following sections, multicriteria optimization is performed to obtain the Pareto optimal sets for profit and sustainable credits. Monte Carlo simulations are used to study the sensitivity of the optimal solution for price variations of raw materials and products in the complex.

6.7 Multiobjective Optimization of the Integrated Biochemical Production Complex

The objective is to find optimal solutions that maximize companies' profits and minimize costs to society. Companies' profits are sales minus economic and environmental costs, as given by Equation 6.4. Economic costs include raw material, utilities, labor, and other manufacturing costs. Environmental costs include permits, monitoring of emissions, fines, etc. The costs to society are measured by sustainable costs. These costs are from damage to the environment by emissions discharged within permitted regulations. Sustainable credits are awarded for reductions in emissions, as shown in Table 6.6, and are similar to emissions trading credits. Detailed discussion regarding the sustainable costs and possible credits can be found in Chapter 7. This section demonstrates the multiobjective optimization results for the superstructure to maximize profit and sustainable credits.

The multicriteria optimization problem can be stated as in terms of profit, P, and sustainable credits/costs, S, for theses two objectives in Equation 6.4 (details on multicriteria optimization algorithm can be found in Appendix B).

$$\text{Max: } P = \sum \text{Product sales} - \sum \text{Economic costs} - \sum \text{Environmental costs}$$

$$= \sum \text{Sustainable (Credits} - \text{Costs)} \tag{6.4}$$

Subject to: Multiplant material and energy balances
　　　　　　　Product demand, raw material availability, plant capacities

Multicriteria optimization obtains solutions that are called efficient or Pareto optimal solutions. These are optimal points where attempting to improve the value of one objective would cause another objective to decrease. To locate Pareto optimal solutions, multicriteria optimization problems are converted to one with a single criterion by the parametric approach method, which is by applying weights to each objective and optimizing the sum of the weighted objectives. The multicriteria mixed-integer optimization problem becomes

$$\text{Max: } w_1 P + w_2 S$$

$$w_1 + w_2 = 1$$

(6.5)

Subject to: Multiplant material and energy balances
Product demand, raw material availability, plant capacities

The chemical complex analysis system was used to determine the Pareto optimal solutions for the weights using $w_1 + w_2 = 1$ given by Equation 6.5; these results are shown in Figure 6.6. The profits for the company are 2 orders of magnitude larger than the sustainable credits/costs. The sustainable credits/costs decline and company's profits increase as the weight, w_1, on company's profit increase. For example, when $w_1 = 1$, the optimal solution is shown in Table 6.14 for $P = \$1660.01$ million per year and $S = \$-9.98$ million per year. The optimal solution with $w_1 = 0$ gave $P = \$1193.45$ million per year and $S = \$26.00$ million per year.

The points shown in Figure 6.6 are the Pareto optimal solutions for w_1 from 0 to 1.0 for increments of 0.001. The values for w_1 equal to 0 and 1.0 and some intermediate ones are shown in Table 6.14. The optimal complex configurations of the Pareto optimal solutions for w1 from 0 to 1.0 for increment of 0.001 are shown in Table 6.15. If a process is selected, the binary variable associated with the process is 1, otherwise it is 0. For each process in Table 6.15, the sums of the binary variable values for the corresponding w1 range are shown, along with the total summation of the times the process was selected.

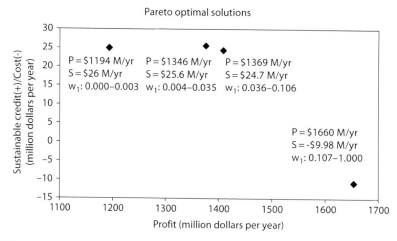

FIGURE 6.6
Optimal solutions generated by multicriteria optimization.

TABLE 6.14

Values of the Pareto Optimal Solutions
Shown in Figure 6.6

Profit (Million Dollars/Year)	Sustainable Credits/ Costs (Million Dollars/Year)	Weight ($w1$)
1660.01	−9.98	1
1660.01	−9.98	0.894
1660.01	−9.98	0.107
1369.32	24.74	0.106
1369.32	24.74	0.036
1346.26	25.60	0.035
1346.26	25.60	0.004
1193.94	26.00	0.003
1193.45	26.00	0

The corn ethanol process was always selected, and the corn stover ethanol process was never selected, in the optimal solution. The acetic acid process from anaerobic digestion and the conventional acetic acid process were never selected, but the acetic acid process from carbon dioxide consumption process was included twice, when sustainable credits had higher weight attached to the process. S and SO_2 recovery from gypsum process ran at lower weights attached to the profit and higher weight for sustainable credits. Hence, the optimal structure is affected, but it did not change significantly (Table 6.15). It is another decision to determine the specific value of the weight that is acceptable to all concerned.

6.8 Sensitivity of the Integrated Biochemical Production Complex

The optimal solution from the superstructure gave a triple bottom line profit of $1650 million per year. The sensitivity of the optimal solution to changes in price for the raw materials and products was studied using the feature in the chemical complex analysis system. The price and standard deviation of the raw materials and products are given in Table 6.6. These prices were collected over a 2 year period and are given in Appendix C. It may be noted that the market conditions were volatile when the price data were collected, and there was wide variability in the price. The average price and standard deviation from the available data were entered in the chemical complex analysis system, and 1000 runs of the optimal solution were obtained. Then the data were exported to Microsoft Excel, and analyzed.

TABLE 6.15

Optimal Structure Changes in Multicriteria Optimization (Number of Times Out of 1000 a Process Is Selected)

Processes	w_1							Total
	0.000–0.149	0.150–0.299	0.300–0.449	0.450–0.599	0.600–0.749	0.750–0.900	0.900–1.000	
Biomass gasification (Y60)	150	150	150	150	150	150	100	1000
Acetic acid anaerobic digestion (Y61)	0	0	0	0	0	0	0	0
Corn stover ethanol (EP1 > 0)	0	0	0	0	0	0	0	0
Corn ethanol (EP2 > 0)	150	150	150	150	150	150	100	1000
Electric furnace phosphoric acid (Y1)	0	0	0	0	0	0	0	0
Acetic acid (Y11)	0	0	0	0	0	0	0	0
New acetic acid (Y12)	2	0	0	0	0	0	0	2
SO$_2$ recovery from gypsum (Y13)	0	0	0	0	0	0	0	0
S and SO$_2$ recovery from gypsum (Y14)	107	0	0	0	0	0	0	107
Methanol (Y16)	150	150	150	150	150	150	100	1000
Haifa process phosphoric acid (Y2)	0	0	0	0	0	0	0	0
Propylene from CO$_2$ (Y23)	147	150	150	150	150	150	100	997
Propylene from propane dehydrogenation (Y24)	147	150	150	150	150	150	100	997
Synthesis gas (Y27)	0	0	0	0	0	0	0	0
Formic acid (Y29)	150	150	150	150	150	150	100	1000
Wet process phosphoric acid (Y3)	150	150	150	150	150	150	100	1000
Methylamines (Y30)	0	0	0	0	0	0	0	0
Methanol (Jun et al., 1998) (Y31)	0	0	0	0	0	0	0	0
Methanol (Bonivardi et al., 1998) (Y32)	0	0	0	0	0	0	0	0
Methanol (Nerlov and Chorkendorff, 1999) (Y33)	0	0	0	0	0	0	0	0
Methanol (Ushikoshi et al., 1998) (Y34)	0	0	0	0	0	0	0	0
New styrene (Y35)	0	0	0	0	0	0	0	0
Ethanol (Y37)	0	0	0	0	0	0	0	0
Dimethyl ether (Y38)	0	0	0	0	0	0	0	0
Graphite (Y39)	147	150	150	150	150	150	100	997
Styrene (Y40)	0	0	0	0	0	0	0	0
Ethyl benzene (Y41)	3	0	0	0	0	0	0	3

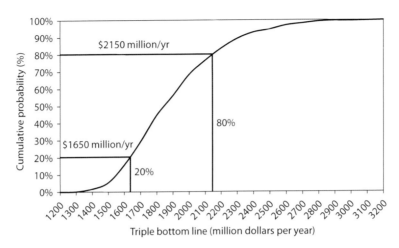

FIGURE 6.7
Cumulative probability distribution for the triple bottom line of the optimal structure.

One of the results from the data analysis was the cumulative probability of the triple bottom line profit, as shown in Figure 6.7. The statistical average triple bottom line profit was $1898 million per year. The standard deviation of the results obtained from Monte Carlo simulations was $312 million per year. The minimum triple bottom line profit was $1251 million per year and the maximum triple bottom line profit was $3134 million per year. From Figure 6.7, it can be said that there is 20% probability of the triple bottom line being less than or equal to $1650 million per year. It can also be inferred that there is 80% probability of the triple bottom line being less than or equal to $2150 million per year.

The optimal solution and the statistical average triple bottom line were not the same. The reason for this is a different configuration of plants is selected for a certain price parameter set in each Monte Carlo run. Some of the configurations may be similar while others will differ. The triple bottom line profit function will change accordingly, for inclusion or exclusion of flow rates from plants in the complex.

The chemical production complex configurations of Monte Carlo simulation solutions for 1000 samples are shown in Table 6.16. If a process is selected, the binary variable associated with the process is 1, otherwise it is 0. For each process in Table 6.16, the sums of the binary variable values for the corresponding iteration range are shown, along with the total summation of the times the process was selected.

The results in Table 6.16 show that the complex was able to curb carbon dioxide emissions for almost all of the case runs (995 out of 1000 times). Corn stover ethanol was selected in 23% of the runs and the corn ethanol process was selected in 77% of the runs. There is 47% probability that the existing

TABLE 6.16

Optimal Structure Changes in Monte Carlo Simulation (Number of Times Out of 1000 a Process Is Selected)

Processes	Monte Carlo Simulation (Iterations)							Total	Probability (%)
	1–150	151–300	301–450	451–600	601–750	751–900	901–1000		
Corn stover ethanol (EP1 > 0)	36	30	33	30	36	37	27	229	23
Corn ethanol (EP2 > 0)	114	120	117	120	114	113	73	771	77
Ethylbenzene (Y41)	71	79	74	72	76	55	46	473	47
Styrene (Y40)	0	0	0	0	0	0	0	0	0
Electric furnace phosphoric acid (Y1)	0	0	0	0	0	0	0	0	0
Acetic acid anaerobic digestion (Y61)	0	0	0	0	0	0	0	0	0
Methanol (Y16)	150	150	150	150	150	150	100	1000	100
Biomass gasification (Y60)	122	134	135	131	124	137	81	864	86
Wet process phosphoric acid (Y3)	150	150	150	150	150	150	100	1000	100
Pure CO_2 emission abatement (S64 = 0)	149	150	150	150	148	149	99	995	100
New acetic acid (Y12)	48	55	50	54	49	62	37	355	36
Acetic acid (Y11)	0	0	0	0	0	0	0	0	0
Propylene from CO_2 (Y23)	130	135	135	142	137	140	95	914	91
Propylene from propane dehydrogenation (Y24)	148	143	144	147	147	145	99	973	97
Synthesis gas (Y27)	6	9	7	5	6	7	4	44	4
Formic acid (Y29)	136	135	132	140	125	142	88	898	90
Methylamines (Y30)	44	51	57	51	43	69	37	352	35
Methanol (Jun et al., 1998) (Y31)	0	0	0	0	0	0	0	0	0
Methanol (Bonivardi et al., 1998) (Y32)	0	0	0	0	0	0	0	0	0
Methanol (Nerlov and Chorkendorff, 1999) (Y33)	0	0	0	0	0	0	0	0	0
Methanol (Ushikoshi et al., 1998) (Y34)	0	0	0	0	0	0	0	0	0
New styrene (Y35)	0	0	0	0	0	0	0	0	0
Ethanol (Y37)	15	13	17	20	18	20	8	111	11
Dimethyl ether (Y38)	0	0	0	0	0	0	0	0	0

ethylbenzene process will be selected, but the styrene processes, both existing and proposed, was never selected in the runs.

The existing acetic acid process and the new acetic acid process from anaerobic digestion of biomass were never selected, but the new acetic acid process from carbon dioxide consumption was selected in 36% of the runs. The existing methanol process in the base case was always selected, and proposed methanol processes from carbon dioxide consumption could never compete with the existing process. The existing phosphoric acid was always selected and the proposed alternatives for phosphoric acid production could never compete with the existing process. The biomass gasification process was selected in 86% of the case runs, while the new synthesis gas process from carbon dioxide consumption was selected in only 4% of the cases. Among the new plants proposed for carbon dioxide utilization, the graphite process was always selected, propylene from CO_2 and formic acid was selected in 90% of the case runs, the methylamines process was selected in 35% of the case runs, ethanol process was selected in 11% of the case runs, and the dimethyl ether process was never selected. Thus, decisions regarding the operation of the plants can be made based on the Monte Carlo sensitivity analysis of the optimal operation of plants.

6.9 Comparison with Other Results

The present research studies the chemical complex optimization for integration of bioprocesses into an existing chemical complex infrastructure. Chemical complex optimization has been studied by Xu (2004) for integrating new carbon dioxide processes into the base case of chemical plants in the lower Mississippi River corridor. Indala (2004) developed HYSYS designs for 14 new processes that converted high-purity carbon dioxide to chemicals. The superstructure developed by Xu (2004) integrated these new processes into the base case and obtained the optimal solution. This research acknowledges the work by Xu (2004) and Indala (2004) for the base case of plants and the carbon dioxide processes developed for integration into the base case. There have been no other reports of chemical complex optimization from a macroscale.

The comparison of results for the base case and superstructure of plants developed by Xu (2004) is given in this section. The chemical complex analysis system models for the base case and superstructure were obtained from www.mpri.lsu.edu. The parameters for price were changed from 2004 to 2010 values, as given in Table 6.6, and the cases were optimized. The solution for the base case is given in Table 6.17. The optimal solution for the superstructure from Xu (2004) is given in Table 6.18.

The base-case flow rates are given in Figure 6.8. From Table 6.17, it is seen that the triple bottom line for the base case increased from $343 million per

TABLE 6.17

Comparison of Results for Base Case
(Million Dollars per Year)

	Base Case (Xu, 2004)	Base Case (Present Research)
Income from sales	1277	2026
Economic costs (Raw materials and utilities)	554	697
Raw material costs	542	685
Utility costs	12	12
Environmental cost (67% of raw material cost)	362	457
Sustainable credits (+)/costs (−)	−18	−18
Triple bottom line	**343**	**854**

Source: Xu, A., Chemical production complex optimization, pollution reduction and sustainable development, PhD dissertation, Louisiana State University, Baton Rouge, LA, 2004 and present research.

TABLE 6.18

Comparison of Results for Optimal Structure
(Million Dollars per Year)

	Optimal Structure (Xu, 2004)	Optimal Structure (Modified for Cost Parameters)
Income from sales	1508	1859
Economic costs (raw materials and utilities)	602	360
Raw material costs	577	334
Utility costs	25	26
Environmental cost (67% of raw material cost)	382	223
Sustainable credits (+)/costs (−)	−15	−14
Triple bottom line	**506**	**1262**

Source: Xu, A., Chemical production complex optimization, pollution reduction and sustainable development, PhD dissertation, Louisiana State University, Baton Rouge, LA, 2004 and present research.

year to $854 million per year. The income from sales increased from $1277 million per year to $2026 million per year—an increase of 58%. The raw material costs increased from $542 million per year to $685 million per year—an increase of 26%.

The optimal structure flow rates for the integration of carbon dioxide processes in the base case are given in Table 6.18. From Table 6.18, it is seen that

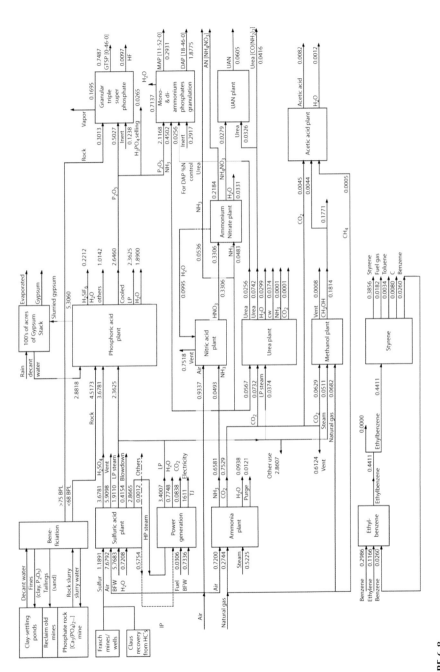

FIGURE 6.8
Base-case flow rates, million metric tons per year.

the triple bottom line increased from $506 million per year to $1262 million per year. The income from sales increased from $1508 million per year to $1859 million per year—an increase of 23%. The raw material costs decreased from $577 million per year to $334 million per year—a decrease of 42%. The major portion of this decrease was due to the exclusion of the ethylbenzene plant in the optimal structure. The ethylbenzene plant was excluded due to increase in the raw material cost of benzene from $303/MT in 2004 to $914/MT in 2010.

A procedure was described by Monteiro et al. (2010) to design an industrial ecosystem for sequestrating CO_2 and consuming glycerol in a chemical complex with 15 integrated processes. Methanol, ethylene oxide, ammonia, urea, dimethyl carbonate, ethylene glycol, glycerol carbonate, β-carotene, 1,2-propanediol, and olefins were produced in a complex that was simulated using UNISIM Design (Honeywell). Environmental impact (EI) was calculated using the waste reduction algorithm, while profit (P) was estimated using classic cost correlations. The objective was granting maximum process sustainability, which involves finding a compromise between high profitability and low environmental impact. Sustainability maximization was a multicriteria optimization problem that was solved by connecting MATLAB® to UNISIM, and Pareto optimization methodology was used for trading off P versus EI. The Pareto frontier was successfully identified, as validated by a sensitivity analysis performed on the solutions found along the Pareto line. In the Pareto frontier, scenarios with varying relevancies of P and EI were tested and the results described.

6.10 Summary

The formulation of the superstructure of chemical and biochemical plants was described in this chapter. Bioprocess models from Chapter 4 were integrated into an existing base case of plants in the lower Mississippi River corridor. The carbon dioxide from the integrated complex was utilized for algae production and other chemicals that consume carbon dioxide. For mixed-integer optimization, relations among the binary variables and the logical constraints were used. The upper bounds and lower bounds of the production capacities of all the plants in the chemical complex are also given.

The economic model for triple bottom line was given, which expanded value-added economic model for profit to include environmental and sustainable costs/credits. The triple bottom line was the objective function for the MINLP problem. The superstructure was optimized with multiplant material and energy balances, plant capacities, availability of raw materials, and demand for products as constraints.

The triple bottom line from the optimal solution was $1650 million per year, an increase of 93% from the base case. The increase was due to sale of new products from the bioprocesses. The utility costs increased due to the addition of new plants. The sustainable costs to the society decreased from the base case as credits were given for carbon dioxide consumption.

The total pure carbon dioxide emission in the optimal structure was reduced to zero from the base case; 84% of the pure carbon dioxide was consumed by algae and used for the production of algae oil. Impure carbon dioxide emissions from the power plant increased by 75% from the base case and contributed to sustainable costs to the society.

The Pareto optimal solutions from multicriteria optimization showed that for maximum weight on sustainability, the complex receives sustainable credits of $26 million per year and a profit of $1194 million per year. The profit is $1660 million per year and $−9.98 million per year for no weights on sustainability. This gives a set of operating conditions for the complex with weights on the criteria to be chosen by the user.

The sensitivity of the optimal solution was studied using Monte Carlo simulations with average price and standard deviation in price of chemicals and raw materials in the complex. The average triple bottom line from sensitivity analysis was $1898. The pure carbon dioxide emission in the 1000 simulation runs of the complex was zero in almost all of the cases.

7

Case Studies Using Superstructure

7.1 Introduction

Chapter 6 described the optimal structure for bioprocess integration into the existing base case with carbon dioxide utilization. This chapter uses the superstructure for studying changes to the optimal solution for the following cases.

The first case studies the modification of the superstructure where carbon dioxide from the integrated complex is not reused in algae oil production or other chemical processes. Currently, there are no incentives or credits for nonemission of carbon dioxide in chemical production processes. However, using renewable raw materials for producing chemicals is a transition from nonrenewable resources as atmospheric carbon dioxide fixed through biomass is used as raw materials. Thus, this case is a study of the effect of using renewable raw materials for chemical production.

The second case is a parametric study of sustainable costs and credits on the optimal structure. These costs are typical of what might apply if the carbon tax was used as sustainable cost that a plant has to pay for carbon dioxide emission. The cost for geological carbon dioxide sequestration is considered as credits that a chemical plant receives for consuming carbon dioxide.

The third case is a parametric study of changes to the optimal structure for costs of algae oil production. The algae oil production unit in the superstructure used new technologies and algae strains for which production costs were zero. There is no cost for carbon dioxide use as raw material and the new technologies with algae strains secreting oil promise to bring down the cost of algae oil production. However, these technologies are currently being developed and industrial-scale production has not been achieved. A parametric study with utility cost parameters for algae oil production was varied in the superstructure to see the effect on the optimal structure and the triple bottom line. Two technologies producing 30% and 50% algae oil were considered for the parametric study.

The fourth case combines the parametric studies from carbon dioxide sustainable costs and credits, and costs for algae oil production. A high cost for carbon dioxide emission and a low credit for carbon dioxide emission were used in this case. Carbon dioxide consumption in the complex included a 30% oil content algae oil production unit (current technology) with high-performance or low-performance plant. Multicriteria optimization was used to obtain Pareto optimal solutions with varying weights on profit and sustainable credits.

The fifth case is a parametric study of biomass feedstock costs. Corn and corn stover were competing feedstock for the ethanol production in the complex. So, these feedstock prices were used in this analysis. There were wide variations in corn prices over the last 10 years. The highest price and lowest price for corn were used in this study along with the highest and lowest price for corn stover. In Chapter 6, all corn ethanol plants were selected in the optimal structure. So, the number of corn ethanol plants for average corn and corn stover prices were varied to see the effect on the triple bottom line.

7.2 Case Study I—Superstructure without Carbon Dioxide Use

This case studies the modification of the superstructure shown in Figure 6.2. The carbon dioxide from the integrated complex is not reused in algae oil production or other chemical processes. Currently, there are no incentives or credits for nonemission of carbon dioxide in chemical production processes. However, using renewable raw materials for producing chemicals is a transition from nonrenewable resources as atmospheric carbon dioxide fixed through biomass is used as raw materials. Thus, this case is a study of the effect of using renewable raw materials for chemicals production, use of carbon dioxide in existing plants in the base case, and the additional carbon dioxide from renewable and nonrenewable resources vented to the atmosphere. This extension of the superstructure without CO_2 utilization will be referred henceforth as "superstructure w/o CO_2 use."

The superstructure was modified to exclude the new plants for CO_2 use as shown in Figure 7.1. The plants included in the superstructure w/o CO_2 use are given in Table 7.1. The stream relations modified to obtain the superstructure w/o CO_2 use are given in Table 7.2.

Equation 1 in Table 7.2 gives the relation for setting the flow rate of CO_2 to zero for new CO_2-consuming processes, including algae. This is denoted by the red lines (signifying zero flow in the carbon dioxide pipeline) in Figure 7.1. The plants that used CO_2 are shown in gray, signifying zero production. Equation 2 sets binary variable for the new acetic acid process consuming carbon dioxide to zero. This means that at most one of the acetic acid plants, the existing plant from base case or the new acetic acid plant from anaerobic digestion process will operate, but the new acetic acid process consuming CO_2 will never operate. Equations 3 and 4 set the binary variables for alternate choices of phosphoric acid plant to zero, as shown in gray in Figure 7.1. Equations 5 and 6 set the binary variables for new processes for sulfur and

(continued)

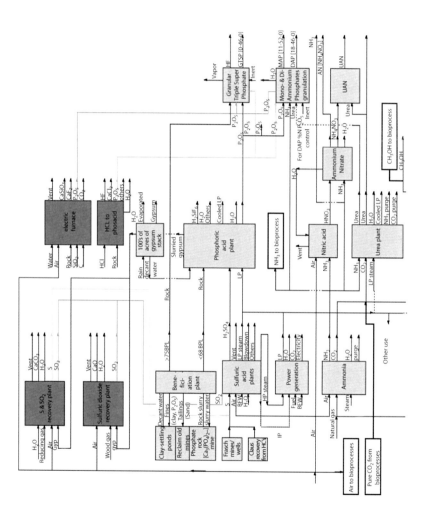

FIGURE 7.1
(See color insert.) Integrated chemical production complex, superstructure w/o CO_2 use.

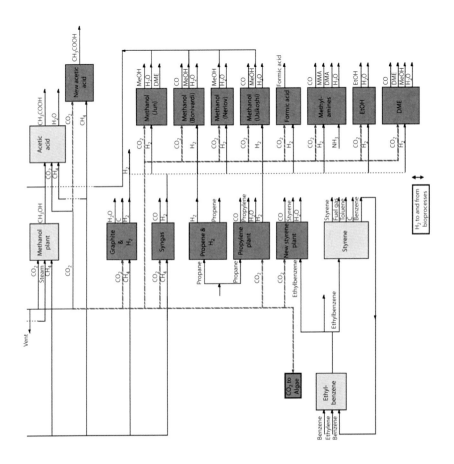

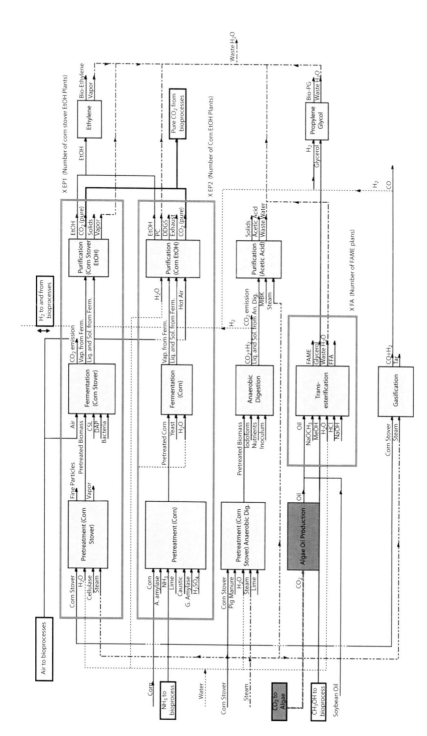

FIGURE 7.1 (continued)

TABLE 7.1

Plants in Base Case and Superstructure w/o CO_2 Use

Plants in Base Case (Xu, 2004)	Plants in Superstructure w/o CO_2 Use
Ammonia	Fermentation ethanol (corn stover)
Nitric acid	Fermentation ethanol (corn)
Ammonium nitrate	Anaerobic digestion to acetic acid (corn stover)
Urea	Algae oil production (not operating)
UAN	Transesterification to FAME and glycerol
Methanol	(soybean oil and algae)
Granular triple super phosphate (GTSP)	Gasification to syngas (corn stover)
MAP and DAP	Ethylene from dehydration of ethanol
Contact process for sulfuric acid	Propylene glycol from glycerol
Wet process for phosphoric acid	
Acetic acid—conventional method	
Ethyl benzene	
Styrene	
Power generation	

TABLE 7.2

Stream Relations Modified for Superstructure w/o CO_2 Use

Relation	Description
1. $F_{922} = 0$	Carbon dioxide to consumption processes, including algae production is zero
2. $Y_{12} = 0$	New acetic acid process from CO_2 is not selected
3. $Y_1 = 0$	Electric furnace phosphoric acid process not selected
4. $Y_2 = 0$	Haifa process phosphoric acid not selected
5. $Y_{13} = 0$	SO_2 recovery from gypsum not selected
6. $Y_{14} = 0$	S and SO_2 recovery from gypsum not selected
7. $Y_{24} = 0$	New propylene from propane dehydrogenation process not selected

sulfur dioxide recovery to zero, shown in gray in Figure 7.1. The upper and lower bounds on the capacity of the plants were same as in Table 6.5.

The superstructure w/o CO_2 use was optimized by maximizing the triple bottom line as given in Equation 6.1, subject to equality and inequality constraints. The plants included in the optimal structure for the case are shown in Table 7.3.

Table 7.4 shows the sources and consumption of carbon dioxide in the base case and optimal structure from the superstructure w/o CO_2 use. The pure carbon dioxide produced from the bioprocesses was 0.16 million MT/year in the optimal structure. This was in addition to the 0.75 million MT/year of pure carbon dioxide produced from the existing ammonia plant in the complex. A total

TABLE 7.3

Plants in the Optimal Structure from the Superstructure w/o CO_2 Use

Existing Plants in the Optimal Structure	New Plants in the Optimal Structure
Ammonia	Fermentation ethanol (corn)
Nitric acid	Algae oil production
Ammonium nitrate	Transesterification to FAME and glycerol
Urea	(soybean oil and algae)
UAN	Gasification to syngas (corn stover)
Methanol	Ethylene from dehydration of ethanol
Granular triple super phosphate (GTSP)	Propylene glycol from glycerol
MAP and DAP	
Contact process for sulfuric acid	
Wet process for phosphoric acid	
Acetic acid—conventional method	
Power generation	
Existing plants not in the optimal structure	New plants not in the optimal structure
Ethyl benzene	Fermentation ethanol (corn stover)
Styrene	Anaerobic digestion to acetic acid (corn stover)

TABLE 7.4

Carbon Dioxide Sources and Consumption in the Base Case and Optimal Structure from Superstructure w/o CO_2 Use (Million Metric Tons per Year)

	Base Case	Optimal Structure w/o CO_2 Use
Pure CO_2 produced by ammonia plant	0.75	0.75
Pure CO_2 produced by bioprocesses	na	0.16
Impure CO_2 emission from power plant	0.08	0.09
Pure CO_2 consumed by existing chemical plants in base case	0.14	0.14
Pure CO_2 vented to atmosphere	0.61	0.76

0.91 million MT/year of carbon dioxide was produced from the addition of the bioprocesses to the base case. Existing plants in the base case (urea, methanol, and acetic acid) utilize 0.14 million MT/year of the pure carbon dioxide. The remaining 0.76 million MT of pure carbon dioxide emission from the biochemical production complex was vented to the atmosphere. This was an increase of 0.15 million MT/year from the base case. The impure carbon dioxide emissions increase by 0.01 million MT/year for the addition of the bioprocesses.

The conventional acetic acid method was selected in the optimal structure w/o CO_2 use over the anaerobic digestion process for acetic acid. The existing ethyl benzene and styrene plants did not operate in the optimal structure.

A comparison of the sales and costs associated with the triple bottom line is shown in Table 7.5 for the base case and the optimal structure w/o CO_2 use.

TABLE 7.5

Sales and Costs Associated with the Triple Bottom
Line for the Base Case and Optimal Structure from
Superstructure w/o CO_2 Use (Million Dollars per Year)

	Base Case	Optimal Structure w/o CO_2 Use
Income from sales	2026	2147
Economic costs (Raw Materials and Utilities)	697	697
Raw material costs	685	671
Utility costs	12	26
Environmental cost (67% of raw material cost)	457	447
Sustainable credits (+)/costs (−)	−17.8	−18.4
Triple bottom line	**854**	**984**

The triple bottom line increased from $854 to $984 million per year for the
optimal operation of the plants. This is approximately 15% increase from
the base case. The product sales increased from $2026 million per year to
$2147 million per year, an increase of 6%. The total economic costs remained
the same in the base case and the optimal structure. The raw material costs
decreased from the base case, because the ethyl benzene plant with high
raw material costs was excluded in the optimal structure. The utility costs
increased from the base case due to the inclusion of bioprocesses in the optimal structure. The sustainable costs to the society increased from the base
case as more carbon dioxide was vented to the atmosphere.

Thus, it can be concluded that the integration of bioprocesses without CO_2
use increases the triple bottom line profit in the optimal structure. The triple
bottom line increases by 15% from the base case. The pure carbon dioxide emissions increase from the base case by 25% as additional corn ethanol plants producing carbon dioxide are included in the optimal structure. This increased
the sustainable costs from the base case to the optimal structure by 3%. Costly
ethyl benzene process was excluded in the optimal structure, and this reduced
raw material costs from the base case to the optimal structure by 2%.

7.3 Case Study II—Parametric Study of Sustainable Costs and Credits

This case study is to determine the effect of sustainable credits and costs on
the triple bottom line. Costs and credits for carbon dioxide were based on
a detailed literature review to construct cases. The triple bottom line, the
sustainability credits and costs, and the carbon dioxide emissions, sources,

and sequestration methods for the optimal solution were compared with the solution of the base case. A reference case was used where the carbon dioxide emission costs and utilization credits were zero.

7.3.1 Carbon Dioxide Costs and Credits

The total cost assessment methodology discusses the sustainable costs to society (Constable et al., 1999). Sustainable costs are costs to society for emissions within regulations. The sustainable cost for carbon dioxide in the report was $3.25 per MT of carbon dioxide emission. This cost was based on the willingness-to-pay of consumers to avoid the occurrence of pollution. Currently, there are no credits for avoiding emissions of greenhouse gases (GHGs), and a hypothetical but conservative credit of twice the sustainable costs was used as shown in Table 6.6. Sustainable credits are the credits that a company may receive for avoiding emissions.

Costs per metric ton of carbon dioxide for the cap-and-trade system, carbon tax, and carbon dioxide sequestration are reported along with a discussion of the perspectives of the government and major companies. Then these costs are used to determine the effect on the optimal structure.

Historically, there had been no governmental regulations on carbon dioxide emissions. However, the increased concerns due to global warming, climate change, and pollution reduction programs prompted the U.S. House of Representatives to pass the American Clean Energy and Security Act of 2009 (ACES, 2010). This bill, if passed, would introduce a cap-and-trade program aimed at reducing GHGs to address climate change. The Environmental Protection Agency (EPA) issued the Mandatory Reporting of Greenhouse Gases Rule in December 2009 (EPA, 2010). The rule requires reporting of GHG emissions from large sources and suppliers in the United States, and it is intended to collect accurate and timely emissions data to inform future policy decisions. Under the rule, suppliers of fossil fuels or industrial GHGs, manufacturers of vehicles and engines, and facilities that emit 25,000 MT or more per year of GHG emissions are required to submit annual reports to the EPA. With these government initiatives and increased global concerns for GHG emissions, alternate pathways for the production of chemicals are required. Renewable resources from biomass give alternative options for producing chemicals.

Companies are examining several options for the reduction of carbon dioxide, and the following section discusses the opinion of the major companies (C&E News, 2009a). Most chemical companies anticipate a limit on carbon emissions. However, there is no unanimous agreement on whether a tax, a cap-and-trade program, or some other mechanism is the most efficient way to limit emissions of carbon dioxide and other man-made GHGs. The other mechanisms may include carbon dioxide sequestration, planting of forests that are carbon dioxide sinks, production of chemicals from carbon dioxide, large-scale production of algae and use of algae as raw material in biochemical processes, etc.

The cap-and-trade program works in the following way. The industry emission limits are set to a certain annual amount. Companies polluting more than the emission limit are required to buy emission allowances from those who pollute less. The cap-and-trade program is effective in the European Union (EU), where ETS (European carbon trading system) carbon allowances varied from $12 to $37 per MT of CO_2. DuPont believes that the cap-and-trade system is more effectively designed to achieve emission reduction targets; a carbon tax cannot guarantee that. The firm also wants cap-and-trade program to take into account already reduced emissions from previous years, like DuPont achieved 72% reductions between 1990 and 2003. DuPont believes that the cap-and-trade should take into account the early achievements by putting a greater reduction burden on those who have yet to make emission cutbacks. Dow chemical favors the cap-and-trade system, but they also believe that any system needs to take into account that hydrocarbons need to be transformed into petrochemicals such as ethylene and polyethylene. Dow proposes that only the energy consumed in the system needs to be considered for carbon emissions. An initially aggressive cap that encourages users to switch from coal to natural gas would drive chemical manufacturers offshore. Shell says that a harmonized global emission trading structure is needed that does not disadvantage any one part of the world. Shell believes that cross-border carbon leakage is going to happen in the absence of a unanimous rule for CO_2.

A tax-based program would require firms to pay a toll on emissions. Exxon Mobil prefers a carbon tax because they believe that cap-and-trade program will bring costly brokerage system subject to volatile price swings and high monitoring costs. Tax is a more direct, transparent, and effective approach, and the user pays for the carbon dioxide emissions associated with a product. It also allows firms to make more accurate decisions to budget the expenses of a firm. The reluctance to pay the higher price for a product due to inclusion of a carbon tax would automatically reduce the demand for a product. The company will produce less of the product and the associated carbon dioxide.

Some firms are developing technologies that capture emissions and sequester them underground. BASF is testing an amine-based solvent to capture CO_2 from coal-burning power plant emissions for injection into wells, and Air Products is adapting their air separation technology to capture CO_2 produced by power plants for sequestration. The costs for sequestration of carbon dioxide are high, and additional energy is required for compressing the gas to liquefy it at 2200 psi pressure.

In the United States, the Montgomery County in Maryland passed the nation's first county-level carbon tax in May 2010 (Faden, 2010). The legislation requires the payment of $5 per ton of carbon dioxide from any stationary source emitting more than 1 million tons of carbon dioxide per year. An 850 MW coal-fired power plant in the state emits carbon dioxide in that range. In May 2008, the Bay Area Air Quality Management District, which covers nine counties in the San Francisco Bay Area, passed a carbon tax on businesses of 4.4 cents per ton of CO_2. The cap-and-trade program in ETS sells carbon allowances for \$12–\$37 per MT of CO_2. Carbon dioxide sequestration costs reported by Katzer (2008) range from \$22 to \$46 per MT of CO_2 avoided. Carbon dioxide sequestration cost considered in NAS (2009) was \$50 per MT of carbon dioxide. An analysis for carbon dioxide sequestration was performed in HYSYS and cost estimation was done in ICARUS in this research. The cost per ton of carbon dioxide sequestered from the simulation was \$30 per MT (Pike and Knopf, 2010).

Banholzer et al. (2008) from the Dow Chemical Company published a comprehensive article on the possibility of bioethanol-based ethylene production. A scenario analysis predicted market penetration of bio-derived ethylene to be around 12% in 2020, provided advances in biotechnology occurs and capital budget is allocated for the bioethylene plant. They also suggested that there exists a trade-off between variable and capital costs. The capital cost and variable cost for competing processes for ethylene production are shown in Figure 7.2. The diagonal lines represent approximate economic-cost-of-production equivalency curves.

As an example, the capital cost versus variable cost of 1000 Gg of olefin production from coal technology required the highest capital investment of \$3250 million but low variable costs. Thus, if cost variability minimization over the life cycle of a project is objective, the coal-to-olefins route is an appropriate choice.

Banholzer et al. (2008) further discuss scenarios where a carbon tax varying between \$25 and \$125 per MT is implemented. They assume that a carbon tax is implemented for a nonrenewable-feedstock-based ethylene production

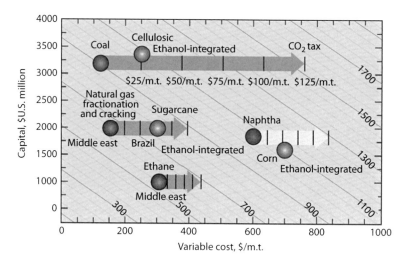

FIGURE 7.2

Contour plot of production cost plus return on investment as a function of capital and variable costs. (Based on 1000 Gg/year of olefin production) (From Banholzer, W.F. et al., *Chem. Eng. Prog.*, 104(3), S7, 2008.)

process, while there is no taxation for cellulosic ethanol, sugarcane ethanol, and corn ethanol plants emitting carbon dioxide. As seen from Figure 7.2, the cellulosic ethanol to bioethylene production becomes comparable in variable costs with a coal-to-olefins plant for a carbon tax of $25 per MT of CO_2, and corn ethanol becomes comparable in variable costs to naphtha cracking at carbon tax of $50 per MT of CO_2.

The ExternE methodology (Bickel and Friedrich, 2005) developed by the EU gives a method to estimate external costs (may be considered as sustainable costs). The first step in the method measures damages to society that are not paid directly by the contributors to the cost. The second step assigns monetary value to the damage and the third step determines how these external costs can be charged to producers and consumers. The main aim of this methodology is to internalize external costs, by taxing the most damaging technologies or subsidizing the cleanest and healthiest ones, which can be the driving force for developing new and energy-efficient processes. The method used pollution reduction to Kyoto target and EU global warming reduction target to 2°C above preindustrial temperatures as standards to calculate avoidance costs. The report estimates costs of €9 per ton of CO_2 as the cost for global warming damage, €5–20 per ton of CO_2 to reach Kyoto standards, and €95 per ton of CO_2 to meet EU target of global warming reduction. This value was unlikely to be accepted by the consumers, and a value of €50 per ton of CO_2 was used in their analysis. These costs are based on the willingness to pay for global warming damage avoidance.

The costs associated with carbon dioxide are summarized in Table 7.6. A carbon tax of $5 per MT has been implemented in Maryland, while the

TABLE 7.6

Costs Associated with Carbon Dioxide for Sustainability Consideration

Type of CO_2 Cost Assessment	Cost ($/MT CO_2)	Acceptance Area	Source
Carbon tax	$5	Implemented in Maryland	Faden (2010)
	$0–$125	Assumed in scenario analysis	Banholzer et al. (2008)
Carbon cap-and-trade program	$12–$37	European carbon trading system	C&E News (2009a)
Carbon dioxide sequestration (geological)	$22–$46	Various advanced power generation technologies from coal	Katzer (2008)
	$50	Assumed in scenario analysis	NAS (2009)
Sustainable cost	0–$3.25 (1998 value)	Damage cost approach based on a willingness-to-pay to avoid adverse human health effects, agricultural effects, and materials damage	TCA methodology, Constable et al. (1999)
	€9–€50 (ca. $11–$64)	Damage cost approach based on a willingness-to-pay	ExternE Methodology (Bickel and Friedrich, 2005)

Dow Chemical Company assumed a carbon tax of \$0–\$125 per ton of CO_2. The carbon cap-and-trade program in the ETS sells carbon permits in the range of \$12–\$37 per ton of CO_2. Carbon dioxide sequestration costs of \$22–\$46 per ton of CO_2 have been reported for coal-power-generating processes. Sustainable costs of \$0–\$3.25 per ton of CO_2 were reported in the TCA methodology and \$11–\$64 per ton of CO_2 were reported in the ExternE methodology based on willingness to pay. Additional costs associated with carbon dioxide are reported in Appendix A.

Thus, costs per metric ton of carbon dioxide for the cap-and-trade system, carbon tax, and carbon dioxide sequestration can be considered for sustainable costs/credits. This is the amount that is paid by the companies to prevent the carbon dioxide from going to the atmosphere. If the carbon dioxide is unregulated, and the company does not pay for the emissions, the society has to pay the same amount for emissions. This is an assumption for developing the cases for sustainability cost and credit parameters used in the base case and the superstructure, as explained in the following section.

7.3.2 Developing the Case for Sustainability Analysis

The costs associated with carbon dioxide were discussed in Section 7.3.1 and given in Table 7.6. The terms associated with CO_2 sequestration will be discussed in this section to develop the case for sustainability.

If atmospheric carbon dioxide, a GHG, is captured and stored in plants by biological processes, it is known as biosequestration (Blanco, 2009). Algae production can be considered a biosequestration process, where carbon dioxide from various sources can be used for the growth of algae.

Chemical sequestration of carbon dioxide involves capturing the carbon dioxide from the air or other sources by means of a chemical reaction (CYPENV, 2010). Conventional chemical sequestration includes converting carbon dioxide to calcium carbonate and dumping it in old coal mines. New methods to convert carbon dioxide have been developed that may use high-purity carbon dioxide for the production of chemicals (Indala, 2004).

Physical sequestration methods involve pumping liquid carbon dioxide in deep saline aquifers. Geologic or terrestrial sequestration procedures involve capturing flue gases from processes, cleaning the carbon dioxide, and applying successive compression, cooling, and separation methods to liquefy carbon dioxide. The liquid is then pumped at around 2200 psi pressure to underground reservoirs, or used for enhanced oil recovery.

From Figure 6.3 it can be seen that in the optimal structure, 1.0 million MT per year of pure carbon dioxide is available in the integrated complex after utilization in the existing plants from the base case (methanol, urea). This means that additional methods to purify the gas are not necessary. There are three options for utilizing the carbon dioxide: geological sequestration, chemical sequestration, or biological sequestration. Among these three methods, the biological sequestration and chemical sequestration methods can use the carbon dioxide directly, while geological sequestration will require additional costs to compress the gas to a liquid, and then use a pump at 2200 psi pressure to store the liquid in underground reservoirs.

Thus, chemical or biological sequestration of carbon dioxide is a possible alternative to geological sequestration, and costly methods to sequester in geological sinks are avoided. This avoidance of geological storage can be considered as a credit for the production units that consume the carbon dioxide.

Table 7.6 shows that the costs for geological storage range from $22 to $47 per ton of CO_2 for advanced carbon capture systems (Katzer, 2008), and a most likely cost of $50 per MT is assumed in an analysis by NAS (2009). The costs of $25 and $50 per MT of CO_2 were used as lower and upper limits for carbon dioxide emission avoidance (or utilization) credits in the economic model to obtain the optimal structure.

The imposition of a carbon tax on a company would mean that if a company emits beyond a certain amount of carbon dioxide, a premium needs to be paid for that emission. The amount of CO_2 has not been decided yet by the government, but the EPA requires mandatory reporting of GHG emissions of more than 25,000 MT per year. The carbon tax is more likely to be an environmental cost to the company if implemented, where noncompliance results in the payment of fines. However, as the tax money will be spent for a sustainable society, the carbon tax may be considered as a sustainable cost that the company has to pay for damages to the society. Table 7.6 shows that a carbon tax of $5 per MT of carbon dioxide has been implemented in Maryland. This is used as the lower limit for carbon dioxide emission cost in the economic model. The carbon tax of $125 per MT of carbon dioxide is considered as the upper limit of carbon tax (based on Banholzer et al., 2008). Two intermediate costs of $25 and $75 per MT of CO_2 are used for sustainable costs in the economic model.

Apart from the costs and credits explained earlier, costs and credits of $0 per MT of CO_2 were used to obtain the optimal solution to provide a reference scale for the results. This is the current industrial scenario, where there are no costs for carbon dioxide emission and no credits are given for carbon dioxide utilization (apart from profit from sales of products for companies using CO_2 as raw material).

7.3.3 Effect of Sustainable Costs and Credits on the Triple Bottom Line

The following cases were used to study the effect of sustainable costs and credits on the triple bottom line for the base case and optimal solution. The following carbon dioxide prices (CP) or costs (based on carbon tax) were used as parameters: $0, $5, $25, $75, and $125 denoted by CP0, CP5, CP25, CP75, and CP125, respectively. The carbon dioxide credits (CCs) (avoidance of geological sequestration) were used as parameters: $0, $25, and $50 denoted by CC0, CC25, and CC50, respectively. Combinations of the aforementioned costs and credits are used to obtain the optimal structure.

TABLE 7.7

Sales and Costs Associated with the Triple Bottom Line for the Optimal Structure for Varying Sustainability Costs and Credits (Million Dollars per Year)

Optimal Solution	CC0– CP0	CC25– CP5	CC25– CP25	CC25– CP75	CC25– CP125	CC50– CP5	CC50– CP25	CC50– CP75	CC50– CP125
CO_2 credit (CC) ($/ton CO_2)	0	25	25	25	25	50	50	50	50
CO_2 price (CP) ($/ton CO_2)	0	5	25	75	125	5	25	75	125
Income from sales	2489	2489	2471	2471	2489	2471	2471	2489	2489
Economic costs (Raw materials and utilities)	516	516	502	502	516	502	502	516	516
Raw material costs	470	470	459	459	470	459	459	470	470
Utility costs	46	46	43	43	46	43	43	46	46
Environmental cost (67% of raw material cost)	313	313	306	306	313	306	306	313	313
Sustainable credits (+)/ costs (−)	−16	9.5	7	1	−8	36	34	26	19
Triple bottom line	1644	1670	1670	1664	1652	1700	1697	1686	1679

The results for sales and costs associated with the triple bottom line for the optimal structure from the superstructure are given in Table 7.7. The emissions, sources, and sequestration of carbon dioxide in the optimal structure are given in Table 7.8. The results for sales and costs associated with the triple bottom line for the base case are given in Table 7.9. The emissions, sources, and sequestration of carbon dioxide in the base case solution are given in Table 7.10.

Figures 7.3 and 7.4 show the triple bottom line for the optimal structure and base case, respectively. The solid lines in the figures denote the reference solution for optimal structure and base case with zero credits for CO_2 consumption and zero cost for CO_2 emission (CC0–CP0). It can be seen that a triple bottom line profit of $855 million per year was obtained in the base case and $1644 million per year was obtained in the optimal solution.

From Figure 7.3 for the optimal solution, it can be seen that the values for triple bottom line for all the scenarios of carbon dioxide cost and credits lie above the zero cost and zero credit reference (CC0–CP0) of $1644 million per year. This is because carbon dioxide consumption occurs in the

TABLE 7.8

Emissions, Sources, and Sequestration for Carbon Dioxide in the Optimal
Structure (Million Metric Tons per Year)

Optimal Solution	CC0– CP0	CC25– CP5	CC25– CP25	CC25– CP75	CC25– CP125	CC50– CP5	CC50– CP25	CC50– CP75	CC50– CP125
Pure CO_2 sources	1.07	1.07	1.07	1.07	1.07	1.07	1.07	1.07	1.07
Pure CO_2 seq. in existing process	0.07	0.07	0.05	0.05	0.07	0.05	0.05	0.07	0.07
Pure CO_2 seq. in algae and new chemicals	1.00	1.00	1.02	1.02	1.00	1.02	1.02	1.00	1.00
Pure CO_2 emission	0	0	0	0	0	0	0	0	0
Impure CO_2 emission (from power plant)	0.15	0.15	0.12	0.12	0.15	0.12	0.12	0.15	0.15

TABLE 7.9

Sales and Costs Associated with the Triple Bottom Line for the Base Case
for Varying Sustainability Costs and Credits (Million Dollars per Year)

Base-Case Solution	CC0– CP0	CC25– CP5	CC25– CP25	CC25– CP75	CC25– CP125	CC50– CP5	CC50– CP50	CC50– CP75	CC50– CP125
CO_2 credit (CC) ($/ton CO_2)	0	25	25	25	25	50	50	50	50
CO_2 price (CP) ($/ton CO_2)	0	5	25	75	125	5	25	75	125
Income from sales	2026	2026	2026	2024	2001	2026	2026	2026	2001
Economic costs (raw materials and utilities)	697	697	697	696	686	697	697	697	686
Raw material costs	685	685	685	685	677	685	685	685	677
Utility costs	12	12	12	11	9	12	12	12	9
Environmental cost (67% of raw material cost)	457	457	457	457	451	457	457	457	451
Sustainable credits (+)/ costs (−)	−16	−16	−31	−65	−91	−13	−27	−62	−87
Triple bottom line	**855**	**855**	**841**	**806**	**774**	**859**	**845**	**809**	**777**

TABLE 7.10

Emissions, Sources, and Sequestration for Carbon Dioxide in the Base Case
(Million Metric Tons per Year)

Base-Case Solution	CC0–CP0	CC25–CP5	CC25–CP25	CC25–CP75	CC25–CP125	CC50–CP5	CC50–CP50	CC50–CP75	CC50–CP125
Pure CO_2 sources	0.75	0.75	0.75	0.75	0.69	0.75	0.75	0.75	0.69
Pure CO_2 seq. in existing process	0.14	0.14	0.14	0.14	0.14	0.14	0.14	0.14	0.14
Pure CO_2 emission	0.61	0.61	0.61	0.61	0.55	0.61	0.61	0.61	0.55
Impure CO_2 emission	0.09	0.09	0.09	0.08	0.07	0.09	0.09	0.09	0.07

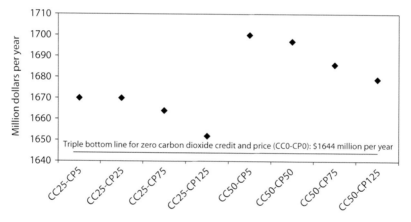

FIGURE 7.3
Optimal structure triple bottom line for CCs of $0, $25, and $50 per MT of CO_2 and carbon
dioxide costs of $0, $5, $25, $75, and $125 per MT of CO_2.

biosequestration process for algae production and in new chemical sequestration processes. The triple bottom line is highest of $1700 million per year
when CCs are highest of $50 per MT and carbon costs are lowest of $5 per
MT (CC50–CP5). The triple bottom line is similar, $1670 million per year, for
carbon credit of $25 per MT of CO_2 sequestered and carbon costs of $5 and
$25 per ton of CO_2 emitted (CC25–CP5 and CC25–CP25). The triple bottom
line is lowest, $1652 million per year, when the carbon costs for emission is
highest of $125 per MT of CO_2 emitted (CC25–CP125). The triple bottom line
for the highest carbon cost of $125 million per MT of CO_2 and carbon credit
of $50 per MT of CO_2 is slightly higher (by about 0.5%) than the low carbon
credit and low carbon tax cases (CC25–CP5, CC25–CP25). This suggests that
at higher credits for CO_2 utilization, the triple bottom line is going to be high
even at high costs of CO_2 emission.

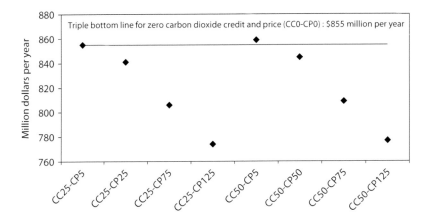

FIGURE 7.4
Base-case triple bottom line for CC of $0, $25, and $50 per MT of CO_2 and carbon dioxide costs of $0, $5, $25, $75, and $125 per MT of CO_2.

From Figure 7.4 for the solution of the base case, it can be seen that the triple bottom line for a carbon credit of $25 per ton of CO_2 consumed and carbon cost of $5 per ton of CO_2 emitted (CC25–CP5) is equivalent to a zero credit–zero cost scenario (CC0–CP0). The triple bottom line for a carbon credit of $50 per ton of CO_2 consumed and carbon dioxide cost of $5 per ton of CO_2 emitted (CC50–CP5) is $859 million per year, about $4 million higher than the zero cost and credit scenario. All the other values for the triple bottom line lie below the CC0–CP0 value of $855 million per year and decrease as the carbon dioxide costs increase. The reason for this decrease is the inclusion of sustainable costs in the triple bottom line for emission of carbon dioxide from the base case. This suggests that without utilization of carbon dioxide, the triple bottom line is going to decrease for any amount of carbon dioxide cost included.

Figures 7.5 and 7.6 show only the sustainable costs/credits for the optimal structure and the base case, respectively, for CCs of $0, $25, and $50 per MT of CO_2 consumed and CP (costs) of $0, $5, $25, $75, and $125 per MT of CO_2 emitted. The solid lines in the figures denote the reference sustainable cost with zero credits for CO_2 consumption and zero cost for CO_2 emission (CC0–CP0). It can be seen that the sustainable cost (cost denoted by negative sign and credit denoted by positive sign in Tables 7.7 and 7.9) was $16 million per year for both the base case and the optimal solution.

From Figure 7.5 it can be seen that the optimal structure has a sustainable cost of $8 million per year for the high CP and low CC case (CC25–CP125). There are sustainable credits in all the other cases with the maximum credits of $36 million per year received for high CC and low CP case (CC50–CP5).

From Figure 7.6 it can be seen that the base case always has a sustainable cost denoted by negative values on the graph. The maximum sustainable cost is $91 million per year for low CC and high price (CC25–CP125).

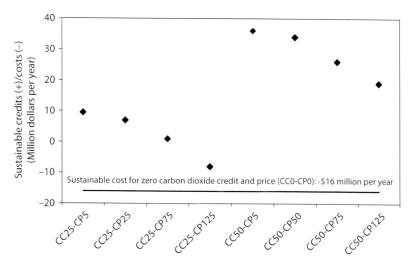

FIGURE 7.5
Optimal structure sustainable costs/credits for CCs of $0, $25, and $50 per MT of CO_2 and carbon dioxide costs of $0, $5, $25, $75, and $125 per MT of CO_2.

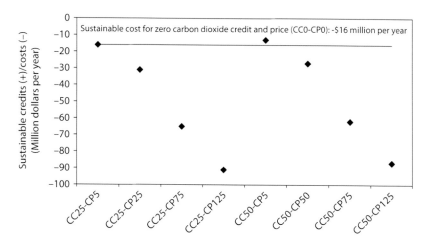

FIGURE 7.6
Base-case sustainable costs/credits for CCs of $0, $25, and $50 per MT of CO_2 and carbon dioxide costs of $0, $5, $25, $75, and $125 per MT of CO_2.

The sustainable costs are lowest of $13 million (lower than the reference case $16 million for CC0–CP0) when the CCs are high and CP is low (CC50–CP5).

The emissions, sources, and sequestration of carbon dioxide for the optimal structure and the base case are shown in Figures 7.7 and 7.8, respectively. Figure 7.7 shows that in the optimal structure, pure carbon dioxide emissions were reduced to zero for all the cases. This was achieved by the utilization of carbon dioxide in algae oil production process and for new carbon

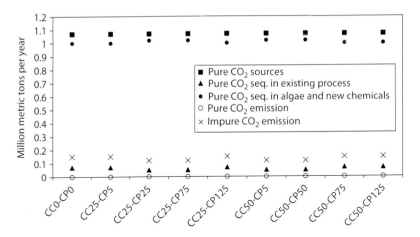

FIGURE 7.7
Optimal structure emissions, sources, and sequestration of carbon dioxide for credits of $0, $25, and $50 per MT of CO_2 and carbon dioxide costs of $0, $5, $25, $75, and $125 per MT of CO_2.

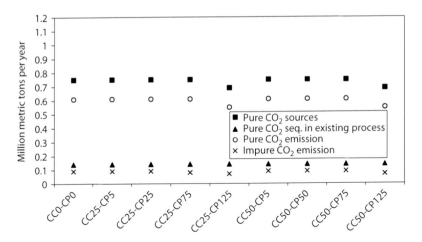

FIGURE 7.8
Base-case emissions, sources, and sequestration of carbon dioxide for credits of $0, $25, and $50 per MT of CO_2 and carbon dioxide costs of $0, $5, $25, $75, and $125 per MT of CO_2.

dioxide–consuming processes. The pure carbon dioxide flow rate increased to 1.07 million MT/year as the sources of carbon dioxide were increased in the bioprocesses. It may be noted here that the pure carbon dioxide sources increased in the optimal structure from the base case (0.75 million MT/year). Thus, it can be concluded that even for higher amounts of carbon dioxide, the biological and chemical sequestration processes were successful in consuming all the carbon dioxide from the chemical complex. The impure carbon dioxide emissions increased to 0.15 million MT/year due to the increase in the energy requirement from the power plant. Overall, it was a profitable process to operate the plants at the optimal structure, as carbon dioxide emissions from pure sources were lowered to zero level and credits were received for this achievement.

From Figure 7.8 for the base case, it can be seen that the pure CO_2 emissions decrease due to a decrease in the pure CO_2 source for high carbon price of \$125 per MT of CO_2 (CC25–CP125 andCC50–CP125). The flow of carbon dioxide from the ammonia plant was decreased as the ammonia plant operated at a lower capacity than in the other cases. This was an important observation and it can be inferred that for very high values of carbon costs (\$125/MT CO_2), the flow rates of production processes are affected to meet the sustainability criteria in the triple bottom line. This case was true even when carbon credits were given (\$25 or \$50/ton CO_2). The carbon credits could not be utilized because the carbon dioxide consumption processes in the base case were limited to methanol, urea, and acetic acid plants, and the surplus pure carbon dioxide was emitted to the atmosphere.

The superstructure w/o CO_2 use discussed in Case Study I was used for the parametric study of carbon dioxide costs and credits. The results are given in Table 7.11 for the sales and costs associated with the triple bottom line. Table 7.12 gives the emissions, sources, and sequestration of CO_2.

Figure 7.9 shows the triple bottom line for changes in credits and costs of CO_2. The triple bottom line is marginally higher than the zero CC and CP (CC0–CP0) case for the low carbon dioxide price of \$5 per MT. All the other cases have decreasing triple bottom line with increasing CP. The minimum triple bottom line was \$884 million per year for low CC of \$25 per MT and high carbon cost of \$125 per MT (CC25–CP125).

Figure 7.10 shows the sustainable costs/credits for varying carbon dioxide costs and credits. The sustainable cost is highest (\$112 million/year) for the low CC and high carbon dioxide cost case (CC25–CP125). This is \$21 million higher than the base case of existing plants for the same CC and CP case. This shows that at high carbon prices, the sustainable costs for emitting pure carbon dioxide from the integrated complex without carbon dioxide use will increase the sustainable costs for the complex.

Figure 7.11 shows the emissions, sources, and sequestration of CO_2 for the optimal solution without CO_2 use. The carbon dioxide produced in the optimal solution was 0.91 million MT per year and an increase of 0.16 million MT from the base case of existing plants except when the carbon dioxide cost was high at \$125 per ton. The ammonia plant produced less ammonia in this case and the CO_2 produced from the ammonia plant was reduced. Thus, the total pure carbon dioxide emission was decreased to 0.85 million MT in the optimal structure without CO_2 use for a high carbon dioxide cost of \$125 for CO_2 emission.

In summary, the optimal solution from the superstructure shows that carbon dioxide emissions are reduced to zero when processes consuming carbon dioxide are available in the complex. These can be algae oil production processes (biosequestration) or chemical processes. The optimal solution

TABLE 7.11

Sales and Costs Associated with the Triple Bottom Line for the Optimal Solution from Superstructure without CO_2 Use for Varying Sustainability Costs and Credits (Million Dollars per Year)

Optimal Solution w/o CO_2 Use	CC0–CP0	CC25–CP5	CC25–CP25	CC25–CP75	CC25–CP125	CC50–CP5	CC50–CP50	CC50–CP75	CC50–CP125
CO_2 credit (CC) ($/ton CO_2)	0	25	25	25	25	50	50	50	50
CO_2 price (CP) ($/ton CO_2)	0	5	25	75	125	5	25	75	125
Income from sales	2147	2147	2147	2146	2122	2147	2147	2147	2122
Economic costs (raw materials and utilities)	697	697	697	696	685	697	697	697	685
Raw material costs	671	671	671	671	663	671	671	671	663
Utility costs	26	26	26	25	23	26	26	26	23
Environmental cost (67% of raw material cost)	447	447	447	447	441	447	447	447	441
Sustainable credits (+)/ costs (−)	−18	−17	−34	−78	−112	−14	−31	−75	−108
Triple bottom line	**984**	**985**	**967**	**924**	**884**	**988**	**971**	**927**	**887**

TABLE 7.12

Emissions, Sources, and Sequestration for Carbon Dioxide in the Optimal Solution from Superstructure without CO_2 Use (Million Metric Tons per Year)

Optimal Solution w/o CO_2 Use	CC0–CP0	CC25–CP5	CC25–CP25	CC25–CP75	CC25–CP125	CC50–CP5	CC50–CP50	CC50–CP75	CC50–CP125
Pure CO_2 sources	0.91	0.91	0.91	0.91	0.85	0.91	0.91	0.91	0.85
Pure CO_2 seq. in existing process	0.14	0.14	0.14	0.14	0.14	0.14	0.14	0.14	0.14
Pure CO_2 emission	0.77	0.77	0.77	0.77	0.71	0.77	0.77	0.77	0.71
Impure CO_2 emission	0.10	0.098	0.098	0.087	0.074	0.098	0.098	0.098	0.074

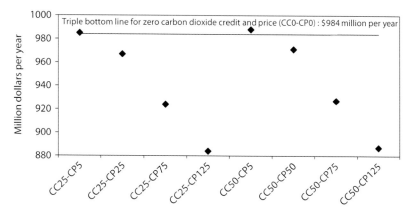

FIGURE 7.9
Triple bottom line of optimal structure w/o CO_2 use for CCs of $0, $25, and $50 per MT of CO_2 and carbon dioxide costs of $0, $5, $25, $75, and $125 per MT of CO_2.

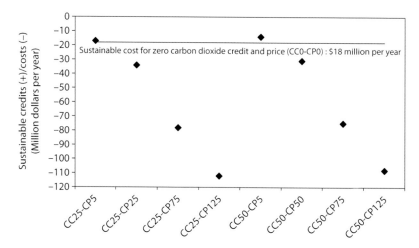

FIGURE 7.10
Sustainable costs/credits in optimal structure w/o CO_2 use for CCs of $0, $25, and $50 per MT of CO_2 and carbon dioxide costs of $0, $5, $25, $75, and $125 per MT of CO_2.

shows that the complex has sustainable credits for carbon dioxide utilization except when sustainable credits are low ($25/MT CO_2 for consumption) and sustainable costs are high ($125/MT CO_2 emitted). The sustainable cost in this case (CC25–CP125) is $8 million per year.

The base case always has a sustainable cost for emission of CO_2. The highest sustainable cost of the base case complex is $91 million per year for sustainable credits of $25 per MT of CO_2 and sustainable costs of $125 per MT of CO_2 (CC25–CP125). The optimal flow rate of ammonia is reduced in the complex to meet the sustainability criteria in the triple bottom line when carbon dioxide emission costs are high.

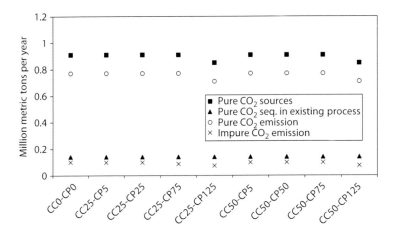

FIGURE 7.11
Emissions, sources, and sequestration of carbon dioxide in optimal structure w/o CO_2 use for credits of $0, $25, and $50 per MT of CO_2 and costs of $0, $5, $25, $75, and $125 per MT of CO_2.

The optimal solution w/o CO_2 use always has a sustainable cost. This is due to emission of pure CO_2 from the optimal structure, which does not have carbon dioxide utilization in biosequestration or chemical sequestration processes. The highest sustainable cost of the optimal solution w/o CO_2 use is $112 million per year for sustainable credits of $25 per MT of CO_2 and sustainable costs of $125 per MT of CO_2 (CC25–CP125). The optimal flow rate of ammonia is reduced in the complex to meet the sustainability criteria in the triple bottom line when carbon dioxide emission costs are high.

7.3.4 Cross-Price Elasticity of Demand for Ammonia

Price elasticity calculations were performed for the demand in chemicals for changes in the price of carbon dioxide. Price elasticity has been discussed in detail in Appendix D with calculations for price elasticity of demand and supply for corn, bioethanol, and ethylene. In this section, cross-price elasticity theory is applied to calculate the PED of ammonia with respect to changes in the price of carbon dioxide.

The change in the flow rate of ammonia in the base case of plants is used to demonstrate how the cross-PED for ammonia can be calculated for changes in carbon dioxide prices. Table 7.13 gives the base case flow rates for carbon dioxide and ammonia obtained from the solution of the base case corresponding to the increasing prices of carbon dioxide. The optimal flow rate can be considered as the demand for ammonia in the chemicals market, guided by the selling price of ammonia. The triple bottom line optimization meets the economic, environmental, and societal criteria for maximum profit. Thus, the selling price of ammonia is not the only component in the profit equation, but has to meet all the three aspects for sustainability.

TABLE 7.13

Cross-PED for Ammonia to Changes in CP (CO_2 Costs)
(CC25–CP25/CC25–CP75/CC25–CP125)

Carbon Dioxide Price	Flow Rate of CO_2 from Ammonia Process	Demand for Ammonia	Change in Price ($\Delta P/P*100$)	Change in Demand for Ammonia ($\Delta Q/Q*100$)	Cross-Price Elasticity of Demand
25	0.753	0.658	−200%	0%	0
75	0.753	0.658	−67%	8.2%	−0.122
125	0.691	0.604			

The optimal flow of ammonia from the base case (0.658 million MT/year) remains the same when CP changes from $25 to $75 per MT of CO_2. This means that the triple bottom line profit meets the sustainability criteria for these costs of CO_2 emission. When the CP changes from $75 to $125 per MT of CO_2 the demand for ammonia reduces from 0.658 million MT/year to 0.604 million MT/year. Thus, a change of 8.2% in demand is caused for ammonia. This means that to meet the sustainability criteria in the triple bottom line, the flow rate of carbon dioxide needed a reduction, and this was achieved by reducing the flow rate of ammonia (source of CO_2 generation). The cross-PED of ammonia with respect to price changes in CO_2 shows that for price changes from $25 to $75 per MT of CO_2, the demand for ammonia is perfectly inelastic (having a value of zero). When the price changes from $75 to $125 per ton of CO_2, it can be seen that the demand for ammonia decreases. The cross-PED for ammonia is −0.122. This means that the demand of ammonia is not perfectly price inelastic and changes in demand occur. The value of −0.122 suggests that the demand for ammonia is relatively price inelastic for changes in the price of CO_2.

In summary, it can be concluded that the triple bottom line for the optimal structure was always greater than the zero carbon dioxide costs and credits scenario. The optimal solution gives sustainable credits for CO_2 utilization in all the cases, except when the CC is low ($25/ton CO_2) and carbon dioxide costs are high ($125/ton CO_2). The pure CO_2 emission is zero for all the cases as the carbon dioxide is consumed in algae and other chemical processes from CO_2.

7.4 Case Study III—Parametric Study of Algae Oil Production Costs

This case study is for the parametric study of algae oil production costs. In this section, algae oil production costs are incorporated in the utility costs of the superstructure from Chapter 6. The changes to the triple bottom line of the optimal structure for various costs of algae production are evaluated in this section.

Currently, for large-scale production of algae oil, there is a significant cost for drying and separation. The superstructure in Chapter 5 considered the algae strain that secreted oil, for example, *Botryococcus braunii*, which has typically lower utility costs for algae drying and oil separation. Also, new methods for algae oil separation have been developed where algae cells are ruptured and oil is liberated without the need for dewatering or solvents (Ondrey, 2009). This process, developed by Origin Oil Inc., has reduced energy costs by 90% and substantial savings have been made to capital cost for oil extraction.

These processes for low-cost algae oil production are not commercialized yet. Thus, it is necessary to study the effect of variations in costs for algae oil production, and how it affects the triple bottom line. Pokoo-Aikins et al. (2010) give the algae oil costs based on oil content and discuss the performance of an algae production plant. A "high performance case" considered low cost of electricity ($0.05/kWh), high production rate (100 ton/day plant), and the use of heat integration in drying using hot flue gases. A "low performance case" assumed high cost of electricity ($20/kWh), low production (1 ton/day), and no heat integration for drying. Oil contents of 30% and 50% (dry basis oil in algae) were considered. The costs for producing algae oil per pound of algae oil given by Pokoo-Aikins et al. (2010) in Table 7.14 were used as price parameters for utility costs in the superstructure.

From Table 7.14, it can be seen that algae oil production costs are highest ($1.14/lb) for low performance (LP) and low oil content of algae (30%). Algae oil production cost was the minimum ($0.07/lb) for high performance (HP) algae oil plant and high oil content (50%). The costs from Pokoo-Aikins et al. (2010) for the LP, HP, and average performance (AP) plant operations and for oil content of 30% and 50% were used to study the effect of algae oil production costs on the triple bottom line. The oil production technologies and oil contents were used to construct the cases in this section as given in the following.

TABLE 7.14

Costs for Producing Algae Oil

	Oil Content			Oil Content		
	30%	50%	Units	30%	50%	Units
Low performance	1.14	0.63	$/lb	2513	1389	$/ton
High performance	0.21	0.07	$/lb	463	154	$/ton
Average	0.68	0.35	$/lb	1500	772	$/ton

Source: Pokoo-Alkins, G. et al., *Clean Technol. Environ. Policy*, 12(3), 239, 2010.

Case studies:

30% algae oil content:

SU30: algae production cost equals zero for low-oil-content algae (30%) (superstructure)

LP30: low-oil-content algae (30%) and low plant performance at $2513 per ton of oil

HP30: low-oil-content algae (30%) and high plant performance at $463 per ton of oil

AP30: low-oil-content algae (30%) and average plant performance at $1500 per ton of oil

SO30: algae production cost for 30% oil content equal to soybean oil purchased ($616/ton)

50% algae oil content:

SU50: algae production cost equals zero for high-oil-content algae (50%)

LP50: high-oil-content algae (50%) and low plant performance at $1389 per ton of oil

HP50: high-oil-content algae (50%) and high plant performance at $154 per ton of oil

AP50: high-oil-content algae (50%) and average plant performance at $772 per ton of oil

SO50: algae production cost for 50% oil content equal to soybean oil purchased ($616/ton)

SU30 is the superstructure with zero production costs for algae oil and 30% oil content. Case 2 SU50 is a modification of the superstructure where oil production costs are zero and the algae yields 50% algae oil. LP30, HP30, and AP30 include the costs of LP, HP, and AP, respectively, for production of algae with 30% oil content. LP50, HP50, and AP50 include the costs of LP, HP, and AP production, respectively, of algae with 50% oil content. SO30 and SO50 are modifications to the superstructure, which considers algae oil production costs equivalent to soybean oil purchased prices of $616 per ton. A carbon dioxide cost of $3.25 per MT of CO_2 emission and CC of $6.50 per MT of CO_2 consumed was used for all the cases.

The superstructure was modified for the aforementioned parameters, and the results from the optimal structure are given in Table 7.15 for 30% oil content algae oil production and in Table 7.16 for 50% oil content algae oil production. The carbon dioxide emission from and consumption by existing and new plants are shown in Figure 7.12 for 30% algae oil content and Figure 7.13 for 50% algae oil content. Figures 7.14 and 7.15 give the costs associated with the triple bottom line for 30% and 50% algae oil content, respectively.

TABLE 7.15

Sales and Costs Associated with the Triple Bottom Line for Varying Algae Oil Content and Production and Costs (Million Dollars per Year)

	SU30	LP30	HP30	AP30	SO30	SU50	LP50	HP50	AP50	SO50
Algae oil production cost ($/ton)	0	2513	463	1500	616	0	1389	154	772	616
Income from sales	2490	2585	2592	2585	2592	2794	2585	2794	2744	2744
Economic costs (raw materials and utilities)	516	782	764	782	821	510	782	636	1013	916
Raw material costs	470	723	531	723	530	456	722	456	468	468
Utility costs	46	59	233	59	291	53	59	180	546	448
Environmental Cost (67% of raw material cost)	313	482	354	482	353	304	482	304	312	312
Sustainable credits (+)/ costs (−)	−10	−14	−10	−14	−10	−10	−14	−10	−10	−10
Triple bottom line	1650	1307	1464	1307	1406	1970	1307	1844	1408	1506

TABLE 7.16

Pure Carbon Dioxide Consumption for Various Configurations of Algae Oil Production and Costs (Million Metric Tons per Year)

	SU30	LP30	HP30	AP30	SO30	SU50	LP50	HP50	AP50	SO50
CO_2 produced by ammonia plant	0.75	0.75	0.75	0.75	0.75	0.75	0.75	0.75	0.75	0.75
CO_2 produced by bioprocesses	0.32	0.16	0.32	0.16	0.32	0.32	0.16	0.32	0.32	0.32
CO_2 consumed (existing chemical plants)	0.07	0.14	0.10	0.14	0.10	0.07	0.14	0.07	0.10	0.10
CO_2 consumed (new chemical plants and algae)	1.00	0.55	0.97	0.55	0.97	1.00	0.55	1.00	0.97	0.97
CO_2 consumed by algae	0.84	0	0.70	0	0.70	0.91	0	0.91	0.70	0.70
CO_2 vented to atmosphere	0	0.23	0	0.23	0	0	0.23	0	0	0

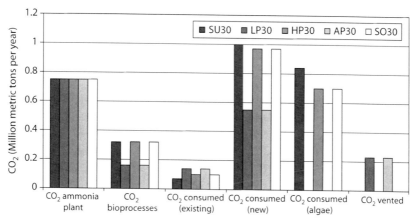

FIGURE 7.12
Carbon dioxide production and consumption for 30% algae oil content (million metric tons per year).

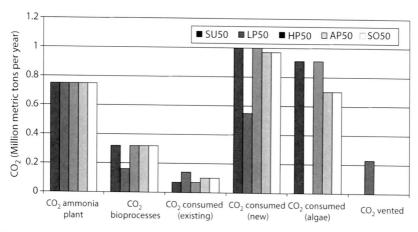

FIGURE 7.13
Carbon dioxide production and consumption for 50% algae oil content (million metric tons per year).

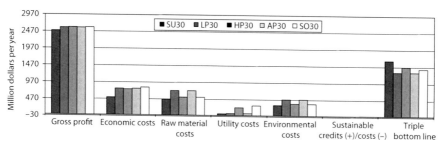

FIGURE 7.14
Costs associated with triple bottom line for 30% algae oil.

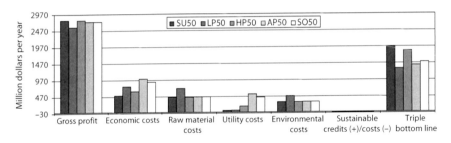

FIGURE 7.15
Costs associated with triple bottom line for 50% algae oil.

From Figure 7.12 for 30% algae oil content, 0.23 million MT of pure carbon dioxide emission occurs from the complex when an LP (LP30) or an AP (AP30) algae plant is selected. The highest carbon dioxide consumption of 1.0 million MT/per year occurs when there is no cost for algae oil production (SU30), as can be expected. There is considerable carbon dioxide consumption of 0.98 million MT/year when an HP algae oil plant operates (HP30) or when the plant operation costs are comparable to soybean oil prices (SO30).

From Figure 7.13 for 50% algae oil content, the pure carbon dioxide emission of 0.23 million MT/year occurs from the complex only when an LP (LP50) algae plant is selected. The maximum carbon dioxide consumption of 1.0 million MT/year occurs when there is no cost for algae oil production (SU50) and for HP algae oil production (HP50). There is considerable carbon dioxide consumption of 0.97 million MT/year when an AP algae oil plant operates (AP50) or when the plant operation costs are comparable to soybean oil prices (SO50).

Figure 7.14 shows the triple bottom line costs for 30% algae oil content plant operations. The triple bottom line for 30% oil production cases is maximum ($1650 million/year) in case of zero algae oil production costs (SU30), as can be expected. There is no difference in the triple bottom line ($1307 million/year) when an LP (LP30) or an AP (AP30) plant operates.

From Figure 7.14, the triple bottom line is $1464 million per year (11% lower than the optimal solution for zero algae oil production cost, SU30) for operating an HP (HP30) algae oil production plant. The triple bottom line is $1406 million per year (15% lower than the optimal solution for zero algae oil production cost, SU30) for algae oil production plant operating at soybean oil prices (SO30). From Table 7.15, the sustainable costs are minimum in these two cases (HP30 and SO30), comparable to the zero algae oil production cost (SU30). The triple bottom line for operating an LP (LP30) or AP (AP30) algae oil production plant is $1307 million per year (21% lower than the optimal solution for zero algae oil production cost, SU30). The sustainable costs are 40% higher in the case of LP or AP algae oil production as emission of pure CO_2 occurs in these cases.

Figure 7.15 shows the triple bottom line costs for 50% oil content algae oil production. The triple bottom line is $1970 million per year for 50% algae oil production (SU50) with zero oil production costs (20% higher than the optimal solution for 30% algae oil content and zero algae oil production cost, SU30).

From Figure 7.15, an LP (LP50) plant for 50% oil has the minimum triple bottom line of $1307 million per year. The triple bottom line for operating an HP algae oil production plant (HP50) is $1844 million per year (12% higher than the optimal solution for zero algae oil production cost, SU30) and for an algae oil production plant operating at soybean oil prices (SO50), the triple bottom line is $1506 million per year (9% lower than the optimal solution for zero algae oil production cost, SU30). The sustainable costs are comparable to the zero production cost case of SU30 in all of the 50% algae oil content cases (BBC50, HP50, AP50, and SO50) except for the low plant performance case (LP50). For the low plant performance case, the sustainable costs are 40% higher than the optimal solution for SU30.

Thus, for 30% algae oil content algae strain, it can be concluded that an HP (HP30) algae oil production plant or a plant with production costs comparable to soybean oil costs (SO30) can reduce the emission of pure CO_2 to zero. The production costs for 30% oil content algae oil included in these two cases reduces the triple bottom line by 15% and 21%, respectively, from the zero production cost case, SU30. So, the operation of an HP algae oil production plant is desirable for sustainability when zero oil production costs are not feasible. If HP production is not achievable, then the oil production costs should be targeted to be reduced to at least the soybean oil purchased costs. This would ensure the selection of algae oil production process and reduce the pure carbon dioxide emissions to zero.

For 50% algae oil content algae strain, it can be concluded that there is emission from the complex only for the LP plant. The high algae oil content and HP plant (HP50) gives a higher triple bottom line than the zero oil production cost and low-oil-content plant (SU30). So, the target should be obtaining the algae oil from the high-oil-content strain with high plant performance.

Figure 7.16 shows the algae oil production costs and the triple bottom line for the cases in Table 7.15. This figure shows the relative triple bottom line changes with respect to algae oil production costs. The triple bottom line for the base case of existing plants in the chemical production complex has been included in the figure for comparison with the optimal solutions for algae oil content of 30% and 50%. It is seen that the triple bottom line is lowest ($1307 million per year) for LP30, AP30, and LP50 cases.

Figure 7.17 shows the triple bottom line changes to algae oil production costs for 30% and 50% algae oil content. From the figure, it is seen that the triple bottom line is not going to be lower than $1307 million per year. The base-case triple bottom line of existing plants is $854 million per year. Thus, it can be concluded that the triple bottom line will be 53% higher than the base case for biomass process integration with carbon dioxide consumption processes.

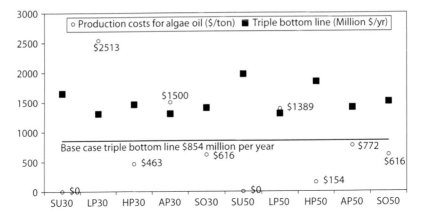

FIGURE 7.16
Algae oil production costs and corresponding triple bottom line.

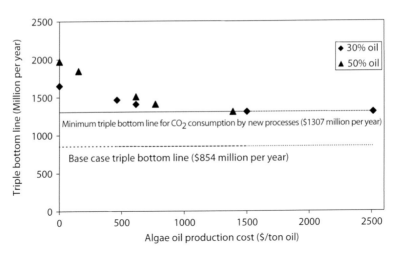

FIGURE 7.17
Triple bottom line (million dollars per year) vs. production costs for algae oil ($/ton of oil).

From Table 7.16, it is seen that the pure carbon dioxide emission is 0.23 million MT/year, and there is no consumption of carbon dioxide by the algae oil production unit. This suggests that the carbon dioxide produced from the complex is being consumed by existing processes and new chemical processes for carbon dioxide consumption. The base case of existing plants emitted of 0.61 million MT/year of pure CO_2 (Table 6.11). Thus, at least a reduction of 66% carbon dioxide is achieved when carbon dioxide–consuming processes are included in the optimal structure.

In summary, variations in algae oil production costs were evaluated for the base case and optimal solution from the superstructure. The triple bottom line for optimal solution is lowest ($1307 million/year) for an LP and AP 30% algae oil content (LP30 and AP30) production plant and an LP 50% algae oil production plant (LP50). In the absence of technology to convert carbon dioxide to algae oil at zero production costs, the best option is to operate HP, 50% oil content algae oil production plant that gives a triple bottom line of $1844 million per year.

The carbon dioxide vented to the atmosphere is zero for all the cases of algae oil production costs considered, except for three of the cases (LP30, AP30, and LP50). These are the LP and AP 30% algae oil content production plants and LP 50% algae oil production plant. In these cases, the pure carbon dioxide is consumed by other processes converting pure carbon dioxide to chemicals.

The sustainable costs for the LP30, AP30, and LP50 cases are 50% higher than the costs for total pure carbon dioxide consumption, but 22% lower than the base case of existing plants. This reduction in sustainable cost is achieved by the utilization of carbon dioxide in carbon dioxide–consuming processes.

7.5 Case Study IV—Multicriteria Optimization Using 30%-Oil-Content Algae and Sustainable Costs/Credits

The objective for multicriteria optimization is to find optimal solutions that maximize companies' profits and minimize costs to society. Companies' profits are sales minus economic and environmental costs as shown in Equation 7.1. Economic costs include raw material, utilities, labor, and other manufacturing costs. Environmental costs include permits, monitoring of emissions, fines, etc. The costs to society are measured by sustainable costs. These costs are from damage to the environment by emissions discharged within permitted regulations. Sustainable credits are awarded for reductions in emissions, and are similar to emissions trading credits.

In the superstructure from Chapter 6, the algae oil production costs were considered negligible. This was in anticipation that new technology will be developed on an industrial scale for algae oil production at minimal costs. Current state of technology is not developed fully to attain this case. Also, in the superstructure, CC of $6.50 and cost of $3.25 per MT of CO_2 was used for sustainable costs and credits.

Case Study III was a parametric study of algae oil content and production costs; 30% algae oil content seems most likely to be attained in the near future. The cost of HP and LP 30% algae oil content oil production costs were $463 and $2513 per MT of oil produced, respectively (Table 7.14, Pokoo-Aikins et al., 2010).

Case study II was a parametric study on CCs and costs (prices). The CC of $25 per MT was considered as a low credit for carbon dioxide consumption by a process. The carbon dioxide emission cost of $125 per MT was considered as the highest cost for carbon dioxide emission.

These parameters were used for multicriteria optimization of the superstructure. The costs for low carbon dioxide utilization credits ($25/MT) and high carbon dioxide emission costs ($125/MT) were used. The LP (LP30, production cost $2513/ MT oil) and HP (HP30, production cost $463/MT oil) algae oil production costs for 30% algae oil content were considered.

The HP algae oil production case with carbon dioxide costs and credits is denoted as CC25–CP125–HP30. The LP algae oil production case with carbon dioxide costs and credits is denoted as CC25–CP125–LP30. Multicriteria analysis was used to determine Pareto optimal sets for the two cases. This section demonstrates the multiobjective optimization results to maximize profit and sustainable credits with production costs consideration for LP and HP algae oil considered in the superstructure with CO_2 use.

The multicriteria optimization problem can be stated in terms of profit, P, and sustainable credits/costs, S, for theses two objectives in Equation 7.1.

$$\text{Max: } P = \sum \text{Product sales} - \sum \text{Economic costs} - \sum \text{Environmental costs}$$

$$S = \sum \text{Sustainable (Credits} - \text{Costs)} \qquad (7.1)$$

Subject to: Multiplant material and energy balances
     Product demand, raw material availability, plant capacities

Multicriteria optimization obtains solutions that are called efficient or Pareto optimal solutions. These are optimal points where attempting to improve the value of one objective would cause another objective to decrease. To locate Pareto optimal solutions, multicriteria optimization problems are converted to one with a single criterion by the parametric approach method, which is by applying weights to each objective and optimizing the sum of the weighted objectives. The multicriteria mixed-integer optimization problem becomes

$$\text{Max: } w_1 P + w_2 S \qquad (7.2)$$

Subject to: Multiplant material and energy balances
     Product demand, raw material availability, plant capacities

The multicriteria optimization was used to determine the Pareto optimal solutions for the weights using $w_1 + w_2 = 1$ given by Equation 7.2. The results for carbon dioxide consumption credit of $25 per MT of CO_2, carbon dioxide emission cost of $125 per MT of CO_2, and HP 30% oil content algae oil production (production cost $463/MTalgae oil) (CC25–CP125–HP30) are shown in Figure 7.18 and the values are given in Table 7.17. The results for

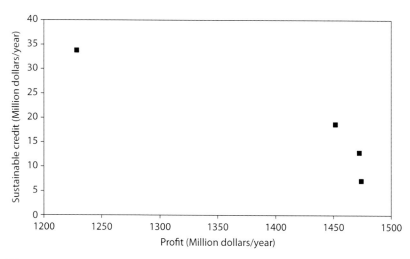

FIGURE 7.18
Pareto optimal solution generated by multicriteria optimization for HP algae oil production (30% oil content) with $25 carbon dioxide consumption credit and $125 carbon dioxide emission cost.

TABLE 7.17

Values of the Pareto Optimal Solutions for CC25–CP125–HP30 Shown in Figure 7.18

Profit (Million Dollars per Year)	Sustainable Credits (Million Dollars per Year)	Weight (w_1)
1228	33.8	0
1452	18.7	0.1
1452	18.7	0.2
1472	12.9	0.3
1472	12.9	0.7
1474	7.1	0.8
1474	7.1	0.9
1474	7.1	1

carbon dioxide consumption credit of $25 per MT of CO_2, carbon dioxide emission cost of $125 per MT of CO_2, and LP 30% oil content algae oil production (production cost $1253/MT algae oil) (CC25–CP125–LP30) are shown in Figure 7.19 and the values are given in Table 7.18.

From Figure 7.18 and Table 7.17 it can be seen that the sustainable credits decline and the company's profits increase as the weight, w_1, on the company's profit increases. For example, when $w_1 = 1$, the optimal solution is shown in Table 7.17 for $P = \$1474$ million per year and $S = \$7.0$ million per year.

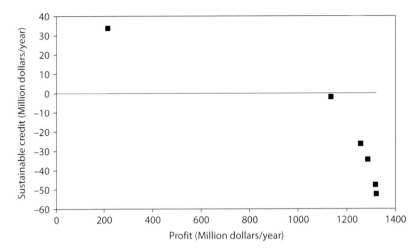

FIGURE 7.19
Pareto optimal solution generated by multicriteria optimization for LP algae oil production (30% oil content) with $25 carbon dioxide consumption credit and $125 carbon dioxide emission cost.

TABLE 7.18

Values of the Pareto Optimal Solutions
for CC25–CP125–LP30 Shown in Figure 7.19

Profit (Million Dollars per Year)	Sustainable Credits (Million Dollars per Year)	Weight (w_1)
215	33.8	0
1136	−1.9	0.1
1257	−26.3	0.2
1286	−34.4	0.3
1318	−47.5	0.7
1318	−47.5	0.8
1321	−52.2	0.9
1321	−52.2	1

The optimal solution with $w_1 = 0$ gave $P = \$1238$ million per year and $S = \$33.8$ million per year.

The points shown in Figure 7.18 are the Pareto optimal solutions for w_1 from 0 to 1.0 for increments of 0.1. The values of profit and sustainable credit for w_1 equal to 0 and 1.0 and some intermediate ones are shown in Table 7.17. The solution always gives a sustainable credit for the complex, when an HP algae production plant is selected (production cost of $463/MT algae oil).

From Figure 7.19 and Table 7.18 it can be seen that the sustainable credits/costs decline and the company's profits increase as the weight, w_1, on the company's profit increases. For example, when $w_1 = 1$, the optimal solution is shown in Table 7.18 for $P = \$1321$ million per year and $S = \$-52.2$ million per year. The optimal solution with $w_1 = 0$ gave $P = \$215$ million per year and $S = \$33.8$ million per year. Thus, when weights on profit are high, the sustainable costs are 2 orders of magnitude lower than the profit, but when the weights on sustainable credits are high, the profit is only 1 order of magnitude higher than the sustainable cost.

The points shown in Figure 7.19 are the Pareto optimal solutions for w_1 from 0 to 1.0 for increments of 0.1. The values of profit and sustainable credit for w_1 equal to 0 and 1.0 and some intermediate ones are shown in Table 7.18. The solution gives a sustainable credit of $33.8 million (profit = $215 million per year) for highest weight on sustainability objective ($w_1 = 0$) in the complex, and this changes to a sustainable cost of 1.9 million (profit = $1136 million per year) for the weight $w_1 = 0.1$. Intermediate points for w_1 may be determined by increasing the increment size to 0.01 or 0.001. Thus, multicriteria optimization of the superstructure shows that the complex operates at sustainable costs for weights on profit at or greater than $w_1 = 0.1$ when an LP algae production plant is selected (production cost of $2513/MT algae oil).

7.6 Case Study V—Parametric Study for Biomass Feedstock Costs and Number of Corn Ethanol Plants

The superstructure in Chapter 6 considers three types of purchased biomass as raw materials, corn (starch-based biomass), corn stover (lignocellulosic biomass), and soybean oil (natural oils). The effect of changes in biomass raw material costs for corn and corn stover are evaluated in this case study.

The costs for corn and soybean oil were available from the U.S. Department of Agriculture (USDA (a), 2010 and USDA(b), 2010). Details are given in Appendix C. Corn stover costs data were very limited with only three data points for 2002, 2007, and projected price for 2012 (Humbird and Aden, 2009). The costs for corn and corn stover used are given in Table 7.19.

TABLE 7.19

Low, High, and Average Price and Standard Deviation in Price for Corn and Corn Stover (Appendix C)

	Cost ($/Ton)		
	Low	High	Average
Corn stover	51	70	61
Corn	72	160	108

From Table 7.19, it is seen that the lowest corn stover cost anticipated was $51/MT (2012 value) and maximum was $70/MT. The corn costs varied from $72 per ton to $160 per ton. PED for corn use as bioethanol and for use as feed and price elasticity of supply for corn are given in Appendix D to account for the wide variation in price. The cases developed for corn and corn stover prices are given in Section 7.6.1.

7.6.1 Options Used in the Parametric Study

The corn and corn stover were competing raw materials for the ethanol process. A combination of corn and corn stover costs were studied to obtain the optimal solution from the superstructure. The prefix CO and CS were used for corn and corn stover, respectively. The suffix HI, LO, and AV are used for high price, low price, or average price, respectively.

The optimal solution for the superstructure in Chapter 6 was obtained with the average price of the raw materials. Then, parameters in the superstructure were modified to obtain the different options. There are nine possible cases with the combination of corn and corn stover prices. Of these nine cases, one is where both corn and corn stover are average price (COAV–CSAV), which is same as the solution obtained in Chapter 6.

Cases are logically picked for the rest of the combinations. There were four extreme cases, COLO–CSLO, COHI–CSLO, COLO–CSHI, and COHI–CSHI. The corn stover and corn prices are likely to be low at the same time, so the case COLO–CSLO was included for analysis. The selection of the next case, COHI–CSLO was guided by the following logic. The cost of corn may not be competitive with gasoline for the production of ethanol as fuel, but it may be competitive for the production of ethanol for chemicals. Corn stover prices on the other hand, are dependent on transportation costs, primarily gasoline. In that case, high corn prices (COHI), and low corn stover price (CSLO) is a possible combination.

The third combination, COLO–CSHI was an unlikely case as the selection of corn stover at high corn stover price (CSHI) was unlikely when corn price was low (COLO). This was a logical deduction from the optimal solution of average corn stover and corn prices. The corn stover process was not selected when both prices were average (COAV–CSAV), so it is unlikely to be chosen when corn price is low and corn stover price is high.

The fourth case was selected where both corn and corn stover prices are high (COHI–CSHI). The remaining four combinations for average cases for corn and corn stover with prices for high and low of the other raw material were not included.

Three other cases were included to study the effect where the number of corn ethanol plants was constrained. The optimal solution for two, three, and four corn ethanol plants for average corn and corn stover ethanol prices was analyzed. The legends 2CO, 3CO, and 4CO were used to denote two, three, and four corn ethanol plants.

Case studies:

COAV–CSAV: average corn ($108/ton) and corn stover ($61/ton) cost (superstructure)

COLO–CSLO: corn cost low ($72/ton) and corn stover cost high ($70/ton)

COHI–CSLO: corn cost high ($160/ton) and corn stover cost low ($51/ton)

COHI–CSHI: corn cost high ($160/ton) and corn stover cost high ($70/ton)

COAV–CSAV–2CO: average corn and corn stover cost, two corn ethanol plants

COAV–CSAV–3CO: average corn and corn stover cost, three corn ethanol plants

COAV–CSAV–4CO: average corn and corn stover cost, four corn ethanol plants

7.6.2 Results of Parametric Study

The aforementioned parameters were incorporated into the superstructure and the optimal solution was obtained for each case. The results are given in Table 7.20. The changes in optimal case triple bottom line are shown in Figure 7.20.

From Figure 7.20 and Table 7.20, it can be seen that the triple bottom line is maximum ($1718 million/year) when both corn and corn stover costs are low

TABLE 7.20

Sales and Costs Associated with the Triple Bottom Line for Varying Costs of Corn and Corn Stover (Million Dollars per Year)

	COAV-CSAV	COLO-CSLO	COAV-CSAV-4CO	COAV-CSAV-3CO	COAV-CSAV-2CO	COHI-CSLO	COHI-CSHI
Corn (CO)/corn stover (CS) process selected	CO	CO	CO	CO	CO+CS	CS	CS
Income from sales	2489	2489	2488	2487	2476	2452	2452
Economic costs (raw materials and utilities)	516	475	516	516	523	520	552
Raw material costs	470	429	470	470	466	441	473
Utility costs	46	46	46	46	57	79	79
Environmental cost (67% of raw material cost)	313	285	313	313	311	294	316
Sustainable credits (+)/Costs (−)	−10.0	−10.0	−10.0	−10.0	−10.5	−11.7	−11.7
Triple bottom line	1650	1718	1649	1648	1632	1627	1572

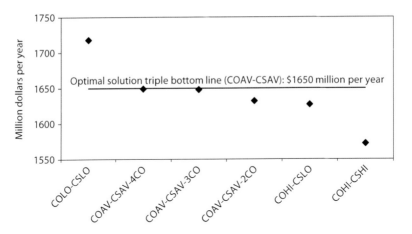

FIGURE 7.20
Triple bottom line for varying corn and corn stover price and plant operation.

(COLO–CSLO). The triple bottom line is almost same as the COAV–CSAV case ($1650 million/year) when four or three corn ethanol plants operate (COAV–CSAV–4CO and COAV–CSAV–3CO) at average corn and corn stover costs. The triple bottom line decreases to $1632 million per year when two corn ethanol plants operate (COAV–CSAV–2CO). When corn costs are high and corn stover costs are low (COHI–CSLO), the corn stover plants operate and the triple bottom line is $1627 million per year, 1.4% lower than the COAV–CSAV case.

From Figure 7.20, the triple bottom line drops further at high costs for both corn and corn stover, when corn stover plants operate and corn plants do not operate (COHI–CSHI). The triple bottom line is $1572 million per year in this case, 5% lower than the COAV-CSAV case.

For high corn costs (COHI–CSLO and COHI–CSHI), it is seen from Table 7.20 that the use of corn stover as raw material decreases the raw material costs from the reference COAV–CSAV case ($470 million/year) for low corn stover cost ($441 million/year, COHI–CSLO) but it is higher than the reference case for high corn stover costs ($473 million/year, COHI–CSHI). The economic costs are higher for both of these cases than the reference case, because the utility costs are increased by 71% from the reference case.

Thus, it can be concluded that for higher corn costs, the operation of corn stover plants may be a possible option for reduction in raw material costs, but the utility costs are increased and the triple bottom line is decreased when operating corn stover plants.

The changes in optimal flow rates for the corn ethanol plants, corn stover ethanol plants, and bioethylene plants are shown in Table 7.21. The number of corn stover and corn ethanol plants operating for the optimum triple bottom line for the different cases is shown in Figure 7.21.

TABLE 7.21

Configuration of Corn and Corn Stover Ethanol Plants and Corresponding Mass Flow Rates (Million Tons per Year)

	COAV-CSAV	COLO-CSLO	COAV-CSAV-4CO	COAV-CSAV-3CO	COAV-CSAV-2CO	COHI-CSLO	COHI-CSHI
No. of CS plants	0	0	0	0	1.33	4.22	4.22
No. of CO plants	5.84	5.84	4	3	2	0	0
CS ethanol each	0	0	0	0	0.079	0.079	0.079
CO ethanol each	0.057	0.057	0.083	0.111	0.114	0	0
Total CS ethanol	0	0	0	0	0.10	0.33	0.33
Total CO ethanol	0.33	0.33	0.33	0.33	0.23	0	0
Bioethylene	0.2	0.2	0.2	0.2	0.2	0.2	0.2

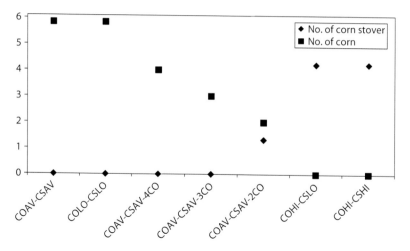

FIGURE 7.21

Number of corn stover and corn ethanol plants operation for maximum triple bottom line.

In Figure 7.21, it can be seen that as the number of corn stover ethanol plants increases, the triple bottom line decreases. A trade-off exists between the number of corn and corn stover plants and the triple bottom line as the number of corn ethanol plants goes from three to zero. This can be explained as when there is possibility of only three or less number of corn ethanol plants, the corn stover plants will operate to meet the demand for ethanol, but the triple bottom line will also decrease. Further parametric studies could be performed for the optimum number of plants and flow rate of ethanol from the corn stover and corn ethanol plants to refine these evaluations.

In Figure 7.22, the flow rates (in million metric tons per year) are given for corn ethanol, corn stover ethanol, and bioethylene produced from the

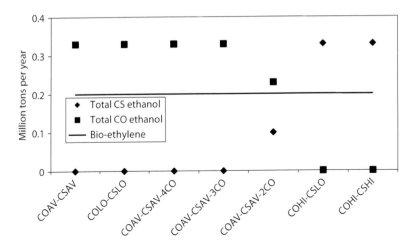

FIGURE 7.22
Corn ethanol, corn stover ethanol, and bioethylene flow rates (Million tons per year).

TABLE 7.22

Pure and Impure Carbon Dioxide Generation from Bioprocesses (Million Metric Tons per Year)

	COAV-CSAV	COLO-CSLO	COAV-CSAV-4CO	COAV-CSAV-3CO	COAV-CSAV-2CO	COHI-CSLO	COHI-CSHI
Pure CO_2	0.317	0.317	0.317	0.317	0.317	0.315	0.315
Impure CO_2	0	0	0	0	0.033	0.106	0.106

bioethanol plants. This shows that the production of bioethylene is not affected by changes in corn or corn stover ethanol flow rates.

The changes in pure and impure carbon dioxide emission from the bioprocesses for the cases are given in Table 7.22 and shown in Figure 7.23. Pure carbon dioxide is the carbon dioxide captured and purified from processes to above 99% purity. Impure carbon dioxide is the carbon dioxide emitted to the atmosphere when carbon dioxide content in streams is low, and the capture and purification of the carbon dioxide is cost intensive. From Figure 7.23, it can be seen that the impure carbon dioxide emissions increase as the number of corn stover ethanol plants increases. This is because there is impure carbon dioxide emission in the bacteria seed generation section for corn stover ethanol plants.

In summary, it can be concluded that for higher costs of corn, the operation of corn stover plants is selected in the optimal structure. However, the triple bottom line decreases as the corn stover ethanol processes are selected. The impure carbon dioxide emissions increase as corn stover plants are included,

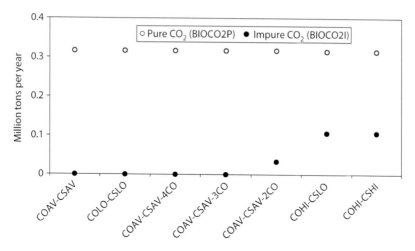

FIGURE 7.23
Pure and impure carbon dioxide emissions from bioprocesses for changes in corn ethanol and corn stover ethanol plant operations.

because it is difficult to capture the carbon dioxide produced in the bacteria seed generation processes. Also, the utility costs are higher when corn stover ethanol plants are operated. These are reasons to be considered when evaluating the triple bottom line.

7.7 Summary

Five case studies are given in this chapter where the optimal structure was used to evaluate the effect of parameter changes. These demonstrate how the chemical complex analysis using the triple bottom line can be used for obtaining the optimal solution and configuration among possible choices.

Case I was a modification of the superstructure to obtain the case for biomass process integration into the base case of existing plants without carbon dioxide utilization. From the optimal solution w/o CO_2 use, it was seen that the triple bottom line increased by 15% from the base-case solution of existing plants (from $854 million per year in the base case to $984 million in the optimal structure). The pure carbon dioxide emissions increased from the base case by 25% as additional corn ethanol plants producing carbon dioxide are included in the optimal structure.

Case II was a parametric study on the superstructure to see the effect of probable sustainable costs and credits on the optimal structure. CCs of $0, $25, and $50 per MT of CO_2 and carbon dioxide costs of $0, $5, $25, $75, and $125 per MT of CO_2 were used in the study. It was concluded that the optimal structure triple bottom line using CO_2 was always greater than the zero carbon dioxide costs and credits scenario. There were no sustainable credits for zero carbon dioxide cost and credit, so the optimal solution had a sustainable cost of 16 million per year, same as the base case. The optimal solution gives sustainable credits for CO_2 utilization in all the cases, except when the CC is low ($25 per ton CO_2) and carbon dioxide costs are high ($125 per ton CO_2). The sustainable costs are $8 million per year in this case. The pure CO_2 emission is zero for all the cases as the carbon dioxide is consumed in algae and other chemical processes from CO_2.

Case III was a parametric study on the superstructure to see the effect of including algae oil production costs on the optimal structure. Two oil contents (30% and 50%) were studied with HP, AP, and LP algae oil production costs. The zero production costs were used as reference. The triple bottom line for optimal solution was for LP and AP 30% algae oil content production plant and LP 50% algae oil production plant. In the absence of technology to convert carbon dioxide to algae oil at zero production costs, the best option is to operate an HP 50% oil content algae oil production plant that gives a triple bottom line of $1844 million per year. The algae oil is not produced in the optimal structure for the LP or AP 30% oil content or the LP 50% oil content algae oil production. The carbon dioxide in the complex is used for the manufacture of other chemicals in the optimal structure, reducing the sustainable costs by 22% from the base case.

Case IV was a multicriteria optimization problem for maximizing profit while minimizing sustainable costs. A most probable case of 30% oil content algae is used in this case. The cost of carbon dioxide at $125 per ton CO_2 emitted and the credit of $25 per MT of CO_2 consumed were used for the sustainable costs/credits. The cost of production in an LP and an HP algae oil production unit for the aforementioned combination of parameters was used to obtain the optimal solution. P = $1474 million per year and S = $7 million per year were obtained for maximum weight on profit and P = $1238 million per year and S = $33.8 million per year for maximum weight on sustainability for the HP plant operation of 30% oil content algae. P = $1321 million per year and S = $–52.2 million per year were obtained for maximum weight on profit and P = $215 million per year and S = $33.8 million per year were obtained for maximum weight on sustainability for the LP plant operation of 30% oil content algae.

Case V was a parametric study for corn and corn stover prices on the superstructure. It was concluded that for higher costs of corn, the operation of corn stover plants is selected in the optimal structure. However, the triple bottom line decreases as the corn stover ethanol processes are selected. The impure carbon dioxide emissions increase as corn stover plants are included, because it is difficult to capture the carbon dioxide produced in the bacteria seed generation processes. Also, the utility costs are higher when corn stover ethanol plants are operated.

Thus, five case studies were presented that demonstrated the use of chemical complex optimization for sustainability analysis. The first case demonstrated how changes to the model of the superstructure can be made for scenario analysis. The second, third, and fifth cases demonstrated the optimization of the triple bottom line by changing parameters for sustainable credits/costs, utility costs (production costs), and raw material costs, respectively. The fourth case demonstrated multicriteria optimization of the complex based on parameter changes for technology and policy changes in the future.

Appendix A: TCA Methodology and Sustainability Analysis

A.1 Introduction

Sustainability or sustainable development is defined as development that meets the needs of the present without compromising the ability of future generations to meet their own needs (United Nations, 1987). There are various methods to evaluate sustainability, some of which are presented in this chapter.

Total Cost Assessment (TCA) was developed at the Tellus Institute for the EPA and New Jersey Department of Environmental Protection in 1991. Later, a detailed methodology was developed based on the TCA concept by an industry group working through the American Institute of Chemical Engineers. The details of TCA methodology are provided in the following section. Some of the other tools like sustainability metrics, indices, life-cycle analysis (LCA), ecoefficiency analysis, etc., used for sustainability analysis are also discussed.

The TCA methodology is a powerful tool for decision making because it incorporates costs associated with the total life cycle of a project. The measurement of the costs associated with a particular process gets increasingly difficult to measure from conventional costs (raw material costs, operating costs, capital costs, etc.) to societal costs (damage to the environment from emissions within regulations). The triple bottom line method described by the TCA methodology incorporates a quantitative measure of sustainability. In contrast, LCA gives a comparative assessment of sustainability using eight impact categories. Evaluation of these categories is based on material and energy balances at stages of the life cycle from collecting raw material from earth and ending when all this material is returned to earth.

The ExternE method developed by the European Commission provides a framework for transforming impacts into monetary values (Bickel and Friedrich, 2005). The method includes activity assessment, definition of impact categories, and externalities. The estimation of impacts or effects of the activity are then performed to find the difference in impact between different scenarios. The monetization of impacts leads to external costs in the next stage of the method. Uncertainties and sensitivity analysis are used to obtain results with the ExternE methodology.

A.2 Total Cost Assessment

TCA is a methodology developed by industry professionals and sponsored by the American Institute of Chemical Engineers (Constable et al., 1999; Laurin, 2007). TCA is a decision-making tool intended for evaluating different alternatives on a cost basis. The tool provides cost information for internal managerial decisions. The TCA methodology identifies five types of cost-associated accounting. These costs are outlined in Figure A.1. Dow Chemical, Monsanto, GlaxoSmithKline, and Eastman Chemical are industrial companies that have made use of the TCA methodology.

Type I and Type II costs cover traditional accounting methods used for decision making and typically focus on expected revenues and direct costs. Type I costs include capital, labor and material, and waste disposal costs, while Type II includes indirect costs such as reporting costs, regulatory costs, and monitoring costs. These costs are calculated based on traditional return on investment model, and uncertainties may be added to these values depending on the user. The TCA model differs from traditional accounting methods when it introduces three other types of costs: Type III, Type IV, and Type V.

Type III, the future and contingent liability costs, includes the costs that a company may have to bear in the case that an event occurs in future. For example, TCA allows for uncertainties like the probability and cost of cleanup of an oil spill to be included in a financial assessment. Thus, if the event occurs in the future, it will become a Type II cost.

Type IV costs include internal intangible costs that are qualitatively defined but difficult to quantify. For example, assigning a dollar value to ideas like

Cost benefit type	Description
I: Direct	Revenues, capital costs, labor, materials, waste disposal. May have uncertainty attached.
II: Indirect	Non-allocated corporate and plant costs (e.g., reporting costs, regulatory costs, monitoring costs)
III: Future and contingent liability	Potential fines, penalties, and future liabilities (e.g., non-compliance, remediation, personal injury, property damage, industrial accident costs, changes in regulations)
IV: Intangible—internal	Difficult to measure costs and benefits borne by the company (e.g., changes in the value of brand value, worker morale, union relations, community relations)
V: Intangible—external	Costs borne by society (e.g., effect of operations on housing costs, degradation of habitat)

FIGURE A.1
Cost benefit types in TCA analysis. (From Laurin, L., *Chem. Eng. Prog.*, 103(6), 44, 2007.)

protecting brand value, improving employee morale, and reducing regulatory scrutiny are difficult, but these affect a company's benefits. TCA allows these benefits to be identified and quantified in decision analysis.

Type V costs include external intangible costs that are not directly borne by the company. This is the cost that society pays as a result of pollution, loss of wetlands, and the effect of employment on human health.

Constable et al. (1999) give a detailed procedure to conduct the TCA for projects. The TCA method can use the life-cycle inventory (LCI) results from LCA (described in a later section) and use the data to conduct a total cost inventory (similar to an LCI). The overview of the process is given in Figure A.2.

The step 4 of the process in Figure A.2 is crucial in the process as it involves the estimation of Type I–V costs that are recurring or nonrecurring (Constable et al., 1999). The uncertainties associated with the occurrence of these types

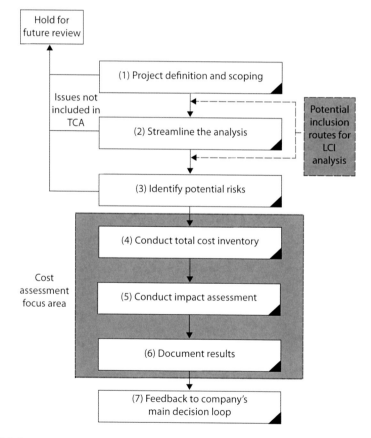

FIGURE A.2
Overview of TCA. (From Constable, D. et al., *Total Cost Assessment Methodology; Internal Managerial Decision Making Tool*, American Institute of Chemical Engineers, AIChE/CWRT, New York, 1999.)

TABLE A.1

Type III, IV, and V Costs in TCA

Cost Type	Categories
Type III—Future and contingent liability	Compliance obligations
	Civil and criminal fines and penalties
	Remedial costs of contamination
	Compensation and punitive damages
	Natural resource damage
	PRP liabilities for off-site contamination
	Industrial process risk
Type IV—Internal intangible	Staff (productivity/morale, turnover, union negotiating time)
	Market share (value chain perception, public perception, consumer perception)
	License to operate
	Relationships with investors, lenders, communities, regulators
Type V—External intangible	Pollutant discharges to air
	Pollutant discharges to surface water
	Pollutant discharges to groundwater/deep well
	Pollutant discharges to land natural habitat impacts: local community, wetlands, wildlife reserves
	Value chain impacts
	Product health impacts

Source: Constable, D. et al., *Total Cost Assessment Methodology; Internal Managerial Decision Making Tool,* American Institute of Chemical Engineers, AIChE/CWRT, New York, 1999.

are also evaluated. Type I and II costs are available from a company's costing analysis data. It is the identifying of the Type III–V scenarios and assigning a cost to them are the challenges faced in using the TCA method. Constable et al. (1999) developed several cost databases and descriptions of how some cost values could be represented. These databases were a result of surveys conducted by the TCA Work Group for costing approaches that have been developed previously. The Type III–V costs can be grouped into costs, as given in Table A.1.

Koch (2002) describes Dow's efforts to develop Total "Business" Cost Assessment, or TBCA, including Type I, Type II, Type III, and Type IV costs from the TCA model, as shown in Figure A.3. Type V, the external benefits and costs, requires additional LCA tools, which can provide an LCI. Dow could not fix the pricing of the various societal impacts, so the Type V elements were excluded. The Dow TBCA pilot is a subset of TCA. Thus, with Type V externalities excluded the model of TCA concerns solely business or the economic aspect for the company and hence derive the name Total "Business" Cost Assessment.

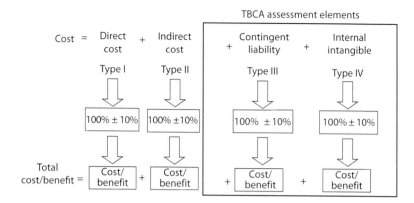

FIGURE A.3
Dow Chemical Company's model of TBCA. (From Koch, D.: Dow chemical pilot of total "business" cost assessment methodology: A tool to translate EH&S "...right things to do" into economic terms (dollars). *Environ. Prog.* 2002. 21(1). 20–28. Copyright Wiley-VCH Verlag GmbH & Co. KGaA. With permission.)

Koch (2002) also gives an example of how Dow implemented TBCA for their 2005 EH&S goals. Dow had a specific set of goals to meet their sustainability standards for 2005. In 2000, Dow conducted a series of TBCA workshops addressing the benefits of their EH&S 2005 goals. The Type I and Type II costs to achieve those goals were determined from these workshop sessions with well-defined costing methods for the company. The Type III and Type IV costs needed identifying and assigning specific metrics. The Type III costs identified included (but not limited to) fines and penalties, legal fees, business interruptions, cost of future environmental cleanup, and future cost to discharge wastewater. Type IV benefits were assessed and dealt with intangible internal issues, such as corporate image, public perception, and worker morale and effectiveness. The cost type identification was accomplished in a series of eight workshops, addressing the following specific EH&S areas: emissions (priority and chemical), waste (kilns and landfills) and wastewater (BOD and hydraulic), energy, loss of primary containment and process safety, personal injury—illness/motor vehicle—and transportation. All of the TBCA workshops conducted within Dow utilized the software tool TCAce™, jointly developed by A.D. Little and Sylvatica. The Dow effort utilized the spreadsheet input/output capability of TCAce to input the data during the workshop sessions. The data entered in the spreadsheets are given in Table A.2.

For the Type III and IV costs, the workshop team at Dow collectively developed scenarios, which had a time line, estimate of benefit or cost if the scenario is fully realized, and the probability of the scenario occurring. These results were analyzed on a time basis time period of 10 years for economic analysis. The results were reviewed after the workshop to include revised projections relative to what might have seemed valid during the workshop discussions. The final results were used to enhance the internal decision-making process

TABLE A.2

TCAce™ Spreadsheet Input for Dow's TBCA Analysis

Alternative—This is the name of the alternative
Alternative description—A textual description of the alternative
Alternative date—The date that this alternative was developed
Scenario type—This is a way to group scenarios together
Scenario—Each unique name in this field will create a new scenario
Description—A textual description of the scenario
Overall probability—This is an estimation of the likelihood that the scenario will occur at all, in some project year between year 0 and the end year of the analysis
Probability of occurrence—This is an estimation of the likelihood that the scenario will occur in the project year listed for this line item
Scenario simulation method—This field indicates how the occurrence of the scenario is simulated or modeled. There are three valid scenario simulation methods that can be entered
    (a) Annual nonrecurrable (AN) ("sample annually, repeats are not possible")
    (b) Annual recurrable (AR) ("sample annually, repeats are possible")
    (c) One possible occurrence, uncertain timing (OU) ("sample once, timing is uncertain")
TCA cost type—This field indicates the cost type under which the particular cost driver will be grouped (e.g., Type III–V)
Activity—This is a way to group cost drivers together within a cost type
Driver—This is the item for which an expense is incurred
Cost kind—Users can enter the following options to indicate the depreciation method or expensed (not depreciated). The four are
    (a) (*OM*) Cost (annualized costs that are not depreciated)
    (b) (*5DD*) 5-year double declining balance
    (c) (*7DD*) 7-year double declining balance
    (d) (*10SL*) 10-year straight line
    (e) (*Exp*) Expensed (not depreciated)
Salvage value—User can enter the salvage value in dollars
Cost model type—Each cost driver can be modeled using one specific cost type
Project year—This is the year of the project that is being estimated. The probability of occurrence is directly related to this value, as is the cost in the year of scenario occurrence
Cost in year of occurrence—This is the cost associated with the line item for cost incurred
Cost 1 (up to 20) years after occurrence (optional)—These fields are optional and can be created to include costs that are incurred as a result of scenario occurrence but in years *after* the year of scenario occurrence

Source: Koch, D.: Dow chemical pilot of total "business" cost assessment methodology: A tool to translate EH&S "…right things to do" into economic terms (dollars). *Environ. Prog.* 2002. 21(1). 20–28. Copyright Wiley-VCH Verlag GmbH & Co. KGaA. With permission.

of the organization by applying these results to EH&S-related Six Sigma projects throughout the company. Thus, improved understanding of the full impact of these EH&S projects in economic metrics was possible, and these results from TBCA analysis are described in Figure A.4.

The specific numeric information for Dow is confidential, but the structure of results is given. The initial summary and assessment confirmed that the Type III and Type IV estimates of value contribution can match or exceed the traditional Type I and Type II economic benefit estimates. Dow had estimated their 2005 resource and productivity goal costs for $1 billion but achieved a return of over $5 billion (Laurin, 2007).

	Type I/II Costs/Benefits		Type III Costs/Benefits		Type IV Costs/Benefits	
	Unit Costs		Unit Costs		Unit Costs	
Emissions Target: 90% Reduction of Primary and 50% Reduction of Chemical	Project Specific	$/ton	X	$/ton	X	$/ton
Waste: Target: 50% Reduction of Waste per Pound of Prod.						
TTUs	Project Specific	$/mt	X	$/mt	X	$/mt
Kilns	Project Specific	$/mt	X	$/mt	X	$/mt
BOD IN WWTP	Project Specific	$/mt	X	$/mt		$/mt
Landfills	Project Specific	$/m3	X	$/m3	X	$/m3
Wastewater: Target: 50% Reduction of Wastewater per Pound Prod.	Project Specific	$/m3	X	$/m3	X	$/m3
Energy: Target: 20% Reduction of Energy per Pound of Production	Project Specific	$/MM Btu	X	$/MM Btu	X	$/MM Btu
LOPC Target: 90% Reduction in Incidents	Project Specific	$/Incident	X	$/Incident	X	$/Incident
Process Safety Target: 90% Reduction in Incidents	Project Specific	$/Event	X	$/Event	X	$/Event
Personal Injury and Target: 90% Reduction in Incidents	Project Specific	$/Incident	X	$/Incident	X	$/Incident

FIGURE A.4
Dow Chemical TBCA results. (From Koch, D.: Dow chemical pilot of total "business" cost assessment methodology: A tool to translate EH&S "… right things to do" into economic terms (dollars). *Environ. Prog.* 2002. 21(1). 20–28. Copyright Wiley-VCH Verlag GmbH & Co. KGaA. With permission.)

A.3 Chemical Complex and Industrial Ecology

"Industrial ecology" refers to the exchange of materials between different industrial sectors where the by-product of one industry becomes the feedstock of another. Integrating the notions of sustainability into environmental and economic systems gives rise to industrial ecology. The concept of industrial ecology can be used to describe chemical complexes throughout the world. A few of these complexes are tabulated in Table 1.1.

The TCA methodology has been incorporated in the chemical complex analysis system developed at Louisiana State University. A base case of chemical plants in the lower Mississippi River corridor was used. The base case contained plants in the existing complex, and the superstructure contained plants with new processes that utilize the CO_2 produced in the complex to make chemicals (Xu, 2004). The new processes were included by using the chemical complex analysis system to form an optimal structure of plants in the chemical production complex.

A complex extension was developed in this research to include biomass feedstock–based chemicals, and the TCA methodology is used to find the optimum plant configuration from the superstructure (Sengupta, 2010).

A.4 Evaluating Sustainability with Metrics and Indices

In this section, some of the methods used to evaluate sustainability are discussed. Most of these methods are aimed at comparing the feasibility of a project based on different scenarios and evaluating which one is the best. The appropriate measure of sustainable development lies in integrating the three aspects in the triple bottom line: economic, environmental, and social.

One way to attempt to evaluate sustainability is using indicators or metrics. Two classes of metrics or indicators are used. These are indicators to indicate the state and the behavior of a system. The indicators that indicate the state of a system are known as *content* indicators and those that measure the behavior of a system are known as *performance* indicators. Sustainability indicators are primarily performance indicators as they focus on the behavior of a system. The goal is to make a system function in a sustainable manner (Sikdar, 2003). The three aspects of sustainability can be denoted in the Venn diagram of Figure A.5. The three circles denote the metrics for each of the sustainability aspect, ecological metrics, economic metrics, and sociological

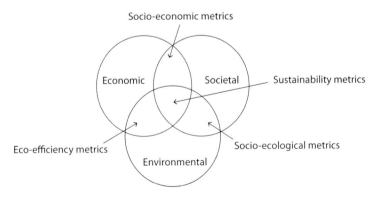

FIGURE A.5
Three intersecting circles to denote sustainability. (From Sikdar, S.K., *AIChE J.*, 49(8), 1928, 2003.)

metrics. The intersections denote four other types of metrics and are sum-marized as follows (Sikdar, 2003):

Group 1 (1-D): economic, ecological, and sociological indicators

Group 2 (2-D): socioeconomic, ecoefficiency, and socioecological indicators

Group 3 (3-D): sustainability indicators and metrics

One-dimensional (1-D) metrics attempt to measure any one aspect of sustainability. Two-dimensional (2-D) metrics attempt to take into account any two aspects. Three-dimensional (3-D) metrics are aimed at measuring all the three aspects of sustainability. 1-D metrics have been well defined by different institutions like the AIChE (United States) and the IChemE (United Kingdom) (Sikdar, 2003). 1-D indicators are grouped into environmental, economic, and social categories. The environmental indicators are further divided into resources or environmental impact categories.

Among resources, the important indicators are energy use, material use, water use, and land use, and among environmental impacts, acidification, global warming, human health, ozone depletion, photochemical smog formation, and ecological health. Economic indicators include value-added measures and R&D expenditures. Social indicators are based on employee benefits, safety, and how the employees are treated in the workplace. 2-D metrics include ecoefficiency analysis, a concept developed by the World Business Council for Sustainable Development (Lehni, 2000) and implemented by BASF (Wall-Markowski et al., 2004). Thus, sustainable process design is said to be viewed as a multiobjective optimization problem in which the cost of manufacture is minimized while improving all 3-D indicators.

Sikdar (2003) described some examples of the indicators. Examples of 3-D indicator include nonrenewable energy usage, material usage, and pollutant dispersion. Energy is the main factor in economic growth. Nonrenewable energy has an ecological impact through the emission of pollutants and greenhouse gases (GHG), and depletion of these resources affects future generations. Material use can have direct ecological impact, is associated with value creation, and can have intergenerational impact. For example, lactic acid produced from bio-based material can be considered to have positive social impact as it can be used as intermediate for the manufacture of many useful products and polymers like polylactic acid, etc., which are otherwise produced synthetically from nonrenewable resources.

Pollutant dispersion is a 3-D indicator, as it represents environmental impact, has economic cost associated with it, and has a bearing on the health of the people and ecosystems in the neighborhood of the manufacturing units.

Examples of 1-D and 2-D indicators include process wastes that are well controlled and contained and water usage. When process wastes account for only economic value losses, they represent 1-D economic indicator. Wastes such as gypsum piles can be 2-D ecoefficiency indicators causing potential pollution problems along with economic loss. Water usage can give rise to residuals from water works which can be an environmental nuisance leading to a 2-D ecoefficiency indicator, or without residuals is a 1-D economic indicator. Example of a 2-D socioeconomic indicator is the cost of manufacturing where nature of technology (economic value creation) and affordability for public consumption (societal value) are two factors. An important example in this case can be the production of ethanol from bio-based material where the ecological aspect is at an advantage for using renewable material but the cost of production (economic) makes its acceptance into the society (societal) a challenge.

Sustainability metrics: Sustainability metrics are intended to improve internal management decision making with respect to the sustainability of processes, products, and services. Sustainability metrics can assist decision makers in setting goals, benchmarking, and comparing alternatives such as different suppliers, raw materials, and improvement options from the sustainability perspective. Development of sustainability metrics was done by BRIDGES to Sustainability™, a not-for-profit organization, by testing, adapting, and refining the sustainability metrics (Tanzil et al., 2003).

There are basic and complementary metrics under six impact categories: material, energy, water, solid wastes, toxic release, and pollutant effects. BRIDGES' sustainability metrics are constructed as ratios with environmental impacts in the numerator and a physically or financially meaningful representation of output in the denominator, the better process being the one with a smaller value for the ratio. The metrics are currently organized into six basic impact categories: material, energy, and water intensities, solid waste to landfills, toxic releases, and pollutant effects. Five of these are the basic metrics, as shown in Table A.3.

The BRIDGES' metrics are unique from other types of metrics in the fact that they are "stackable" along the supply chain, thereby avoiding local optimization while affecting the overall ecoefficiency of life cycle in a negative manner. The BRIDGES' metrics have been piloted in various manufacturing facilities like Formosa Plastics (petrochemical), Interface Corporation (carpeting), and Caterpillar Inc. (tool manufacturing).

Sustainability indices: Sustainability indices for countries provide a 1-D metric to valuate country-specific information on the three dimensions of sustainable development: economic, environmental, and social conditions. The indices give a measure of sustainable development within a community, within a country, or among different places across the world. Bohringer and Jochem (2007) discuss 11 sustainability indices along with their scales, normalization, weighting, and aggregating methods, as summarized in Table A.4.

TABLE A.3

Basic Sustainability Metrics Developed by BRIDGES

Output: Mass of product or sales revenue or value-added	Material intensity
	$$\frac{\text{Mass of raw materials} - \text{Mass of products}}{\text{Output}}$$
	Water intensity
	$$\frac{\text{Volume of freshwater used}}{\text{Output}}$$
	Energy intensity
	$$\frac{\text{Net energy used as primary fuel equivalent}}{\text{Output}}$$
	Solid waste to landfill
	$$\frac{\text{Total mass of solid waste disposed}}{\text{Output}}$$
	Toxic release
	$$\frac{\text{Total mass of recognized toxics released}}{\text{Output}}$$

Source: Tanzil, D. et al., Sustainability metrics, innovating for sustainability, *11th International Conference of Greening of Industry Network*, San Francisco, CA, October 12–15, 2003.

A.5 Sustainable Process Index

The concept of Sustainable Process Index (SPI) is based on the sustainable flow of solar energy (Krotscheck and Narodoslawsky, 1996). The utilization of the solar energy is based on area available. The area can be defined according to its usage on land, in water, and in air. The production in these areas is denoted by production factors. Thus, with the dual function of area as a recipient of solar energy and as a production factor, the SPI can measure and relate the ecological impact of a process with respect to the quantity and the quality of the energy and mass flow it induces.

Processes needing more area for the same product or service are less competitive under sustainable economic conditions. SPI is the ratio of two areas in a given time period. One area is needed to embed the process to produce the service or product unit sustainably in the ecosphere, and another is the area available for the sustainable existence of the product.

The SPI is a number, which is based on the ratio of two areas in a given time period (usually per year) in order to provide one inhabitant with a certain service or product. One area is needed to embed the process sustainably into the ecosphere. The other is the area available (on a statistical base) for every inhabitant. The SPI, thus, is the fraction of the area per inhabitant

TABLE A.4

Sustainability Indicators with Adequate Requirements
for Sustainability Indicators

Index	Scale	Normalization	Weighting	Aggregation
Living planet index	RNC	$\left(\dfrac{x_{i,t}}{x_{i,t-1}}\right)$	Equal	$\sqrt[N]{\displaystyle\prod_{i=1}^{N}\dfrac{x_{i,t}}{x_{i,t-1}}}$
Ecological footprint	RNC	Transformation in square km	Equal	$\displaystyle\sum_{i=1}^{N} x_i$
City development index	RNC	$\left(\dfrac{x_i - \underline{x}}{\bar{x} - \underline{x}}\right)$	Two steps PCA/experts	$\dfrac{1}{N}\displaystyle\sum_{i=1}^{N} w_i x_i$
Human development index	RNC	$\left(\dfrac{x_i - \underline{x}}{\bar{x} - \underline{x}}\right)$	Equal	$\dfrac{1}{N}\displaystyle\sum_{i=1}^{N} x_i$
Environmental sustainability index 2005	RNC	Standard deviation	Equal/experts	$\dfrac{1}{N}\displaystyle\sum_{i=1}^{N} x_i$
Environmental performance index	RNC	Best = 100 Worst = 0	PCA and experts	$\displaystyle\sum_{i=1}^{N} w_i x_i$
Environmental vulnerability index	RNC/INC	Aim = 1 worst = 7	Equal	$\dfrac{1}{N}\displaystyle\sum_{i=1}^{N} x_i$
Index of sustainable economic welfare	RNC	Monetized	Equal	$\displaystyle\sum_{i=1}^{N} x_i$
Well-being index	RNC	Best = 100 Worst = 0	Subjective (not derived)	$\dfrac{1}{N}\displaystyle\sum_{i=1}^{N} (w_i) x_i$
Genuine savings index	RNC	Monetized	Equal	$\displaystyle\sum_{i=1}^{N} x_i$
Environmentally adjusted domestic product (EDP)	RNC	Monetized	Equal	$\displaystyle\sum_{i=1}^{N} x_i$

Source: Bohringer, C. and Jochem, P.E.P., *Ecol. Econ.,* 63(1), 1, 2007.
With variables represented by x_i, weights by w_i, and countries by i and years by t.
RNC, ratio-scale noncomparability; INC, interval-scale noncomparability.

TABLE A.5

Equations for the Calculation of SPI

$A_{\text{tot}} = A_{\text{R}} + A_{\text{E}} + A_{\text{I}} + A_{\text{S}} + A_{\text{P}}$ (m²)	A_{tot} = Total area assigned to embed a process sustainably
	A_{R} = Area requirement to produce raw materials
	A_{E} = Area necessary to provide process energy
	A_{I} = Area to provide the installations for the process
	A_{S} = Area required for the staff
	A_{P} = Area to accommodate products and by-products
$a_{\text{tot}} = \dfrac{A_{\text{tot}}}{S_{\text{tot}}} = \dfrac{1}{y_{\text{tot}}}$ (m² year unit⁻¹)	a_{tot} = Specific (sustainable) service area
	y_{tot} = Specific yield (inverse specific service area)
	S_{tot} = Number of unit services (e.g., product units) supplied by the process in question
$\text{SPI} = \dfrac{a_{\text{tot}}}{a_{\text{in}}}$ (cap unit⁻¹)	a_{in} = Area per inhabitant in the region being relevant to the process

Source: Krotscheck, C. and Narodoslawsky, M., *Ecol. Eng.*, 6(4), 241, 1996.

related to the delivery of a certain product or service unit. The calculation of the SPI is based on the computation of the following equations given in Table A.5. The area for raw materials, energy, process installation, staff, and products has separate equations and is detailed by (Krotscheck and Narodoslawsky, 1996).

A.6 Ecoefficiency Analysis

Ecoefficiency analysis is a life-cycle tool that allows data to be presented in a concise format for use by decision makers (Wall-Markowski et al., 2004). Ecological indicators are combined to provide an "ecological fingerprint," as shown in Figure A.6, which is plotted against the life-cycle cost of process options. The process that has the lowest of both measures is judged to have superior ecoefficiency.

Ecoefficiency analysis starts with identifying viable alternatives of a process or product. Data are collected for the production, use, and disposal phases of life cycle and impacts considering all the alternatives in the following environmental categories are determined: resource consumption, energy consumption, emissions, risk potential, health effect potential, and land use. These categories are shown in Figure A.6.

These results are then weighted and aggregated into a total environmental impact in each of the categories, and then further consolidated into one overall relative environmental impact. The weighting factors include a societal weighting factor depending upon perceived relative importance of the environmental categories (as shown in Figure A.7), a relevance weighting factor giving a relative environmental impact for alternatives to the total

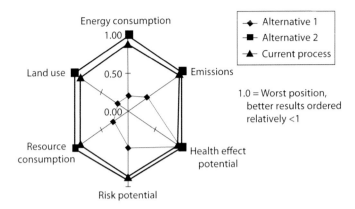

FIGURE A.6
Ecoefficiency environmental categories "ecological fingerprint." (From Wall-Markowski, C.A., Kicherer, A., Saling, P.: Using eco-efficiency analysis to assess renewable-resource–based technologies. *Environ. Prog.* 2004. 23(4). 329–333. Copyright Wiley-VCH Verlag GmbH & Co. KGaA. With permission.)

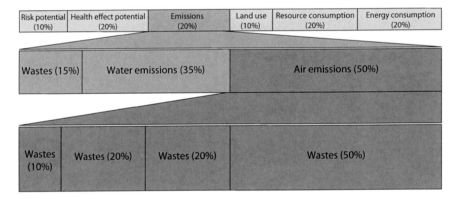

FIGURE A.7
Ecoefficiency analysis societal weighting factors. (Adapted from Wall-Markowski, C.A. et al., *Environ. Prog.*, 23(4), 329, 2004.)

regional impact, and an impact weighting factor reflecting the impacts at the chemical level.

Wall-Markowski et al. (2004) from BASF give a case study of renewable-resource- versus petroleum-based polymers. BASF conducted an ecoefficiency analysis comparing production of polymer granules for four petroleum-based polymers to two renewable-resource-based polymers: two petroleum-based polyamides, two petroleum-based polyesters, a biopolymer based on both petroleum and renewable resources, and a biopolymer based completely on renewable resources.

The ecoefficiency case study was a "cradle-to-gate" assessment and stops at the point that the polymer granules leave the production facility.

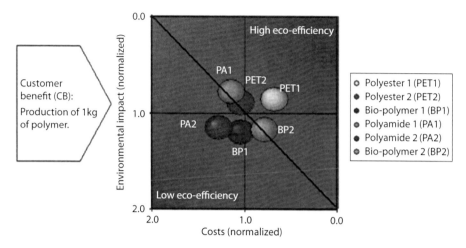

FIGURE A.8
Ecoefficiency analysis for nonrenewable versus bio-based polymer. (From Wall-Markowski, C.A., Kicherer, A., Saling, P.: Using eco-efficiency analysis to assess renewable-resource–based technologies. *Environ. Prog.* 2004. 23(4). 329–333. Copyright Wiley-VCH Verlag GmbH & Co. KGaA. With permission.)

Thus, it was not a true LCA that includes "cradle-to-grave" analysis. The results of this analysis showed how to compare the overall cradle-to-gate ecoefficiency of renewable-resource- and petroleum-based polymers. The cost for production of 1 kg of polymers is plotted versus environmental impact due to that kg of polymer, as shown in Figure A.8. From the figure, the following conclusions were made:

- The petroleum-based polyester 1 was the most ecoefficient, based on its low cost, and had less overall environmental impact than that of the biopolymers.

- The 100% renewable-resource-based polymer (biopolymer 2) had an ecoefficiency similar to that of the petroleum-based polyamide 1 and polyester 2 because its lower cost counterbalanced its higher environmental impact.

- The biopolymer 1 alternative, which was partially based on renewable resources, had a lower ecoefficiency than that of the biopolymer 2. Polyamide 2 was slightly less ecoefficient than the two biopolymers due to higher cost. The unit costs for the raw materials impacted the total cost and hence affected the ecoefficiency.

- Polyester 1 had the best economic position because of its low raw material costs and less processing energy, and thus low utilities costs. Biopolymer 2 had the highest utilities costs arising from the high processing energy, although this was counterbalanced by the low raw material costs.

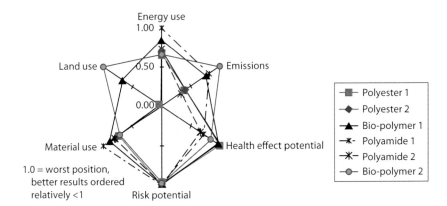

FIGURE A.9
Ecological fingerprint for renewable-resource- and petroleum-based polymers. (From Wall-Markowski, C.A., Kicherer, A., Saling, P.: Using eco-efficiency analysis to assess renewable-resource–based technologies. *Environ. Prog.* 2004. 23(4). 329–333. Copyright Wiley-VCH Verlag GmbH & Co. KGaA. With permission.)

- Polyamide 2 had the highest raw material costs, resulting in the lowest total ecoefficiency.
- The environmental axis demonstrated that the renewable-resource-based polymers had greater overall environmental impact than that of three of the four petroleum-based alternatives.

The ecological fingerprint for the process given in Figure A.9 provided additional details for the environmental categories considered. Each polymer has advantages and disadvantages in the six different categories, as shown in Figure A.9. In the material use category, biopolymers used renewable-resource-based raw materials, but in the energy use category, significant processing energy was necessary to convert plants into material suitable for durable goods manufacture, resulting in no net advantage in energy or material use. In emission category, advantages such as carbon dioxide uptake by plants were counterbalanced by factors in the emission category when water emissions resulted from the starch-manufacturing process. Plant-based products required agriculture, which had an impact in the land use category.

This case study did not include a full LCA from cradle to grave. A different result can be expected by including such an analysis.

A.7 Shear Zones

A company's ability to successfully compete in the marketplace in a sustainable manner depends on the "triple bottom line" (Elkington, 1998). The triple bottom line involves economic, environmental, and sustainable costs.

There is instability in constant flux due to social, economic, political, and environmental pressures, cycles, and conflicts. These conflicts lead to shear zones, where ecological equivalents of tremors and earthquakes occur. Taking care of these shear zones will result in sustainable growth of the company. These three shear zones are the following:

- *Economic-environmental shear zone*: This is the easiest zone for business to manage. These costs are tangible, and new agendas such as ecoefficiency, environmental cost accounting, and ecological tax reforms have emerged to take care of these costs.
- *Social-environmental shear zone*: This zone is the most challenging zone, where concepts of environmental justice and the effects to the society due to environmental imbalance are studied. Environmental literacy and training and intergenerational equity are some other aspects of research in this area.
- *Economic-social shear zone*: This zone relates to direct relationship of the company with the society and addresses issues like downsizing, unemployment, minority rights, and business ethics.

The current methods of evaluating the merits or demerits of a certain project depend upon the direct return on investment for investors. This means that economic aspect is given importance for the company's profit. The environmental aspects are dealt with to keep a company running within regulations set forth by the environmental agencies. However, social aspects need to be addressed for future well-being of the society as a whole. The problem of addressing social issues lies in the fact that there are limited ways to economically assess the impact of these issues. In the following sections, methods to evaluate the feasibility of a project based on the aspects of economic, environmental, and societal aspects are discussed. The processes are also supported by a literature review of case studies.

A.8 Life-Cycle Assessment

Life-cycle assessment is a "cradle-to-grave" approach for assessing industrial systems (SAIC, 2006). "Cradle to grave" begins with the gathering of raw materials from the earth to create the product and ends at the point when all materials are returned to the earth. LCA evaluates all stages of a product's life from the perspective that they are interdependent, meaning that one operation leads to the next. LCA enables the estimation of the cumulative environmental impacts resulting from all stages in the product life cycle, often including impacts not considered in more traditional analyses (e.g., raw material extraction, material transportation, ultimate product disposal, etc.). By including the impacts throughout the product life cycle, LCA provides a comprehensive view of the environmental aspects of the product or process and a more accurate picture of the true environmental trade-offs in product and process selection.

An LCA allows a decision maker to study an entire product system, hence avoiding the suboptimization that could result if only a single process were the focus of the study. For example, when selecting between two rival products, it may appear that option 1 is better for the environment because it generates less solid waste than option 2. However, after performing an LCA, it might be determined that the first option actually creates larger cradle-to-grave environmental impacts when measured across all three media (air, water, and land) (e.g., it may cause more chemical emissions during the manufacturing stage). Therefore, the second product (that produces solid waste) may be viewed as producing less cradle-to-grave environmental harm or impact than the first technology because of its lower chemical emissions.

The life-cycle assessment is a systematic approach to evaluate a process and contains the four stages, as shown in Figure A.10 (SAIC, 2006). These stages are as follows:

- Goal definition and scoping: Define and describe the product, process, or activity. Establish the context in which the assessment is to be made and identify the boundaries and environmental effects to be reviewed for the assessment.
- Inventory analysis: Identify and quantify energy, water, and material usage and environmental releases (e.g., air emissions, solid waste disposal, and wastewater discharges).

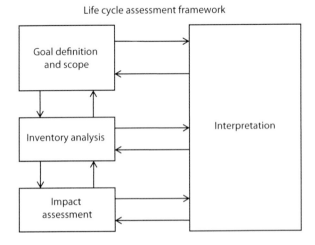

FIGURE A.10
Life-cycle assessment framework. (From SAIC, *Life Cycle Assessment: Principles and Practice*, Scientific Applications International Corporation, National Risk Management Research Laboratory, Cincinnati, OH, EPA/600/R-06/060, 2006.)

- Impact assessment: Assess the potential human and ecological effects of energy, water, and material usage and the environmental releases identified in the inventory analysis.

- Interpretation: Evaluate the results of the inventory analysis and impact assessment to select the preferred product, process, or service with a clear understanding of the uncertainty and the assumptions used to generate the results.

The goal definition and scoping stage defines goal and scope of the entire process to ensure that the most meaningful results are obtained. Every decision made throughout the goal definition and scoping phase impacts either how the study will be conducted, or the relevance of the final results. This process includes defining the goal(s) of the project, determining the type of information needed for the decision makers, determining the specificity required, determining how the data should be organized and the results displayed, defining the scope of the study, and determining the ground rules for performing the work.

The LCI analysis is conducted to collect and organize all data that quantify energy and raw material requirements, atmospheric emissions, waterborne emissions, solid wastes, and other releases for the entire life cycle of a product, process, or activity. In this stage, a flow diagram of the processes being evaluated and a data collection plan are developed. The data are then collected and evaluated to give reported results. In the flow diagram, a "system" or "system boundary" is identified having unit processes inside the system boundary with inputs and outputs (material and energy) to the processes. The output of the process includes categorizing and quantifying products and coproducts along with three types of emissions (atmospheric, waterborne, and solid wastes).

The life-cycle impact assessment (LCIA) phase of an LCA is the evaluation of potential human health and environmental impacts of the environmental resources and releases identified during the LCI. Impact assessment addresses ecological and human health effects and resource depletion. An LCIA attempts to establish a linkage between the product or process and its potential environmental impacts. The results of an LCIA show the relative differences in potential environmental impacts for each option. For example, an LCIA could determine which product/process causes more global warming potential. The steps for conducting LCIA include selection and definition, classification, characterization, normalization, grouping and weighting of impact categories, and evaluating and reporting LCIA results. Impacts are defined as the consequences that could be caused by the input and output streams of a system on human health, plants, and animals, or the future availability of natural resources.

Typically, LCIAs focus on the potential impacts to three main categories: human health, ecological health, and resource depletion. The commonly used impact categories for LCIA are given in Table A.6. The characterization of impact factors provides a way to directly compare the LCI results within

TABLE A.6

Impact Categories Used in LCIA

Impact Category	Scale	Examples of LCI Data (i.e., Classification)	Common Possible Characterization Factor	Description of Characterization Factor
Global warming	Global	Carbon dioxide (CO_2) Nitrogen dioxide (NO_2) Methane (CH_4) Chlorofluorocarbons (CFCs) Hydrochlorofluorocarbons (HCFCs) Methyl bromide (CH_3Br)	Global warming potential	Converts LCI data to carbon dioxide (CO_2) equivalents Note: global warming potentials can be 50, 100, or 500 year potentials
Stratospheric ozone depletion	Global	Chlorofluorocarbons (CFCs) Hydrochlorofluorocarbons (HCFCs) Halons Methyl bromide (CH_3Br)	Ozone depleting potential	Converts LCI data to trichlorofluoromethane (CFC-11) equivalents
Acidification	Regional local	Sulfur oxides (SOx) Nitrogen oxides (NOx) Hydrochloric acid (HCL) Hydrofluoric acid (HF) Ammonia (NH_4)	Acidification potential	Converts LCI data to hydrogen (H+) ion equivalents
Eutrophication	Local	Phosphate (PO_4) Nitrogen oxide (NO) Nitrogen dioxide (NO_2) Nitrates Ammonia (NH_4)	Eutrophication potential	Converts LCI data to phosphate (PO_4) equivalents

Photochemical smog	Local	Nonmethane hydrocarbon (NMHC)	Photochemical oxidant creation potential	Converts LCI data to ethane (C_2H_6) equivalents
Terrestrial toxicity	Local	Toxic chemicals with a reported lethal concentration to rodents	LC50	Converts LC50 data to equivalents; uses multimedia modeling, exposure pathways
Aquatic toxicity	Local	Toxic chemicals with a reported lethal concentration to fish	LC50	Converts LC50 data to equivalents; uses multimedia modeling, exposure pathways
Human health	Global regional local	Total releases to air, water, and soil	LC50	Converts LC50 data to equivalents; uses multimedia modeling, exposure pathways
Resource depletion	Global regional local	Quantity of minerals used Quantity of fossil fuels used	Resource depletion potential	Converts LCI data to a ratio of quantity of resource used versus quantity of resource left in reserve
Land use	Global regional local	Quantity disposed of in a landfill or other land modifications	Land availability	Converts mass of solid waste into volume using an estimated density
Water use	Regional local	Water used or consumed	Water shortage potential	Converts LCI data to a ratio of quantity of water used versus quantity of resource left in reserve

Source: SAIC, *Life Cycle Assessment: Principles and Practice*, Scientific Applications International Corporation, National Risk Management Research Laboratory, Cincinnati, OH, 2006.

each impact category, known as impact indicators, with the use of character-ization factors, as shown in Equation A.1. The choice of the characterization factor determines appropriate impact indicator result. For some impact cat-egories, such as global warming and ozone depletion, there is a consensus on acceptable characterization factors. For other impact categories, such as resource depletion, a consensus is still being developed.

$$\text{Inventory data} \times \text{Characterization factor} = \text{Impact indicators} \qquad (A.1)$$

EPA's Tool for the Reduction and Assessment of Chemical and Other Environmental Impacts (TRACI) is an impact assessment tool that supports consistency in environmental decision making (SAIC, 2006). TRACI allows the examination of the potential for impacts associated with the raw mate-rial usage and chemical releases resulting from the processes involved in producing a product. It allows the user to examine the potential for impacts for a single life-cycle stage, or the whole life cycle, and to compare the results between products or processes. The purpose of TRACI is to allow a determi-nation or a preliminary comparison of two or more options on the basis of the environmental impact categories given in Table A.6.

Normalization is an LCIA tool used to express impact indicator data in a way that can be compared among impact categories by dividing the data by a selected reference value which can be the total emissions or resource use for a given area (global, regional, or local), the total emissions or resource use for a given area on a per capita basis, the ratio of one alternative to another (i.e., the baseline), or the highest value among all options. The normalized data can be compared only within an impact category. The grouping step sorts or ranks impact categories on the basis of characteristics like emissions (e.g., air and water emissions) or location (e.g., local, regional, or global) or on the basis of a ranking system, such as high, low, or medium priority. The weighting step (also referred to as valuation) of an LCIA assigns weights or relative values to the different impact categories based on their perceived importance or relevance. The last step in LCA includes the identification of the significant issues based on the LCI and LCIA stages; evaluation of the results considering completeness, sensitivity, and consistency checks; and reporting conclusions and recommendations.

Singh et al. (2008) give an environmental impact assessment for potential carbon nanotube processes developed by Agboola (2005). Two CNT processes were considered using a plug flow reactor and a fluidized bed reactor. The environmental impact data obtained from the LCI and LCIA stages of LCA are given in Tables A.7 and A.8. Table A.7 gives the data where CO_2 is released into the atmosphere (Case 1), and Table A.8 gives the data where CO_2 is uti-lized within the system (Case 2). Both of the CNT process designs are concep-tual, and it is clearly evident from the data in Case 1 that the CNT processes cannot release the CO_2 into the atmosphere. The Case 2 is also an ideal case where all the CO_2 is utilized in the process. The Case 2 reduces the global

TABLE A.7

Environmental Impact Data for Base Design of CNT-PFR
and CNT-FBR Process

Impact Category	Unit	CNT-PFR Process	CNT-FBR Process
Global warming	CO_2 eq.	6.29	5.82
Acidification	H+ moles eq.	0.1570	0.0799
HH noncancer	Toluene eq.	0.0583	0.0695
Smog	NOx eq.	0.00252	0.00939
HH criteria air-mobile	PM2.5 eq.	0.000421	0.000182
HH criteria air-point source	PM2.5 eq.	0.00040	0.00017
Eutrophication	N eq.	0.000102	0.0000621
HH cancer	Benzene eq.	0.0000559	0.0000644
Ecotoxicity	2,4-D eq.	0.0000373	0.0000137

TABLE A.8

Environmental Impact Data for New Design

Impact Category	Unit	CNT-PFR Process	CNT-FBR Process
Global warming	CO_2 eq.	1.81	1.24
Acidification	H+ moles eq.	0.157	0.0799
HH cancer	Benzene eq.	0.0000559	0.0000644
HH noncancer	Toluene eq.	0.0583	0.0695
HH criteria air-point source	PM2.5 eq.	0.00040	0.00017
HH criteria air-mobile	PM2.5 eq.	0.000421	0.000182
Eutrophication	N eq.	0.0001020	0.0000621
Ecotoxicity	2,4-D eq.	0.0000373	0.0000137
Smog	NOx eq.	0.00252	0.00154

warming impact category from Case 1. Thus, this demonstrates the need to use LCA in determining the environmental impact of a particular process.

A comparative analysis of various design schemes for a process or complex is performed to isolate rank contribution of units to harm to the environment during the life cycle. Results from applying life-cycle assessment to the base case and optimal configuration of plants in the chemical production complex is described using TRACI, a program developed by the USEPA (Singh et al., 2007).

Singh et al. (2007) give the environmental impact assessment for the chemical production complex in the lower Mississippi River corridor. The base case contains plants in the existing complex, and the superstructure contains plants with new processes that utilize the CO_2 produced in the complex (Xu, 2004). The new processes were included by using the Chemical Complex Analysis System to form an optimal structure of plants in the chemical production complex. The environmental impact factors based on the base case and the superstructure are given in Table A.9.

TABLE A.9

Comparison of Impact Category for Base Case and
Optimal Superstructure for Chemical Production Complex
in Lower Mississippi River Corridor

Base Case		New Design Scheme	
Process	Value	Process	Value
Acidification (moles of H⁺ equivalent)			
Nitric acid	20	Nitric acid	19
Sulfuric acid	879	Sulfuric acid	879
Urea	21	Urea	21
Total	920	Total	919
Fossil fuel (MJ)			
Ammonia	1,480	Ammonia	1,480
Methanol	368	Methanol	368
Power generation	10,973	Power generation	20,191
		Acetic acid	12
		Graphite	198
		Synthesis gas	299
Total	12,820	Total	22,547
Global warming (kg CO₂)			
Sulfuric acid	9	Sulfuric acid	9
Nitric acid	1	Nitric acid	1
Power generation	310	Power generation	571
Urea	0	Urea	0
Methanol	1		
Ammonia	1,350		
Total	1,672	Total	581
Water (gal)			
Ammonia	138	Ammonia	138
Sulfuric acid	1,752	Sulfuric acid	1,752
Urea	10	Urea	10
Methanol	14	Methanol	14
Phosphoric acid	2,213		
		Power generation	599
Total	4,126	Total	2,512
Eutrophication (kg N)			
Nitric acid	0.02	Nitric acid	0.02
Urea	0.03	Urea	0.03
Total	0.05	Total	0.05

TABLE A.9 (continued)

Comparison of Impact Category for Base Case and
Optimal Superstructure for Chemical Production Complex
in Lower Mississippi River Corridor

Base Case		New Design Scheme	
Process	Value	Process	Value
Human health noncancer (lbs of C_7H_7 equivalent)			
Sulfuric acid	0.01	Sulfuric acid	0.01
Urea	0.70	Urea	0.70
		Propylene	0.65
Total	0.71	Total	1.36
Photochemical smog (g NO_x equivalent)			
Methanol	0.00	Propene	279.93
Human health criteria (DALY)			
Sulfuric acid	0.24	Sulfuric acid	0.24

Source: Singh, A. et al., *Resour. Conserv. Recycl.*, 51(2), 294, 2007.

Niederl and Narodoslawsky (2004) reported a life-cycle assessment for
the transesterification of tallow (TME) and used vegetable oil (UVO) to fatty
acid methyl and ethyl esters, commonly known as biodiesel. They use SPI
and Centrum Milieukunde Leiden (CML) method as the impact assessment
methods to evaluate the process. The process began with goal definition or
function of the biodiesel from UVO and TME, and the functional unit used
to quantify the goal was the combustion energy (calorific value) of biodiesel.
The reference flow was set at 1 MJ of combustion energy.

Three different scenarios were analyzed for the LCA. These are discussed
as follows:

- *Scenario I:* This scenario addresses the production of biodiesel from
  UVO. The first step in the life cycle of the production of biodiesel
  from UVO is the collection of waste cooking oil and transportation
  caused for the collection. The raw material collected is then processed in a transesterification step to produce biodiesel. The biodiesel thus produced is then transported to facilities before being burnt
  as a fuel. A similar scenario is also applicable for tallow starting at
  the collection of tallow from meat rendering facilities.

- *Scenario II:* This scenario is included for the tallow as raw material.
  The meat rendering waste is further processed into meat and bone
  meal and tallow. The tallow is then transported to the biodiesel
  manufacturing facility and transesterified.

- *Scenario III*: This scenario further analyzes the slaughtering process before the production of tallow. The tallow is a by-product of the meat production process, and the returns from the meat sale and slaughtering process influence the production of tallow.

Based on the system boundaries, the SPI was obtained between −1.2 and $4.8\,m^2a/MJ$ for UVO and between 0.85 and $8.3\,m^2a/MJ$ for TME as compared to $26.1\,m^2a/MJ$ for fossil diesel. The lower SPI for bio-based and waste feedstock showed a positive impact when compared to higher values for fossil diesel.

A.9 Total Value Proposition

Pater (2006) gives a method to evaluate the total value proposition (TVP) of technologies described as clean energy technologies. These technologies use renewable sources of energy and hence are called clean energy. The method for using clean energy can also be used to evaluate the total value for evaluating the use of biomass as feedstock. TVP includes the evaluation of five categories to capture the total value and financial benefits. These categories include risk management, emission reduction, direct policy incentives, reduced resource use, and corporate social responsibility. Most of the benefits included in these categories fit within the conventional definition of value, which includes revenue enhancement, cost reduction, and brand value.

Apart from the benefits to the owners of intellectual property, manufacturers, or distributors of the technology, these categories also take into account the end users. Investors often cannot predict how much end users will pay for the benefits. As a result, investors feel uncomfortable developing a business plan around such uncertain values, resulting in an undervaluation of clean energy technologies. A sixth category of benefits, societal economic benefits, accrues to a wide range of beneficiaries including individual firms through tax breaks and other incentives. These categories and their benefits are discussed in detail in Table A.10.

A.10 Carbon Dioxide Costs

This section gives the costs for carbon dioxide reported in the literature. Some of these were included in Chapter 7 for Case Study II.

Carbon offsets: One of the methods of assigning sustainable costs is carbon offsets (LaCapra, 2007). Carbon offset programs require consumers to pay certain amounts which are used for programs that create renewable energy

TABLE A.10

Categories in TVP Applied in Using Clean Energy Technologies

Category	Benefits/Values
Risk management	• Hedge against fuel-price volatility • Hedge against grid outages • Getting ahead in the futures markets • Prepare for regulatory change • Reduce insurance premiums • Reduce future risks of climate change
Emission reductions	• Generate emission reduction credits/offsets • Reduce fees for emissions • Avoid remediation costs
Policy incentives	• Production tax credit • Accelerated depreciation • Preferential loan treatment • Renewable portfolio standard • Renewable energy certificates • System benefit funds • Rebates • Feed-in tariffs • Net metering • Property tax break • Sales-tax exemption • Local R&D incentives • Other financial incentives
Reduced resource use	• Reduce water use and consumption • Reduce energy use • Decrease production costs
Corporate social responsibility	• Improve stakeholder relations • Satisfy socially responsible investing (SRI) portfolio criteria
Societal economic benefits	• Rural revitalization • Jobs • Economic development • Avoided environmental costs of fuel extraction/transport • Avoided costs of transmission and distribution infrastructure expansion

Source: Pater, J.E., *A Framework for Evaluating the Total Value Proposition of Clean Energy Technologies*, National Renewable Energy Laboratory, Golden, CO, 2006.

or absorb carbon dioxide to counteract emissions. The average global citizen emits 4.5 tons of carbon dioxide per year, and for a U.S. citizen, this amount is 21 t/year. A grown tree uses 3–15 lb carbon dioxide per year. Carbon offset programs are offered by companies like Carbonfund.org, DrivingGreen.com, TerraPass, NativeEnergy, myclimate, CarbonNeutral, etc. These companies have projects aimed at utilizing carbon dioxide, and consumers pay on the basis of such programs. Some of the companies allow the calculation of

TABLE A.11

Sample Carbon Offset Prices Offered by Companies

Organization/Site	Sample Project	Price per Ton of CO_2 Offset
Carbonfund.org	Providing solar energy for low-income families in the area	$5.50
DrivingGreen.com	Converting methane from farm animal waste to renewable energy	$8.00
TerraPass	Purchasing carbon credits on the CCX	$10.00
NativeEnergy	Funding wind turbine projects in Native American and Alaska Native communities	$12.00
Myclimate	Constructing solar greenhouses in the Himalayas so that produce does not have to be flown there	$18.00
CarbonNeutral	Promoting energy-efficient lighting in tourism sector, particularly hotels	$18.40
Climate Trust	—	$12.00

Source: LaCapra, L.T., *The Wall Street Journal*, March 1, 2007; Hileman, B., *Chem. Eng. News*, 31, February 19, 2007.

carbon offsets based on carbon emissions of road trips, flights, or homes, and then donate the cost to the companies which funds the projects based on the donations. There are several costs associated with different carbon dioxide usage. A transatlantic flight costs $9, and a year's driving costs amount to $50. The price per ton of carbon offset for different sites is given in Table A.11.

Emissions trading: The Chicago Climate Exchange (CCX) is the world's first GHG-based emissions trading system (CCX, 2007). CCX emitting members make a voluntary but legally binding commitment to meet annual GHG emission reduction targets. Those who reduce below the targets have surplus allowances to sell or bank; those who emit above the targets comply by purchasing CCX Carbon Financial Instrument® (CFI™) contracts. The CFI contract represents 100 MT of CO_2 equivalents. CFI contracts are comprised of exchange allowances and exchange offsets. Exchange allowances are issued to emitting members in accordance with their emission baseline and the CCX emission reduction schedule. Exchange offsets are generated by qualifying offset projects. The CFI price for October with the totally traded and anticipated carbon for future years is given in Table A.12.

Chicago Climate Futures Exchange (CCFE), a wholly owned subsidiary of the CCX, is a U.S. Commodity Futures Trading Commission designated contract market which offers standardized and cleared futures contracts on emission allowances and other environmental products (CCFX, 2007). The products offered by the CCFE are given in Table A.13 along with sample data for December 2007 settlement prices of the products.

A futures contract is a standardized contract, traded on a futures exchange, to buy or sell a certain underlying instrument at a certain date in the future,

TABLE A.12

CFI Monthly Summary from CCX

Product	Vintage	High[a]	Low[a]	Close	Change	Volume[b]
CFI	2003	$3.00	$2.50	$2.60	−0.50	76,700
CFI	2004	$3.00	$2.50	$2.50	−0.55	61,700
CFI	2005	$3.00	$2.50	$2.50	−0.50	86,600
CFI	2006	$3.00	$2.50	$2.50	−0.50	88,900
CFI	2007	$3.00	$2.50	$2.50	−0.50	97,100
CFI	2008	$3.00	$2.50	$2.50	−0.50	47,300
CFI	2009	$3.00	$2.50	$2.50	−0.50	30,500
CFI	2010	$3.00	$2.50	$2.50	−0.50	35,600
Total electronically traded volume						524,400

Source: CCX, Chicago Climate Exchange, http://www.chicagoclimatex.com/, accessed October 8, 2007.
[a] Price units: per metric ton of CO_2.
[b] Volume: electronically traded volume reported in metric tons CO_2.

at a specified price (CFTC, 2012). The future date is called the delivery date or final settlement date. The preset price is called the futures price. The price of the underlying asset on the delivery date is called the settlement price. Thus, this can account for universally accepted Type V costs identified in AIChE/TCA report.

Carbon dioxide sequestration cost: Another method to account for sustainability is to evaluate the cost to sequester carbon dioxide. Carbon dioxide sequestration is the removal of atmospheric carbon dioxide and storage under the ground or in the deep sea. Technology for capturing of CO_2 is commercially available for large CO_2 emitters, such as power plants, but storage systems of CO_2 have not been developed yet on commercial scale.

Carbon capture and storage (CCS) applied to a modern conventional power plant can reduce CO_2 emissions to the atmosphere by approximately 80%–90% compared to a plant without such facility (Metz, B. et al., 2005). However, capturing and compressing CO_2 requires energy and would increase the fuel needs of a plant with CCS by about 10%–40%. These and other system costs are estimated to increase the cost of energy from a power plant with CCS by 30%–60% depending on the specific circumstances. Intergovernmental Panel on Climate Change (Metz et al., 2005) reports the costs associated with carbon dioxide capture and storage. Three types of costs include capture (including compression), transport, and storage.

Capture costs for different types of power plants are represented as an increase in the electricity generation cost (U.S.$ MWh^{-1}) (Metz et al., 2005). For most large sources of CO_2 (e.g., power plants), the cost of capturing CO_2 is the largest component of overall CCS costs. In this report, capture costs include the cost of compressing the CO_2 to a pressure suitable for pipeline transport (typically about 14 MPa). The design and operation of both the CO_2 capture system

TABLE A.13

CCFE Products and Futures Settlement Prices

Symbol	Product	Settlement Price for December 2007
CER	Certified emission reduction futures CERs are compliant GHG emission reduction credits issued by the United Nations for approved and verified projects undertaken in developing countries	$22.03/CER allowance 1 Contract is equivalent to 1000 mt CER
CFI	CFI futures	$2.61/mt CO_2 1 Contract is equivalent to 1000 mt of CO_2
ECFI	European CFI futures	$0.11/mt CO_2 1 Contract is equivalent to 1000 mt of CO_2
ECO-index	ECO-clean energy index futures	$254.00 1 Contract is equivalent to 50 USD times the value of the ECO-clean energy index
IFEX-ELF	IFEX event linked futures	$1.4–$6.6/U.S. wind event from $50 billion to $10 billion 100 USD multiplied by the event loss trigger index value
NFI-A	Nitrogen financial instrument (annual) futures	$2400–$4000/ton NOx (December 2008) Range depends on 1–4 years deferred vintage value 1 Contract is equivalent to 1 U.S. EPA CAIR annual NOx emission allowance
NFI-OS	Nitrogen financial instrument- ozone season futures	$651–$704/ton NOx Range depends on various vintage year values 1 Contract is equivalent to 5 U.S. EPA NOx emission allowances
SFI	Sulfur financial instrument futures and options	$573.60/ SO_2 emission allowance 1 Contract is equivalent to 25 U.S. EPA SO_2 emission allowances

Source: CCFX, Chicago Climate Futures Exchange, http://www.ccfe.com/, accessed October 8, 2007.

and the power plant or industrial process to which it is applied influence the overall cost of capture. The studied systems are new power plants based on coal combustion or gasification. These processes with percentage reduction of emissions and increase in cost of electricity generation are given in Table A.14.

Transport costs are given in U.S.$/t$CO_2$ per kilometer. The economical method reported is to transport large amounts of CO_2 through pipelines. A cost competitive transport alternative for longer distances at sea is the use of large tankers. Cost elements for pipelines are construction costs (e.g., material, labor, and possible booster station), operation and maintenance costs (e.g., monitoring, maintenance, and possible energy costs), and other costs (e.g., design, insurance, fees, and right-of-way). Special land conditions, like

TABLE A.14

Cost of CO_2 Capture for Representative Processes

Type of Plant	Technology Used	Cost of Electricity U.S.\$ MWh^{-1} Increase (%)	CO_2/Kilowatt-Hour (kWh) Decrease (%)
Modern (high-efficiency) pulverized coal-burning power plant (PC)	Amine-based scrubber	40–70	85
New natural gas combined cycle (NGCC)	Amine-based scrubber	37–69	83–88
New coal based plant employing an integrated gasification combined cycle (IGCC) system	Water gas shift reactor followed by a physical absorption system	20–55	81–91
New hydrogen plant	CO_2 compression	5–33	72–96

Source: Metz, B. et al., *IPCC Special Report on Carbon Dioxide Capture and Storage*, Cambridge University Press, New York, 2005.

heavily populated areas, protected areas such as national parks, or crossing major waterways, may have significant cost impacts. Figure A.11 shows the cost for CO_2 transport for "normal" terrain conditions. Tankers could also be used for transport. Here, the main cost elements are the tankers themselves (or charter costs), loading and unloading facilities, intermediate storage facilities, harbor, and special purpose CO_2 tankers. On the basis of preliminary designs, the costs of CO_2 tankers are estimated at U.S.\$ 34 million for ships

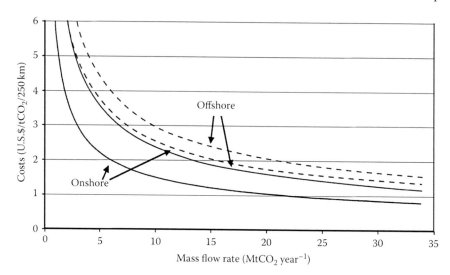

FIGURE A.11

CO_2 transport costs range for onshore and offshore pipelines per 250 km, "normal" terrain conditions. (From Metz, B. et al., *IPCC Special Report on Carbon Dioxide Capture and Storage,* Cambridge University Press, New York, 2005.)

TABLE A.15

Estimates of CO_2 Storage Costs

Option	Representative Cost Range (U.S.$/ton CO_2 Stored)	Representative Cost Range (U.S.$/ton C Stored)
Geological—storage	0.5–8.0	2–29
Geological—monitoring	0.1–0.3	0.4–1.1
Ocean pipeline	6–31	12–16
Ocean ship (platform or moving ship injection)	22–114	44–59
Mineral carbonation	50–100	180–370

Source: Metz, B. et al., *IPCC Special Report on Carbon Dioxide Capture and Storage,* Cambridge University Press, New York, 2005.

of 10,000 t, U.S.$ 58 million for 30,000 t vessels, and U.S.$ 82 million for ships with a capacity of 50,000 t. To transport 6 Mt CO_2 per year, a distance of 500 km by ship would cost about 10 U.S.$/t$CO_2$ (37 U.S.$/tC) or 5 U.S.$/tCO_2/250 km (18 U.S.$/tC/250 km). However, since the cost is relatively insensitive to distance, transporting the same 6 Mt CO_2 a distance of 1250 km would cost about 15 U.S.$/t$CO_2$ (55 U.S.$/tC) or 3 U.S.$/tCO_2/250 km (11 U.S.$/tC/250 km). This is close to the cost of pipeline transport, illustrating the point that ship transport becomes cost competitive with pipeline transport if CO_2 needs to be transported over larger distances.

The storage costs are stated in U.S.$/t$CO_2$ stored. The storage can be geological storage, ocean storage, and storage via mineral carbonation. Table A.15 gives a cost range for the three types of storage costs. The total cost based on the three aspects, capture, transport, and storage, for the power plants are given in Table A.16.

Summary for sustainable costs: There is limited information for sustainable and societal costs, and estimation of these costs can be made using data such as carbon offsets and cost to sequester carbon dioxide. The AIChE/TCA report gives some costs due to emissions, and some of these costs are based on damage cost approach based on a willingness to pay to avoid adverse human health effects, agricultural effects, and material damage by the Minnesotans for an Energy-Efficient Economy website. The carbon offset method discussed is a way to compensate for carbon dioxide usage by individuals. These offset prices can be extended to emissions by a chemical plant. Offset prices offered range from $5 to $12. These prices are not controlled by any governing body, though the Federal Trade Commission has announced that they will investigate the prices. The CCX is the world's first emissions-based trading system, and the cost for a CFI contract costs $2.50–$3.00. The CCFE sells futures contracts, which can be considered for the future costs that a company pays for sustainability. The CCS costs are estimated at $50/ton CO_2, and this cost is paid by the company to fulfill the sustainability criteria. These costs can be used in the triple bottom line to account for sustainability costs.

TABLE A.16

Total Costs for CO_2 Capture, Transport, and Geological Storage Based on Current Technology for New Power Plants

	Pulverized Coal Power Plant	Natural Gas Combined Cycle Power Plant	Integrated Coal Gasification Combined Cycle Power Plant
Cost of electricity without CCS (U.S.$ MWh^{-1})	43–52	31–50	41–61
Power plant with capture			
Increased fuel requirement (%)	24–40	11–22	14–25
CO_2 captured (kg MWh^{-1})	820–970	360–410	670–940
CO_2 avoided (kg MWh^{-1})	620–700	300–320	590–730
% CO_2 avoided	81–88	83–88	81–91
Power plant with capture and geological storage			
Cost of electricity (U.S.$ MWh^{-1})	63–99	43–77	55–91
Electricity cost increase (U.S.$ MWh^{-1})	19–47	12–29	10–32
% increase	43–91	37–85	21–78
Mitigation cost (U.S.$/tCO$_2$ avoided)	30–71	38–91	14–53
Mitigation cost (U.S.$/tC avoided)	110–260	140–330	51–200
Power plant with capture and enhanced oil recovery			
Cost of electricity (U.S.$ MWh^{-1})	49–8	37–70	40–75
Electricity cost increase (U.S.$ MWh^{-1})	15–29	6–22	(−5) to 19
% increase	12–57	19–63	(−10) to 46
Mitigation cost (U.S.$/tCO$_2$ avoided)	9–44	19–68	(−7) to 31
Mitigation cost (U.S.$/tC avoided)	31–160	71–250	(−25) to 120

Source: Metz, B. et al., *IPCC Special Report on Carbon Dioxide Capture and Storage*, Cambridge University Press, New York, 2005.

Transport costs range from 0 to 5 U.S.$/tCO$_2$.

Geological storage cost (including monitoring) ranges from 0.6 to 8.3 U.S.$/tCO$_2$.

Costs for geological storage including EOR range from −10 to −16 U.S.$/tCO$_2$ stored.

A.11 Summary

The TCA methodology was used in the present research. TCA gives a quantitative method for evaluating the triple bottom line profit. All the other methods consider separate criteria for sustainability analysis, whereas the TCA compares costs in the triple bottom line. Possible methods to estimate sustainable costs are given in this chapter, which are based on cost of carbon dioxide emitted to the atmosphere. These were the basis for sustainable cost/credit evaluation in the case studies.

Appendix B: Optimization Theory

This section describes the optimization algorithm used for solving the mixed-integer nonlinear programming problem (MINLP) and the multicriteria optimization problem for this research. These problem statements are given in this section. A brief review of the various types of optimization problems and how they can be solved is given. These include local optimization problems like linear programming (LP), nonlinear programming (NLP), mixed-integer linear programming (MILP), and MINLP followed by multicriteria or multiobjective optimization. A discussion on global optimization accompanied with solution techniques and GAMS (general algebraic modeling system) solvers is given. The chemical complex optimization was done with the help of the chemical complex analysis system developed at Minerals Processing and Research Institute at LSU, which is discussed in a separate chapter.

B.1 Algorithm for Optimization of Chemical Complex

In the present research, an input–output block model was developed from each of the process designs for mathematical representation of the biochemical processes. For fermentation and anaerobic digestion processes, the complexity of the models required three blocks each to describe the overall processes. Then, these models were included into a base case of chemical plants to form the superstructure. Additional processes for utilizing carbon dioxide were added to the superstructure. Alternative choices were provided in the superstructure for production of chemicals like acetic acid production from the base case, the anaerobic digestion process, or the process utilizing CO_2. Also, the choice of at least one hydrogen process for production of chemicals from CO_2 was provided. MINLP methods were applied to the models with an objective function incorporating economic, environmental, and sustainable criteria (based on TCA methodology) with material and energy balance equations and capacity of plants as constraints.

The statement for the optimization problem in the chemical production complex can be given as follows:

Optimize: Objective function
Subject to: Constraints from plant model

The objective function is a profit function for complex economic optimization shown as follows:

$$\text{Triple bottom line} = \text{Profit} - \Sigma \text{ Environmental costs}$$
$$+ \Sigma \text{ Sustainable (Credits} - \text{Costs)} \qquad \text{(B.1)}$$

$$\text{Profit} = \Sigma \text{ Product sales} - \Sigma \text{ Raw material costs} - \Sigma \text{ Energy costs} \qquad \text{(B.2)}$$

$$\text{Triple bottom line} = \Sigma \text{ Product sales} - \Sigma \text{ Raw material costs} - \Sigma \text{ Energy costs}$$
$$- \Sigma \text{ Environmental costs} + \Sigma \text{ Sustainable (Credits} - \text{Costs)}$$
$$\text{(B.3)}$$

The constraint equations describe relationship among variables and parameters in the processes, and they are material and energy balances, chemical reaction rates, thermodynamic equilibrium relations, and others. There have been no reports on methods to evaluate the sustainable development of chemical production complexes except by Xu (2004). Recent trends in biomass feedstock utilization for production of chemicals were relevant for this research to study the integration of bioprocesses into existing nonrenewable feedstock-based chemical production complex. This was a key component of this research. The base case of existing chemical plants in the lower Mississippi River corridor was used to demonstrate the results of integrating bioprocesses in the nonrenewable feedstock based chemical production complex.

The multicriteria algorithm in chemical complex analysis system is given in the following text. The objective of optimization is to find optimal solutions that maximize industry' profits and minimize costs to society. This multicriteria optimization problem can be stated as in terms of industry's profit, P, and society's sustainable credits/costs, S, and these two objectives are given in Equation B.4. To locate Pareto optimal solutions, multicriteria optimization problems are converted to a single criterion by applying weights to each objective and optimizing the sum of the weighted objectives (Equation B.5):

$$\text{Max: } P = \Sigma \text{ Product sales} - \Sigma \text{ Economic costs} - \Sigma \text{ Environmental costs} \qquad \text{(B.4)}$$

$$S = \Sigma \text{ Sustainable (Credits} - \text{Costs)}$$

Subject to: Multiplant material and energy balances, product demand, raw material availability, plant capacities

$$\text{Max: } w_1 P + w_2 S \qquad \text{(B.5)}$$

$$w_1 + w_2 = 1$$

Subject to: Multiplant material and energy balances, product demand, raw material availability, plant capacities

The GAMS was used for optimization in the Chemical Complex Analysis System. With this method, the relationships among the economic, environmental, and sustainable costs were evaluated by maximizing the triple bottom line for the chemical production complex.

B.2 Optimization of Chemical Process Systems

Synthesis of chemical process systems have been discussed in details by Grossmann et al. (1999). The mathematical programming approach to design and integration problems involves three steps. First, a set of alternatives is developed from which optimum solution is selected. Second, the formulation of a mathematical problem that involves discrete and continuous variables for the selection of the configuration and operating levels is required. The third step involves solving the optimization model from which the optimal solution is determined.

The mathematical form for mixed-integer optimization problems expressed in algebraic form is given by Grossmann et al. (1999) as shown in Equation B.6:

$$\text{Minimize: } Z = f(x, y) \tag{B.6}$$

$$\text{Subject to: } h(x, y) = 0$$

$$g(x, y) \leq 0$$

$$x \in X, \quad y \in \{0, 1\}$$

where
   $f(x, y)$ is the objective function (e.g., cost)
   $h(x, y) = 0$ are the equalities that describe the performance of the system (mass and heat balances, design equations)
   $g(x, y) \leq 0$ are inequalities that define the specifications or constraints for feasible choices

The variables x are continuous and generally correspond to the state or design variables, while y are the discrete variables, which generally are restricted to take binary values to define the selection of an item or an action. The aforementioned equation is a mixed-integer programming model formulation. If any of the functions involved in Equation B.1 is nonlinear, the problem corresponds to a mixed-integer nonlinear program. If all functions are linear, it corresponds to a mixed-integer linear program. If there are no binary variables (0–1), then the problem reduces to a nonlinear program or linear program depending on whether the functions are linear.

The solution of these problems can be effectively performed in modeling systems such as GAMS. GAMS can be interfaced with codes (solvers) for optimizing various types of problems. The chemical complex analysis system was developed at LSU Minerals Processing Research Institute, which interfaces with GAMS to determine the best configuration of plants in a chemical complex based on economic, energy, environmental, and sustainable costs. The details of the chemical complex analysis system are given in a later chapter.

In the application of mathematical programming techniques to design and synthesis problems, it is always necessary to postulate a superstructure of alternatives (Grossmann et al., 1999). This is true whether one uses a high-level aggregated model or a detailed model. There are two major issues that arise in postulating a superstructure. The first is, given a set of alternatives that are to be analyzed, the determination of major types of representations that can be used with the implications for the modeling is necessary. The second is, for a given representation that is selected, the feasible alternatives must be included to guarantee that the global optimum is found.

The selection of the level of detail of the optimization model is closely related to selection of the superstructure. Mathematical models can be classified into three main classes: aggregated models, shortcut models, or rigorous models (Grossmann et al., 1999). Aggregated models are high-level representations in which the design or synthesis problem is greatly simplified by an aspect or objective that tends to dominate the problem at hand. Shortcut models refer to fairly detailed superstructures that involve cost optimization (investment and operating costs) but in which the performance of the limits is predicted with relatively simple nonlinear modes in order to reduce the computational cost and/or for exploiting the algebraic structure of the equations, especially for global optimization. Rigorous models rely on detailed superstructures and involve rigorous and complex models for predicting the performance of the units.

B.3 Local Optimization

LP is the simplest type of optimization problem where the objective function and constraint equations are all linear. The general equation for LP can be written as given in Equation B.7. LP requires all constraint equations be written as equalities. Inequalities need to be converted to equality constraints using slack and surplus variables.

$$\text{Optimize: } \sum_{j=1}^{n} c_j x_j$$

$$\text{Subject to: } \sum_{j=1}^{n} a_{ij} x_j = b_i \tag{B.7}$$

$$x_j \geq 0 \quad j = 1, 2, \ldots, n, \quad i = 1, 2, \ldots, m$$

NLP refers to multivariable optimization procedures where the equation for objective function and constraint equations are nonlinear functions of variables. The general representation of the NLP problem is given as in Equation B.8. There are n independent variables, $x = (x_1, x_2, \ldots, x_n)$, and m constraint equations, h of them being equality constraints. The values of x_j's can have lower and upper bounds specified.

$$\text{Optimize: } \qquad y(\mathbf{x})$$

$$\text{Subject to: } \quad \begin{aligned} f_i(\mathbf{x}) &= 0 \quad \text{for } i = 1, 2, \ldots, h \\ f_i(\mathbf{x}) &\geq 0 \quad \text{for } i = h+1, \ldots, m \end{aligned} \tag{B.8}$$

MILP refers to an extension of LP problem where some variables are required to be integers. The use of integer variables makes possible the formulation of models with discrete selection of variables or constraints. The problem statement of MILP is given in Equation B.9. When all the variables are integers, the problem is referred to as integer programming. A special case of integer programming is binary integer programming (BIP), where a variable takes only 1 or 0 as values. BIP is applied to problems where "yes-or-no decisions" are important.

$$\text{Optimize: } \mathbf{c}^{\mathsf{T}}\mathbf{x} + \mathbf{h}^{\mathsf{T}}\mathbf{y} \tag{B.9}$$

$$\text{Subject to: } \mathbf{A}\mathbf{x} + \mathbf{G}\mathbf{y} \leq \mathbf{b}$$

$\mathbf{A}$ and $\mathbf{G}$ are $m \times n$ and $m \times p$ matrices
$\mathbf{b}$ is m-dimensional vector
$\mathbf{x}^{\mathsf{T}} = (x_1, \ldots, x_n)$ $\mathbf{x}$ is n-dimensional vector of positive continuous variables
$\mathbf{y}^{\mathsf{T}} = (y_1, \ldots, y_p)$ $\mathbf{y}$ is p-dimensional vector of positive integer variables

MINLP problem refers to that class of optimization problems where the variables are a combination of binary, integer, and continuous variables. $\mathbf{x}$ is a vector of continuous variables, which represent the process variables such as flow rates, temperature, pressures, etc., and y is a set of binary variables, which can be used to define the topology of the system representing the existence or nonexistence of different processing units. The nonlinearities in the economic and process models appear in the terms $f(x)$, $g(x)$, and $h(x)$. The problem statement can be written as in Equation B.10:

$$\text{Optimize:} \quad \mathbf{c}^{\mathsf{T}}\mathbf{y} + f(\mathbf{x})$$

$$\text{Subject to:} \quad \begin{aligned} \mathbf{A}\mathbf{y} + h(\mathbf{x}) &= 0 \\ \mathbf{B}\mathbf{y} + g(\mathbf{x}) &\leq 0 \end{aligned} \tag{B.10}$$

$\mathbf{x}^{\mathsf{T}} = (x_1, \ldots, x_n)$ $\mathbf{x}$ is n-dimensional vector of positive continuous variables
$x^L \leq x \leq x^U$ lower and upper bounds on each variable
$\mathbf{y}$ is a vector taking only the values 0 and 1
$\mathbf{A}\mathbf{y} \leq \mathbf{a}$ constraint on y

B.4 Multiobjective Optimization

Multiobjective optimization, also called multicriteria optimization, is the simultaneous optimization of more than one objective function. The general multiobjective problem (MOP) is defined as in Equation B.11:

$$\text{Optimize: } F(\mathbf{x}) = [f_1(\mathbf{x}), f_2(\mathbf{x}), \ldots, f_k(\mathbf{x})]^{\mathsf{T}} \tag{B.11}$$

$$\text{Subject to: } g_i(\mathbf{x}) \geq 0 \quad i = 1, 2, \ldots, m$$

$$h_j(\mathbf{x}) = 0 \quad j = 1, 2, \ldots, p$$

$$x^L \leq x \leq x^U$$

There are various methods to solve multicriteria optimization problems like utility function, hierarchical methods, and goal programming. Of these, using the utility function or weighted objective method is the most commonly used. In this method, weights are assigned to the different objective functions, and the sum of the weights equals 1.0. This can be represented as in Equation B.12:

$$\text{Optimize: } F'(\mathbf{x}) = \sum_{i=1}^{n} w_i f_i(x)$$

$$\sum_{i=1}^{n} w_i = 1 \tag{B.12}$$

The multicriteria problem can be a MINLP problem where the multiple objective functions and the constraints are nonlinear and the variables are continuous or integer. The MINLP problem in this research was formulated into a multicriteria problem by maximizing the profit and the sustainability credits simultaneously.

A detailed review of multicriteria optimization in sustainable energy decision making was given by Wang et al. (2009). Technical criteria, economic criteria, environmental criteria, and social criteria were discussed in the paper along with weighted objective methods.

B.5 Methods for Solving Optimization Problems

LP problems are solved using simplex algorithm for local optimization. NLP problems and MILP problems can be solved using branch and bound techniques.

MINLP problems can be solved using several algorithms including branch and bound, generalized benders decomposition (GBD), the alternative dual approach, the outer approximation/equality-relaxation (OA/ER), and the feasibility technique (Grossmann et al., 1999). Each of these methods has advantages and disadvantages. For example, the branch and bound methods can require solution of a large number of NLP subproblems unless the NLP relaxation is very tight. GBD, on the other hand, requires many major iterations in successfully solving the NLP and MILP master problems, and it allows exploitation of special structures in the NLP subproblems. In OA/ER, the number of required major iterations is small but the size of MILP master problem is quite large. Moreover, the MILP master problem in the OA/ER algorithm predicts stronger lower bounds and also provides good initial guesses for the NLP subproblems. The feasibility technique is the least expensive, and all methods find only local optima.

All the methods described earlier find local optima for a problem. The next section introduces the concept of global optimization and various methods that are used to solve the maximum or minimum.

B.6 Global Optimization

Significant research has been spent developing algorithm that finds the global optimum of a problem directly. This would eliminate using the procedure of finding all the local optima and then comparing these local optima to find the largest one "global optima."

Global optimization is the task of finding the absolutely best set of values of variables to optimize an objective function (Gray et al., 1997). Global optimization problems are typically difficult to solve. Global optimization problems are solved by extension of ideas from local optimization. These algorithms are integrated into computer programs for solving the problems.

The GAMS is a high-level modeling language for mathematical programming and optimization. It consists of a language compiler and integrated high-performance solvers. GAMS is tailored for complex, large-scale modeling applications and allows building of large maintainable models that can be adapted quickly to new situations. The GAMS offers a wide range of solvers, which allow the optimization based on type of problem. These include LP, NLP, MILP, MINLP, and global optimization solvers. The following section describes the global optimization solvers used in this research.

B.7 Optimization Solvers

GAMS can interface with various solvers developed on the algorithms mentioned earlier for solving various types of problems. An extensive list of solvers can be found at GAMS website (GAMS, 2010) for solving LP, NLP, MIP, MILP, and MINLP problems. Combinations of these solvers are required to solve the optimization problems. The solvers used to solve the problem in the Chemical Complex Analysis System were simple branch and bound (SBB) solver for MINLP, CONOPT for NLP, and CPLEX for MIP.

The NEOS server for optimization hosted by the Argonne National Laboratory is an open and free to use server for solving optimization problems (NEOS, 2010). The optimization solvers at NEOS represent the state-of-the-art in optimization software. Optimization problems are solved automatically with minimal input from the user. The users only need a definition of the optimization problem, and all additional information required by the optimization solver is determined automatically by the server. For example, the solver choice for MINLP is required, but the subchoices for LP and NLP need not be specified in the server.

The superstructure was solved using five different solvers from the NEOS server. These were DICOPT, SBB, BARON, ALPHAECP, and LINDOGLOBAL. Two of these solvers were listed exclusively under global solvers, which accepted GAMS input (BARON and LINDOGLOBAL), and the other three were listed under MINLP solvers (DICOPT, SBB, and ALPHAECP). The results for computation time and solver status from the NEOS server solution are given in Table B.1. The SBB, DICOPT, and BARON gave a normal completion

TABLE B.1

Comparison of Solvers in NEOS Server for Optimal Solution

Solver	SBB (MINLP)	DICOPT (MINLP)	ALPHAECP (MINLP)	BARON (Global)	LINDOGLOBAL (Global)
Objective value	16.500316	16.500313	NA	16.49418566	NA
Solver status	Normal completion	Normal completion	Normal completion	Normal completion	Iteration interrupt
Model status	Integer solution	Integer solution	Infeasible—no solution	Integer solution	No solution returned
Additional solvers chosen by NEOS	CONOPT 3 (NLP)	XPRESS (MIP) CONOPT 3 (NLP)	—	ILOG CPLEX (LP) MINOS (NLP)	—
Iteration count	246/10,000	318/10,000	47/10,000	0/10,000	0/10,000
Resource usage	0.340/1,000.000	0.370/1,000.000	62.110/1,000.000	40.000/1,000.000	10.336/1,000.000
Compilation time	0.037 s	0.034 s	0.036 s	0.034 s	0.037 s
Generation time	0.024 s	0.025 s	0.014 s	0.025 s	0.014 s
Execution time	0.026 s	0.027 s	0.016 s	0.027 s	0.016 s

TABLE B.2

Comparison of Solvers in NEOS Server and Local Machine

Solver	SBB (MINLP) (NEOS Server)	SBB (MINLP) (Local Machine)
Objective value	16.500316	16.500316
Solver status	Normal completion	Normal completion
Model status	Integer solution	Integer solution
GAMS version	GAMS Rev 228 x86/Linux	GAMS Rev 232 WIN-VIS 23.2.1 x86/MS Windows
Additional solvers chosen by NEOS	CONOPT 3 (NLP)	CONOPT
Iteration count	246/10,000	214/2,000,000,000
Resource usage	0.340/1,000.000	0.359/1,000.000
Compilation time	0.037 s	0.015 s
Generation time	0.024 s	0.063 s
Execution time	0.026 s	0.063 s

with identical solutions for the objective value. The LINDOGLOBAL was unable to solve because of an iteration interrupt. The ALPHAECP gave a normal completion with infeasible solution. Table B.2 gives the comparison of the solution using SBB in the Neos server and the local machine.

Appendix C: Prices of Raw Materials and Products in the Complex

C.1 Price of Renewable Raw Materials

The renewable raw materials are corn (for fermentation to ethanol), corn stover (for fermentation to ethanol, anaerobic digestion to acetic acid, and gasification to syngas), and soybean oil (for transesterification to fatty acid methyl ester [FAME]). Algae are raw materials for producing algae oil used in transesterification to FAME. The raw materials are discussed in detail in the following section.

C.1.1 Corn

The historical price, demand, and supply of corn in the United States were obtained from Feedgrains Yearbook 2010 from USDA (2010b). The data were used in this section for computing the average price and standard deviation of corn and for use in the figures and tables. In the database, data "disappearance" of corn was given, which is interpreted as "demand" for all calculations in this section.

The historical demand and supply of corn for 1976–2011 are shown in Figure C.1 (USDA, 2010b). The values for 2010 and 2011 are projected values estimated by the USDA. The primary use for corn historically has been for feed use. The other uses for corn include alcohol for fuel additive, high-fructose corn syrup (HFCS), glucose and dextrose, starch, alcohol for beverages and other manufacturing, cereals, and other products and seeds.

Figure C.2 shows the use of corn for the aforementioned categories except feed grain use. Comparing to Figure C.1, it can be seen that more than 5000 million bu/year of corn was used for feed in 1981 and approximately 800 million bu/year was for these other uses.

From Figures C.1 and C.2, it can also be seen that there was an increase in the use of corn from 2000 to 2010. In the period from 2005 to 2010, the demand for corn increased by 115%. This was attributed to the production of alcohols as fuel additives from corn.

The historical price (1867–2011) of corn is shown in Figure C.3 (USDA, 2010b). The recent price (2000–2011) of corn is shown in Figure C.4. The price

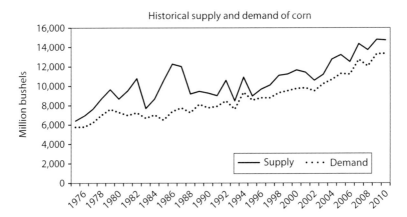

FIGURE C.1
Historical supply and demand for corn. (From USDA, Feed grains database: Yearbook tables, http://www.ers.usda.gov/data/feedgrains/, accessed April 8, 2010, 2010b.)

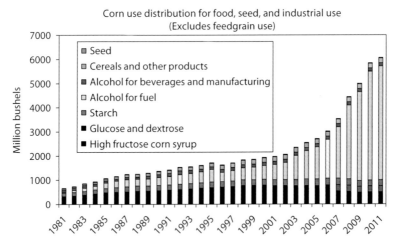

FIGURE C.2
Distribution of corn for food, seed, and industrial use. (From USDA, Feed grains database: Yearbook tables, http://www.ers.usda.gov/data/feedgrains/, accessed April 8, 2010, 2010b.)

of corn increased to $4.20/bu in 2008. The increase in demand as seen in Figures C.1 and C.2 is one of the reasons for the increase in price of corn.

Table C.1 shows the average price and standard deviation in the price of corn for the whole period for which data are available (1867–2011) and the last 10 years (2000–2011). U.S. corn bushel weighs 56 lb/bu. It is seen from Table C.1 that the deviations in price for the whole time period and for the last 10 years are the same.

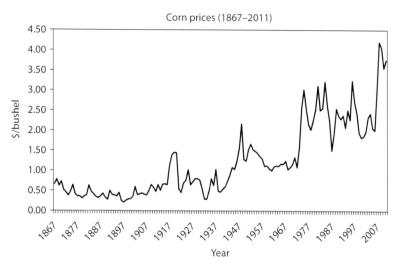

FIGURE C.3

Historical price of corn. (From USDA, Feed grains database: Yearbook tables, http://www.ers.usda.gov/data/feedgrains/, accessed April 8, 2010, 2010b.)

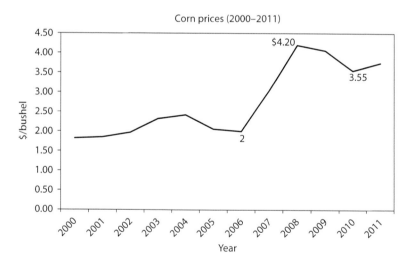

FIGURE C.4

Recent price of corn. (From USDA, Feed grains database: Yearbook tables, http://www.ers.usda.gov/data/feedgrains/, accessed April 8, 2010, 2010b.)

C.1.2 Corn Stover

Perlack and Turhollow (2002) discuss the logistics and estimate the delivered costs for collecting, handling, and hauling corn stover to an ethanol conversion facility. The costs for two conventional baling systems (large round bales and large rectangular bales), a silage-harvest system, and an unprocessed

TABLE C.1

Average Price and Standard Deviation
in the Price of Corn

	Price ($/bu)	Price ($/MT)
Average (1867–2011)	1.23	48.40
Standard deviation (1867–2011)	0.91	35.86
Average (1981–2011)	2.75	108.26
Standard deviation (1981–2011)	0.91	35.86

pickup system are discussed in this paper. The results indicate that stover can be collected, stored, and hauled for about $43.60–$48.80/dry ton ($48.10–$53.80/dry Mg), using conventional baling equipment for conversion facilities ranging in size from 500 to 2000 dry ton/day (450–1810 dry Mg/day). These estimates are inclusive of all costs including farmer payments for the stover. The results also suggest that costs might be significantly reduced with an unprocessed stover pickup system provided more efficient equipment is developed.

Aden et al. (2002) estimated direct costs of baling and staging stover at the edge of the field and the cost of transportation from the farm to the plant gate. Total delivered cost of corn stover was $62/dry MT ($56/dry ST) of stover. Of this, 23% of the cost was for transportation, 47% was for baling and staging the stover, 12% was for fertilizers, and 18% was farmer premium. A nominal cost of $33/dry MT ($30/dry ST) was estimated through improved collection (e.g., single pass) techniques to calculate the cost for cellulosic ethanol. An updated report from the NREL in 2009 (Humbird and Aden, 2009) stated that the cost basis used in their technology assessments for washed and milled corn stover delivered to the throat of the pretreatment reactor was $60/dry ton through 2008 and $46/dry ton in the years 2009–2012.

Idaho National Laboratory (INL) and Oak Ridge National Laboratory (ORNL) estimated feedstock harvest and logistics and grower payments (Humbird and Aden, 2009). They report feedstock cost estimated at $69.60/dry ton in 2007 and expect to decrease each year to reach $50.90/dry ton in 2012. This cost increase over the previous assumption by Aden et al. (2002) showed that a cost reduction to $33/dry ton was not likely to be achieved in the near future.

In summary, the average price and standard deviation using the data points for price of corn stover delivered to the plant gate are $60.83/MT (dry) and $9.40/MT (dry), respectively.

C.1.3 Soybean Oil

The historical price, demand, and supply of soybean oil in the United States were obtained from Oil Crops Yearbook 2010 from USDA (2010a). The datasets were used to obtain the graphs and calculate the average price and standard deviation in the price for soybean oil. Such deviations need to be accounted for while using the oil as feedstock. Apart from food use, soybean oil is currently

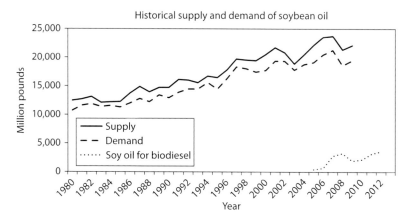

FIGURE C.5

Historical supply and demand of soybean oil and use for biodiesel. (From USDA, Oil crops year-book, http://usda.mannlib.cornell.edu/MannUsda/viewDocumentInfo.do?documentID=1290, accessed April 8, 2010, 2010a.)

used for the production of FAME, also known as biodiesel. Biodiesel is a substitute for diesel and blends of 20% biodiesel and 80% petrodiesel are used.

The historical demand and supply of soybean oil for 1980–2009 are shown in Figure C.5 (USDA, 2010a). The use of soy oil in the production of FAMEs (biodiesel) for the period 2005–2009 and projected usage (medium usage) for 2010–2012 is also shown in Figure C.5 (AGMRC, 2010). The historical price (1980–2009) of soybean oil is shown in Figure C.6 (USDA, 2010a). The recent price (2000–2009) of soybean oil in Figure C.7 shows the recent trend in usage of soybean oil for biofuels. Table C.2 shows the average price and standard deviation in the price of soybean oil for the last 30 years (1980–2009) and last 10 years (2000–2009).

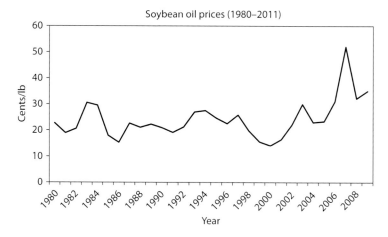

FIGURE C.6

Historical price of soybean oil. (From USDA, Oil crops yearbook, http://usda.mannlib.cornell.edu/MannUsda/viewDocumentInfo.do?documentID=1290, accessed April 8, 2010, 2010a.)

FIGURE C.7
Recent price of soybean oil. (From USDA, Oil crops yearbook, http://usda.mannlib.cornell.edu/MannUsda/viewDocumentInfo.do?documentID=1290, accessed April 8, 2010, 2010a.)

TABLE C.2

Average Price and Standard Deviation in the Price of Soybean Oil

	Price (cents/lb)	Price ($/MT)
Average (1980–2009)	24.17	532.85
Standard deviation (1980–2009)	7.45	164.24
Average (2000–2009)	27.92	615.52
Standard deviation (2000–2009)	10.88	239.86

Source: USDA, Oil crops yearbook, http://usda.mannlib.cornell.edu/MannUsda/viewDocumentInfo.do?documentID=1290, accessed April 8, 2010, 2010a.

From Figure C.7, it can be seen that the price of soybean oil reached a record high of 52 cents/lb of oil in 2007–2008. The average price of soybean oil from 2000–2009 was 28 cents/lb, with a standard deviation of 11 cents/lb. This is a high deviation of 39%.

C.2 Chemical Prices for Bioprocess Catalysts and Nutrients

The raw materials for bioprocesses were discussed in the previous section. In this section, the price of biocatalysts, catalysts, and nutrients used for the bioprocesses are discussed. The prices used in the model are shown in Table C.3. The description of the prices and the respective references are given in the following section.

TABLE C.3

Raw Material and Product Prices in Chemical and Biochemical Complex

Raw Materials	Cost ($/MT)	Standard Deviation ($/MT)	Products	Cost ($/MT)	Standard Deviation ($/MT)
Corn stover	60.83	9.4	Bioethylene	930	—
Corn	108.26	36	FAME	968	213
Soybean oil	616	240	Biopropylene glycol	1636	84
Cellulase	146	—	Acetic acid	515	35
Corn steep liquor	177	—	DDGS	99	—
Bacteria	146	—	Ammonia	424	237
Sodium methylate	980	—	Methanol	435	211
HCl	215	—	Acetic acid	515	35
NaOH	617	—	GTSP	370	—
Lime	90	—	MAP	423	—
Iodoform	3300	—	DAP	457	7.89
MIBK	1290	—	Ammonium nitrate	373	—
α-Amylase	3300	—	Urea	354	17.4
Caustic	12	—	UAN	237	—
Glucoamylase	3300	—	Phosphoric acid	772	—
Sulfuric acid	110	—	Hydrogen	1490	460
Yeast	5510	—	Ethylbenzene	1543	—
Steam	9.83	—	Styrene	1260	—
Water	0.02	—	Propylene	1207	442
Natural gas	382	105	Formic acid	735	—
Phosphate rock			MMA	1610	—
Wet process	27	—	DMA	1610	—
Electric furnace	34	—	DME	946	—
Haifa process	34	—	Ethanol	1224	108
GTSP process	32	—	Toluene	813	222
HCl	215	—	Graphite	2500	—
Sulfur			Fuel gas	1274	—
Frasch	53	9.5	CO	70	19
Claus	21	3.55			
Coke electric furnace	124	—			
Propane	180	—			
Benzene	914	337			
Ethylene	1071	378			
Reducing gas	75	—			
Wood gas	88	—			

The corn stover fermentation section required cellulase enzymes, corn steep liquor, diammonium phosphate (DAP), and bacteria to seed fermentors. The cost of cellulase enzymes was set at $0.10/gal ethanol produced by Humbird and Aden (2009). An update to the report in 2009 reported the price of cellulase enzymes to be $0.12/gal of ethanol and was not expected to go below that price. So, the price of cellulase enzymes was taken as $0.12/gal of ethanol, which converted to $0.06624/lb (using data from Aden et al., 2002 and Humbird and Aden, 2009) or $146/MT. The price of corn steep liquor was reported as $0.0804/lb ($177/MT) in Aden et al., (2002). The price of DAP is taken as $457/MT from Zaworski (2010). Bacteria for corn stover fermentation section were generated in the fermentation facility with seed bacteria. The cost for seed bacteria was assumed to be same as that for cellulase enzymes at $146/MT. The products from the corn stover fermentation section were ethanol, fine particles, steam, and residual solids from centrifuge. The solids were considered waste. The price of bioethylene computed from HYSYS and ICARUS was $930/MT.

For the transesterification section, the raw materials include algae oil and soybean oil. The costs for algae oil reported by Pokoo-Aikins et al. (2010) are $0.68/lb ($1500/MT) oil (with 30% oil content) and $0.35/lb ($772/MT) oil (with 50% oil content), as given in Table 7.14. The average price for soybean oil is $616/MT and standard deviation is $240/MT. The price for sodium methylate catalyst was $980/MT (Haas et al., 2006). The price for hydrochloric acid was $209–$220/MT (Zaworski, 2010). The average cost of $215/MT was used for HCl. The price for sodium hydroxide was $617/MT (Haas et al., 2006). The selling price of FAME as computed from HYSYS and ICARUS was $2.26/gal or $688/MT. A wide variation in the price of FAME has been reported. Haas et al. (2006) reported a biodiesel cost (FAME) of $2.00/gal, and Myint and El-Halwagi (2009) assumed biodiesel costs of $2.75/gal and $3.00/gal for production of FAME from soybean oil. Pokoo-Aikins et al. (2010) considered biodiesel selling price of $3.69/gal and $4.20/gal for algae oil as raw material. These data points were averaged and that price for FAME was used for the complex. The average price for FAME was $3.18/gal ($968/MT), and the standard deviation was $0.70/gal ($213/MT). The average price for propylene glycol was $1636/MT, and the standard deviation was $84/MT (Zaworski, 2010).

For the anaerobic digestion process, the inputs were corn stover, lime, iodoform, nutrients, terrestrial inoculum through pig manure, and methyl isobutyl ketone as solvent for extraction. The terrestrial inoculum and pig manure did not have any price associated, as they were cheap and no transportation cost was required for the raw material. Corn stover cost was already incorporated in the price for corn stover fermentation. The price of lime is reported as $43/MT (Holtzapple et al., 1999). Lime is also used in the pretreatment for corn in the corn fermentation section, and an updated price of $0.09/kg ($90/MT) for lime was given by Kwiatkowski et al. (2006). The price of iodoform was reported as $3.3/kg ($3300/MT) in Holtzapple et al. (1999). The price of solvent (MIBK) was reported as $1290/MT in ICIS (2008a).

The outputs from the process were waste solids. The acetic acid price reported in Zaworski (2010) was $480–$550/MT. The average price of $515/MT and deviation of $35/MT were used in the complex.

For the corn ethanol fermentation section, the inputs were corn, α-amylase, glucoamylase, and yeast. Lime, caustic, and sulfuric acid were also used in the pretreatment process. The price of corn over the last 10 years was used in the complex and was computed as shown in the previous section. The average price for corn was $108.26/MT and the standard deviation was $36/MT. The price for amylase and yeast was obtained from Kwiatkowski et al. (2006) as $3.310/kg ($3310/MT) and $5.510/kg ($5510/MT), respectively. The price of sulfuric acid was $0.11/kg ($110/MT) (Kwiatkowski et al., 2006). The caustic was diluted in water in the process, and the price of the diluted mixture was included in the complex as $0.012/kg ($12/MT). The products from the process were ethanol, which was used for ethylene, and distillers dry grain solids (DDGS) sold at $99/MT (Kwiatkowski et al., 2006).

The biomass gasification plant used corn stover and produced hydrogen and carbon monoxide, which were existing chemical in the complex. These costs are explained in the next section where the prices of raw materials and products are updated from the base case by Xu (2004). The average price of carbon monoxide was $70/MT and the standard deviation was $19/MT.

C.3 Prices of Chemicals in Existing Complex and Chemicals from Carbon Dioxide

This section describes the price of raw materials and products in the existing base case of plants. This section also includes the price of chemicals that were produced from carbon dioxide in the complex. The prices are given in Table C.3. The data were collected from various sources (ICIS 2006a,b, 2007, 2008a,b,c,d,e, 2009a,b,c,d,e,f,g, 2010a,b,c) and explained in the following.

C.3.1 Natural Gas

The price of natural gas in the United States was obtained from Energy Information Administration (EIA, 2010d). The price for natural gas from January 2001 to January 2010 was used for Figure C.8. The average price and standard deviation of natural gas price over the period were $6.73 and $2.15 per thousand cubic feet, respectively.

Figure C.9 shows the price of natural gas over the period of 2005–2009. This is shown in a separate graph as this period had major fluctuations in the price of natural gas. The average price for the 5 year period of 2005–2010

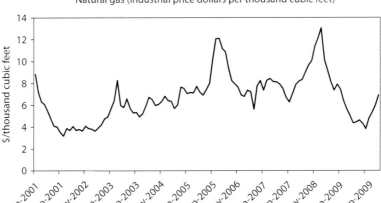

FIGURE C.8
Price of natural gas. (From EIA, United States Natural Gas Industrial Price (dollars per thousand cubic feet), http://tonto.eia.doe.gov/dnav/ng/hist/n3035us3m.htm, accessed April 10, 2010, 2010d.)

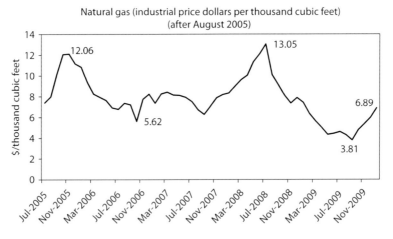

FIGURE C.9
Price of natural gas (after August 2005). (From EIA, United States Natural Gas Industrial Price (dollars per thousand cubic feet), http://tonto.eia.doe.gov/dnav/ng/hist/n3035us3m.htm, accessed April 10, 2010, 2010d.)

was $7.85 per thousand cubic feet and the standard deviation was $2.15 per thousand cubic feet. This price was used for the economic model.

The cost per metric ton of natural gas was computed based on the conversion 1.0 MT LNG = 48.7 thousand cubic feet (Hofstrand, 2010). With this relation, the cost for natural gas was $382/MT natural gas and the standard deviation was $105/MT.

C.3.2 Hydrogen

The price of hydrogen depends on the price of natural gas. The price is shown in Figure C.10 and explained in the following. Using the price of natural gas as $7.5 per thousand cubic feet or million BTUs, the formula given by Kuehler (2003) to compute the hydrogen price is as follows:

Hydrogen price ($/thousand SCF)

$$= \frac{0.9 \times (\text{Natural gas price in } \$/\text{MBTU or thousand cubic feet})}{2} + 0.45$$

where SCF is standard cubic feet

$$= 0.45 \,(\text{natural gas price in } \$/\text{MBTU}) + 0.45$$

$$= (0.45 \times 7.5 + .45)\$/1000 \text{ ft}^3$$

$$= (3.825)\ \$/28.316 \text{ m}^3$$

$$= 0.135 \ \$/\text{m}^3$$

Thus, 1 m^3 of hydrogen costs $0.135.

Kuehler (2003) reported that the energy content (heat of combustion) of natural gas was 310 BTU/SCF. The density of hydrogen at standard state

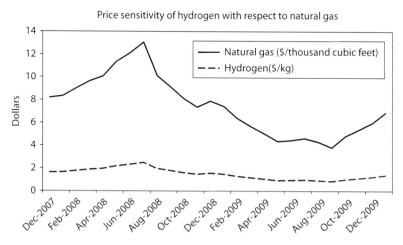

FIGURE C.10
Price of hydrogen and natural gas. (From EIA, United States Natural Gas Industrial Price (dollars per thousand cubic feet), http://tonto.eia.doe.gov/dnav/ng/hist/n3035us3m.htm, accessed April 10, 2010, 2010d; Kuehler, G.P., Exxon Mobil Baton Rouge Refinery, Private Communication, October 10, 2003.)

taken from *Perry's Chemical Engineers' Handbook* is $0.0898 \, kg/m^3$. Using the density of hydrogen, the price of hydrogen can be represented in terms of $\$/kg$ of H_2.

Thus, the price of hydrogen = $0.135/0.0898 \, \$/kg \, H_2$
$$= 1.504 \, \$/kg \, H_2$$
$$= 0.68 \, \$/lb \, H_2$$

The previous equation determines that the price of natural gas is the variable in the cost for hydrogen. Thus, to compute the true price for hydrogen over the past 2 years, the price of natural gas is considered, and the sensitivity of hydrogen to the price of natural gas is determined, and this is shown in Table C.4. Figure C.10 shows the price of natural gas and the price of

TABLE C.4

Price of Hydrogen with Respect to Natural Gas

	Natural Gas Price	Hydrogen Price $\$/1000 \, ft^3$	Hydrogen Price $\$/m^3$	Hydrogen Price $\$/kg$
December 2007	$8.18	$4.13	$0.15	$1.62
January 2008	$8.33	$4.20	$0.15	$1.65
February 2008	$9.00	$4.50	$0.16	$1.77
March 2008	$9.64	$4.79	$0.17	$1.88
April 2008	$10.06	$4.98	$0.18	$1.96
May 2008	$11.36	$5.56	$0.20	$2.19
June 2008	$12.11	$5.90	$0.21	$2.32
July 2008	$13.05	$6.32	$0.22	$2.49
August 2008	$10.11	$5.00	$0.18	$1.97
September 2008	$9.13	$4.56	$0.16	$1.79
October 2008	$8.11	$4.10	$0.14	$1.61
November 2008	$7.36	$3.76	$0.13	$1.48
December 2008	$7.89	$4.00	$0.14	$1.57
January 2009	$7.43	$3.79	$0.13	$1.49
February 2009	$6.37	$3.32	$0.12	$1.30
March 2009	$5.65	$2.99	$0.11	$1.18
April 2009	$5.03	$2.71	$0.10	$1.07
May 2009	$4.35	$2.41	$0.09	$0.95
June 2009	$4.45	$2.45	$0.09	$0.96
July 2009	$4.62	$2.53	$0.09	$0.99
August 2009	$4.31	$2.39	$0.08	$0.94
September 2009	$3.81	$2.16	$0.08	$0.85
October 2009	$4.80	$2.61	$0.09	$1.03
November 2009	$5.37	$2.87	$0.10	$1.13
December 2009	$5.97	$3.14	$0.11	$1.23
January 2010	$6.89	$3.55	$0.13	$1.40
		Average		$1.49
		Standard deviation		$0.46

hydrogen from December 2007 to January 2010. The average price of hydrogen computed over the period was $1.49/kg H_2 and the standard deviation was $0.46. The price of $1490/MT and standard deviation of $460/MT for hydrogen were used in the complex.

C.3.3 Benzene

The average price of benzene was $3.04/gal ($914/MT) and the standard deviation was $1.12/gal ($337/MT). The price of benzene is shown in Figure C.11. The data for benzene were obtained from key indicator prices in ICIS (2008–2010).

C.3.4 Ethylene

The average price of ethylene was 48.58 U.S. cents/lb ($1071/MT) and the standard deviation was 17.16 U.S. cents/lb ($378/MT). The price of ethylene is shown in Figure C.12. The data for ethylene were obtained from key indicator prices in ICIS (2008–2010).

C.3.5 Ammonia

The average price of ammonia was 424.20 $/t and the standard deviation was 237.19 $/t for the period January 2008–2010. The price of ammonia is shown in Figure C.13. The data for ammonia were obtained from key indicator prices in ICIS (2008–2010).

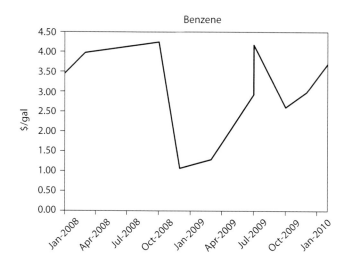

FIGURE C.11
Benzene price (January 2008–2010).

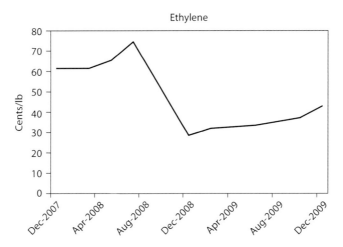

FIGURE C.12
Ethylene price (December 2007–2009).

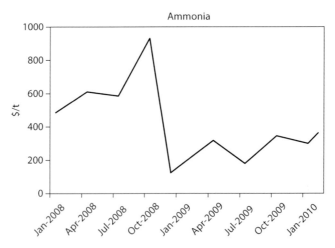

FIGURE C.13
Ammonia price (January 2008–February 2010).

C.3.6 Methanol

The average price of methanol was 130.78 cents/gal and the standard deviation was 63.34 cents/gal in the January 2008–February 2010 period. Using a
conversion rate of 332.6 gal/MT for ethanol (Methanex, 2010), the price for
methanol was $435/MT and standard deviation was $211/MT. The price of
methanol is shown in Figure C.14. The data for methanol were obtained from
key indicator prices in ICIS (2008–2010).

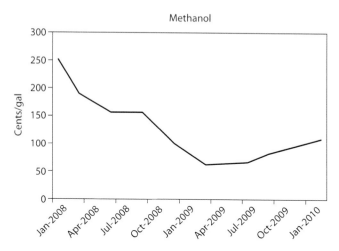

FIGURE C.14
Methanol price (January 2008–February 2010).

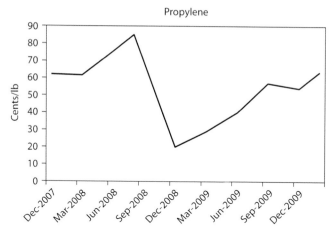

FIGURE C.15
Propylene price (December 2007–February 2010).

C.3.7 Propylene

The average price of propylene was 54.75 cents/lb ($1207/MT) and the standard deviation was 20.04 cents/lb ($442/MT) in the December 2007–February 2010 period. The price of propylene is shown in Figure C.15. The data for propylene were obtained from key indicator prices in ICIS (2008–2010).

C.3.8 Ethanol

The average price of ethanol was $3.39/gal ($1224/MT) and the standard deviation was $0.3/gal ($108.34/MT). The price of ethanol is shown

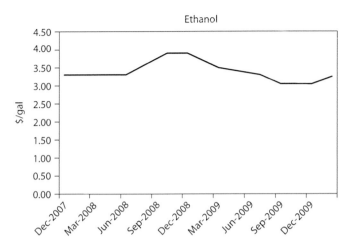

FIGURE C.16
Ethanol price (December 2007–February 2010).

in Figure C.16. The data for ethanol were obtained from key indicator prices in ICIS (2008–2010).

C.3.9 Toluene

The average price of toluene was $813/MT and the standard deviation was $222/MT. The price of toluene is shown in Figure C.17. The data for toluene were obtained from key indicator prices in ICIS (2008–2010).

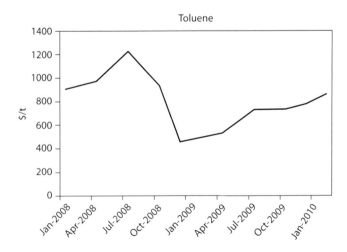

FIGURE C.17
Toluene price (January 2008–February 2010).

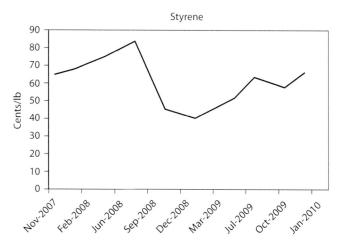

FIGURE C.18
Styrene price (December 2007–January 2010).

C.3.10 Styrene

Spot prices of styrene were reported to hit a record high of $1810–$1830/t in July 2008, but the price dropped in the fourth quarter of 2008 (ICIS, 2009g). The prices reported for February 2009 was $690–$710/t. The average price of $1260/MT of styrene was used in the complex. The price of styrene is shown in Figure C.18. The data for styrene were obtained from key indicator prices in ICIS (2008b–e, 2009c–f, 2010a–b).

C.3.11 Fertilizers: Urea, UAN, Ammonium Nitrate, GTSP, MAP, and DAP

The GTSP price for February 2010 was $350–$390/MT (Zaworski, 2010). The average price of $370/MT was used for GTSP. The price of $423/MT for monoammonium phosphate (MAP) was obtained from linear extrapolation with price for DAP as reported in Xu (2004) and price of DAP from Zaworski (2010). The DAP price for February 2010 was $452–$461/MT (Zaworski, 2010). The average price of $457/MT and standard deviation of $7.89/MT (from Xu, 2004) were used in the complex. The price of $373/MT for ammonium nitrate was obtained from linear extrapolation with price for DAP as reported in Xu (2004) and price of DAP from Zaworski (2010). Urea price reported for February 2010 ranged from $350.53 to $358.25/MT. The average price of $354/MT was used for urea and the standard deviation of $17.4/MT was used from Xu (2004). The price of $237/MT for UAN was obtained from linear extrapolation with price for urea as reported in Xu (2004) and price of urea from Zaworski (2010).

C.3.12 Ethylbenzene

The price of ethylbenzene was reported as 70 cents/lb (ICIS, 2010). The price was converted to $1543/MT and used in the complex.

C.3.13 Phosphoric Acid

The price for phosphoric acid was reported as $35/cwt in Hunt (2004). This was converted to $772/MT for using in the complex.

C.3.14 Formic Acid

Contract price for June 2006 in the United States for formic acid was reported as $0.70–$0.77/kg (Burridge, 2006). The average price of $0.735/kg ($735/MT) was used in the complex.

C.3.15 Methylamines (MMA and DMA) and Dimethyl Ether

The ICIS (2006) reported a hike of €20–€200/MT for methylamine and its derivatives. Due to lack of any other data, the price of $1610/MT reported by Xu (2004) was used in the complex. The price of $946/MT for dimethyl ether (DME) was used from Xu (2004).

C.3.16 Graphite

The price of graphite was obtained from a graphite producer company, and it was sold at $2500/MT in the complex (WGPL, 2010). The projections for graphite price did not change, so standard deviation was not considered for the price.

C.3.17 Fuel Gas

The price of fuel gas was linearly extrapolated using values for fuel gas and natural gas price from Xu (2004) and current natural gas price (EIA, 2010d). It was sold at $1274/MT in the complex.

C.3.18 Carbon Monoxide

The price for carbon monoxide was estimated based on the fuel value of carbon monoxide and the cost and heat of combustion for methane since the price of carbon monoxide was not available (Indala, 2004). The price of methane, or natural gas, was $382/MT and the standard deviation was $105/MT. Using these values and method outlined in Indala (2004), the average price of carbon monoxide was computed to be $70/MT and the standard deviation was $19/MT.

Appendix D: Supply, Demand, and Price Elasticity

D.1 Introduction

Describing supply and demand of goods and services in the economy is complicated. To provide insight, two measured parameters are used: price elasticity of demand (PED) and price elasticity of supply (PES). In economics, elasticity is defined as the ratio of the percent change in one variable to the percent change in another variable.

PED is a measure of the responsiveness of quantity demanded to changes in price (Arnold, 2008) as shown in Equation D.1. Mathematically, it is the ratio of percent change in the quantity of goods or services, Q, to the percent change in price, P. It shows the response of a quantity demanded for goods or services to a change in price. PED is almost always negative, that is, an increase in price will cause a reduction in demand. A value of PED between zero and minus one is considered inelastic:

$$\text{PED} = \frac{\Delta Q/Q \cdot 100}{\Delta P/P \cdot 100} \tag{D.1}$$

Economists often use the absolute value of price elasticity for analysis. The value of PED can be interpreted as follows:

- If $\text{PED} < -1$, then demand is price elastic (quantity demanded changes proportionately more than price changes).
- If $\text{PED} = -1$, then demand is unit elastic (quantity demanded changes proportionately to price changes).
- If $\text{PED} > -1$, then demand is price inelastic (quantity demanded changes proportionately less than price changes).
- If $\text{PED} = 0$, then demand is perfectly inelastic (quantity demanded does not change as price changes).
- If $\text{PED} = -\infty$, then demand is perfectly elastic (quantity demanded is extremely responsive to even small changes in price).

PES is the ratio of percent change in the quantity supplied, S, to the percent change in price, P, as shown in Equation D.2. It measures the sensitivity of the quantity of goods and services to the change in market price for those goods or services. PES is almost always positive, that is, an increase in price will cause an increase in supply. A value of PES less than one is considered inelastic:

$$PES = \frac{\Delta S / S \cdot 100}{\Delta P / P \cdot 100} \tag{D.2}$$

The value of PES can be interpreted as follows:

- If PES > 1, then supply is price elastic (supply quantity changes proportionately more than price changes).
- If PES = 1, then supply is unit elastic (supply quantity changes proportionately to price changes).
- If PES < 1, then supply is price inelastic (supply quantity changes proportionately less than price changes).
- If PES = 0, then supply is perfectly inelastic (supply quantity does not change as price changes).
- If PES = ∞, then supply is perfectly elastic (supply quantity is extremely responsive to even small changes in price).

Other parameters measured to describe economic interactions are given in the following. They have definitions similar to PED and PES. More information is provided in "Elasticity (economics)," Wikipedia (2010). Table D.1 gives some values of PED and PES reported in "Elasticity (economics)," Wikipedia (2010):

- Income elasticity of demand
- Cross PED

TABLE D.1

Some Typical Values of PED and PES

PED		PES	
Oil (world)	−0.4	Heating oil	1.57
Gasoline	−0.25 to −0.64	Gasoline	1.61
Transportation	−0.20 (bus) to −2.8 (car)	Housing	1.6–3.7
Steel	−0.2 to −0.3	Steel	1.2
Rice	−0.47	Cotton	0.3–1.0
Livestock	−0.5 to −0.6	Tobacco	7.0
Airline travel	−0.3 to −1.5		

Source: Wikipedia, Elasticity (economics), http://en.wikipedia.org/wiki/Elasticity_%28economics%29, accessed June 21, 2010, 2010.

- Cross elasticity of demand between firms
- Elasticity of intertemporal substitution
- Elasticity of scale

The method for determining PES and PED, as defined earlier, can be applied to raw materials, products, and intermediates in chemical manufacturing. The market for fermentation ethanol (also known as bioethanol) as a chemical is not yet established in the United States. However, the potential for bioethanol as a chemical exists. The PED and PES of corn (raw material), the PED and PES for bioethanol (intermediate for future petroleum ethanol substitute), and the PED for ethylene (current petroleum feedstock based ethylene) are estimated in this section, so that an insight can be gained for the requirements to evaluate these parameters.

D.2 Price Elasticity of Supply and Demand for Corn

Historically, corn has been used for food use and feed grain. With the ongoing efforts to substitute fossil fuels with biofuels, there has been a rise in the importance of fuel use of cereals (Banerjee, 2010). To study the effect of the rise in demand of corn, price elasticity was used. The data for price, supply, and demand for corn were obtained from USDA (2010b). The price of corn for 1981–2011 is given in Figure D.1. The total supply of corn, the total demand for corn, and the demand for fuel and feed use are given in Figure D.2.

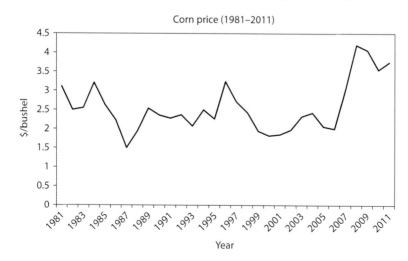

FIGURE D.1

Corn price 1981–2011. (From USDA, Feed grains database: Yearbook tables, http://www.ers.usda.gov/data/feedgrains/, accessed April 8, 2010, 2010b.)

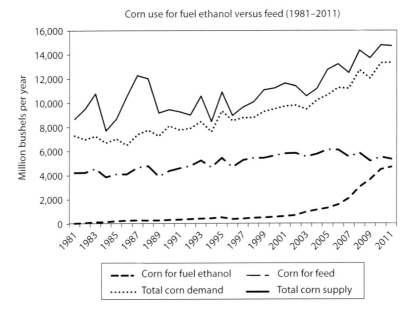

FIGURE D.2
Total corn demand and supply, corn demand for fuel ethanol, and demand for feed. (From USDA, Feed grains database: Yearbook tables, http://www.ers.usda.gov/data/feedgrains/, accessed April 8, 2010, 2010b.)

The short-term and long-term price elasticity for the supply of corn was computed for the period 1981–2011 in intervals of 5 years for the short-term price elasticities. The PED for corn in alcohol fuel use and for feed use were computed for the period 1981–2011 in intervals of 5 years for the short-term price elasticities. Also, for PED, the annual price elasticity data are given for the period of 2000–2010.

D.2.1 Price Elasticity of Supply for Corn

The supply for corn was distributed primarily for the production of fuel ethanol and the production of feed. Other uses of corn include food uses, alcoholic beverages, and seed uses. The long-term and short-term average PES are given in Table D.2. The complete data are given in Table D.10 at the end of this chapter.

Historically, corn has been used as a major feed grain, and as with all other feed grain crops, it is expected to be price inelastic. This means that with an increase or decrease in the supply of corn, the price is not expected to change. The long-term PES of corn calculated over the period 1981–2010 was −0.16, and the short-term elasticities averaged over each 5 year period between the period 1981 and 2010 varied between −1.29 and 0.59. The results for average PES for corn computed using Equation D.2 is given in Table D.2. The value

TABLE D.2

PES for Corn

Year	Average PES
1981–1985	0.59
1986–1990	−0.24
1990–1995	−1.29
1996–2000	0.12
2001–2005	−0.31
2006–2010	0.18
1981–2010 (long-term PES)	−0.16

of PES suggests that the supply was inelastic over the short terms, except for 1990–1995 when the supply was −1.29, a value slightly greater than −1. The rest of the short-term PES conforms with the long-term PES of corn supply being insensitive to price changes.

D.2.2 Price Elasticity of Demand for Corn in Fuel Use

The PED for corn in fuel use was computed using Equation D.1. The short-term average price elasticities are shown for each of the 5 year periods from 1981 to 2010 in Table D.3. The complete data are given in Table D.11 at the end of this chapter.

The long-term PED of corn converted to alcohol for fuel use has a value 0.51 as shown in Table D.3. The short-term average PED values vary between −1.08 and 4.18. The PED suggests that corn demand for conversion to fuel alcohol is price inelastic, particularly in the last two time periods between 2001–2005 and 2006–2010.

TABLE D.3

Average PED for Corn Used for Alcohol

Corn Used for Fuel Alcohol Production	
Year	Average PED
1981–1985	4.18
1986–1990	−0.89
1990–1995	0.22
1996–2000	0.64
2001–2005	0.01
2006–2010	−1.08
1981–2010 (long-term PED)	0.51

TABLE D.4

Annual PED for Corn Used for Alcohol from 2001 to 2010

Year	Production (Million Bushels per Year)	$\Delta Q/Q * 100$	Price ($/Bushel)	$\Delta P/P * 100$	PED
2001	629.83	12.29	1.85	6.49	1.89
2002	707.24	40.76	1.97	17.77	2.29
2003	995.50	17.28	2.32	4.31	4.01
2004	1,167.55	13.33	2.42	−14.88	−0.90
2005	1,323.21	21.17	2.06	−2.91	−7.27
				Average PED	0.01
2006	1,603.32	32.19	2.00	52.00	0.62
2007	2,119.49	43.87	3.04	38.16	1.15
2008	3,049.21	20.58	4.20	−3.33	−6.18
2009	3,676.88	22.39	4.06	−12.56	−1.78
2010	4,500.00	4.44	3.55	5.63	0.79
				Average PED	−1.08

To study the PED in these two time periods, the demand, price, and PED for each year are shown in Table D.4. The annual price elasticities reveal that the PED varied from −7.27 to 4.01 in the year range 2001–2005 and −6.18 to 1.15 in the year range 2006–2010. These high values in PED show that the demand in corn uses for alcohol production has increased steadily. Also, the change in demand was never negative for the corn use in fuel. The price for corn has fluctuated over time, varying from $1.85/bu in 2001, increasing to $2.42/bu in 2004, decreasing to $2.00/bu in 2006, and again increasing to $4.20/bu in 2008. These results show that the demand of corn for alcohol use has been price elastic over the last 10 years.

D.2.3 Price Elasticity of Demand for Corn in Feed Use

The PED for corn in feed use was computed using Equation D.1. The short-term average price elasticities are shown for each of the 5 year periods from 1981 to 2010 in Table D.5. The complete data are given in Table D.12 at the end of this chapter.

The long-term PED of corn used as feed has a value −0.01 as seen from Table D.5. This confirms the general notion that the price of feed grains is perfectly inelastic, meaning that the demand for feed is not going to decrease even for changes in price (this can happen due to several factors, one of them being the unavailability of alternatives) (Banerjee, 2010). From Table D.5, it is seen that the value of PED ranges from −0.80 to 0.59.

Table D.6 gives the result for annual price elasticity for the 10 year period from 2001 to 2010. This table also shows that the corn used for feed is price inelastic for each year except for the year 2008. A closer look at the quantity

TABLE D.5

Average PED for Corn Used for Feed

Corn Used for Feed	
Year	Average PED
1981–1985	0.59
1986–1990	−0.80
1990–1995	−0.46
1996–2000	0.06
2001–2005	0.09
2006–2010	0.45
1981–2010 (long-term PED)	−0.01

TABLE D.6

Annual PED for Corn Used for Feed from 2001 to 2010

Year	Production (Million Bushels per Year)	$\Delta Q/Q * 100$	Price ($/Bushel)	$\Delta P/P * 100$	PED
2001	5822.05	0.46	1.85	6.49	0.07
2002	5848.75	−5.14	1.97	17.77	−0.29
2003	5548.31	4.20	2.32	4.31	0.97
2004	5781.24	6.12	2.42	−14.88	−0.41
2005	6135.08	−0.33	2.06	−2.91	0.11
				Average PED	0.09
2006	6115.06	−9.40	2.00	52.00	−0.18
2007	5540.13	5.73	3.04	38.16	0.15
2008	5857.74	−11.14	4.20	−3.33	3.34
2009	5205.28	6.14	4.06	−12.56	−0.49
2010	5525.00	−3.17	3.55	5.63	−0.56
				Average PED	0.45

demanded in the year 2009 shows that the quantity demanded for feed dropped in 2009 for a 38% increase in price from 2007 to 2008 for corn. This is the only instance in the time period when the price of corn reached a record high of $4.20/bu of corn and that reflected in the PED for corn used as feed.

An insight into this period reveals that the demand for corn used for alcohol production was gaining impetus during this period, and the demand for feed production remained fairly constant. The cost for corn remained low, enabling a higher market for alcohol production from corn. The results from the price elasticity analysis suggest that corn production and demand for ethanol are highly elastic to changes in corn prices, whereas the market for feed is generally inelastic to price changes in corn (Banerjee, 2010).

D.3 Price Elasticity of Supply and Demand for Ethanol

Supply and demand elasticities in the U.S. ethanol fuel market have been evaluated by Luchansky and Monks (2009) with data from various sources for the periods shown in Table D.7. For their demand model, ethanol demand elasticity was −1.605 to −2.915 during the period January 1984–December 1987 and −0.417 to −1.503 for January 1988–May 1993. They offered the explanation of very price elastic for demand being caused by the changing availability of gasoline additives such as MTBE. Based on their supply model, ethanol PES ranged from 0.224 to 0.258 during the period January 1984–December 1987 and was 0.044 for January 1988–May 1993, essentially inelastic. They reported that ethanol production was running largely at capacity during these periods. Effects of corn and gasoline supply and demand were discussed in relation to these price elasticities.

Ethanol elasticity of supply, PES, was estimated for 2009 using data from the Commodities Report of *Ethanol Producer Magazine* (Kment, 2009), and the results are given in Table D.8. The average was 0.425 with all of the values less than 1.71, implying inelasticity. During this period, the Commodities Report describes the market in terms like "gasoline and ethanol markets continue to

TABLE D.7

Ethanol Supply and Demand Price Elasticity

	PED	PES
January 1984–December 1987	−1.605 to −2.915	0.224–0.258
January 1988–May 1993	−0.417 to −1.503	0.043–0.044

Source: Luchansky, M.S. and Monks, J., *Energy Econ.*, 31, 403, 2009.

TABLE D.8

Estimation of the Elasticity of Supply for Ethanol in 2009

Date	Production		Price		
	(Bbl/day)	$\Delta Q/Q$	($/gal)	$\Delta P/P$	PES
March 2009	669,000	0.0000	1.67	−0.012	0.000
April 2009	669,000	0.0000	1.65	0.055	0.000
May 2009	669,000	0.0374	1.74	0.034	1.084
June 2009	694,000	0.0476	1.8	0.028	1.712
July 2009	727,000	−0.0028	1.85	−0.027	0.102
August 2009	725,000	0.0000	1.8	−0.028	0.000
September 2009	725,000	0.0221	1.75	0.286	0.077
				Average PES	0.231

Source: Kment, R., *Ethanol Producer Mag.*, p. 21, May 2009; p. 21 June 2009; p. 23, July 2009; p. 19, September, 2009; p. 19, October, 2009; p. 19 November, 2009, 2010.

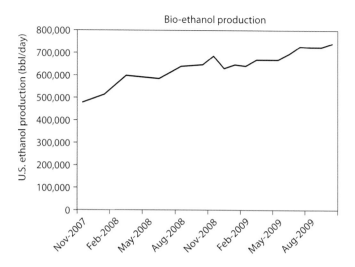

FIGURE D.3
U.S. bioethanol production. (From Kment, R., *Ethanol Producer Mag.*, p. 21, May 2009; p. 21 June 2009; p. 23, July 2009; p. 19, September, 2009; p. 19, October, 2009; p. 19 November, 2009, 2010.)

gain light support"; "current production level of ethanol is enough to handle current and expected demands"; "prices bounced 20–30 cents higher from mid-April to mid-May and have the potential to increase an additional 20–30 cents throughout the summer"; "ethanol prices have weakened significantly through the first half of the summer"; "overall demand is expected to remain stable to strong over the near future"; and "over the past several months, corn prices have had a great impact on the price of ethanol, giving the more of a 'cost-plus' feel than a true companion market to gasoline." In Figure D.3, the production of ethanol is shown from November 2007 to October 2009, and in Figure D.4, the price of ethanol is shown from March 2009 to January 2010 from *Ethanol Producer Magazine* (Kment, 2009).

D.4 Price Elasticity of Demand for Ethylene

The PED for ethylene was estimated using limited data available from C&E News (2009b) for ethylene production shown in Figure D.5 and ICIS (2009b) for ethylene prices in the United States shown in Figure D.6, as shown in Table D.9. The value of −0.416 for the PED was the result of a decrease in price that resulted in an increase in demand. The ICIS Chemical Business (2009b) reported that during this period, "buyers pushed for a decrease in price on the heels of ample supply and lower production costs."

In summary, PED and PES are useful measures of economic activity, but there is very limited data available to evaluate these parameters. Having values

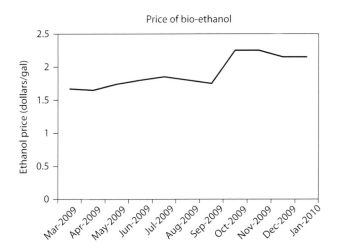

FIGURE D.4

U.S. ethanol price. (From Kment, R., *Ethanol Producer Mag.*, p. 21, May 2009; p. 21 June 2009; p. 23, July 2009; p. 19, September, 2009; p. 19, October, 2009; p. 19 November, 2009, 2010.)

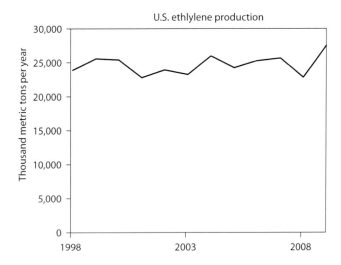

FIGURE D.5

Annual ethylene production (thousand metric tons). (From C&E News, *Chem. Eng. News*, 87(13), 16, 2009a; C&E News, *Chem. Eng. News*, 87(27), 51, 2009b.)

for chemicals from biomass, such as ethanol and ethanol derivatives, glycerol and glycerol derivatives, acetic acid, and fatty acid methyl and ethyl esters (FAME and FAEE), would aid in determining their potential to enter the market place in competition to those from petroleum. The only detailed evaluation by Luchansky and Monks (2009) was for bioethanol, and it was for the period from 1988 to 1994 where ethanol price elasticity ranged from 0.224 to 0.258. To obtain more recent values for ethanol, very limited data were

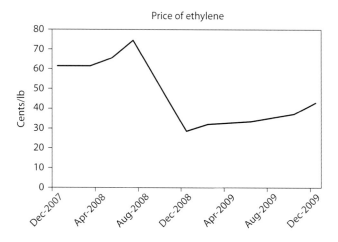

FIGURE D.6
Ethylene price (cents/lb). (From ICIS, *ICIS Chemical Business*, 276(6), 36, August 24, 2008a; ICIS, *ICIS Chemical Business*, 273(3), January 21, 2008b; ICIS, *ICIS Chemical Business*, 274(13), October 6, 2008c; ICIS, *ICIS Chemical Business*, 273(14), April 7, 2008d; ICIS, *ICIS Chemical Business*, 274(2), July 14, 2008e.)

TABLE D.9

Estimation of the Elasticity of Demand for Ethylene
for the Period 2008–2009

Year	Ethylene Production (Thousand Metric Tons)	Ethylene Price (Cents/lb)	$\Delta P/P$	$\Delta Q/Q$	PED
2008	22,554	65.75			
2009	27,252	32.81	0.2083	−0.500	−0.416

available in 2009 to evaluate the PES for ethanol. The average was 0.425 with all of the values less than 1.71, implying inelasticity. The PED for ethylene was estimated using limited data available for ethylene prices in the United States during 2008–2009. The value of −0.416 for the PED for ethylene was the result of a decrease in price that resulted in an increase in demand. All of these values imply inelasticity, which is probably typical for commodity chemicals.

Estimates of PED and PES can be incorporated as constraints in optimization of a chemical complex. Estimates of supply and demand changes will move upper and lower limits on availability of raw materials and demand for products. With price elasticity values, corresponding changes in prices are estimated for raw materials and products. Using PED and PES is given in Section 7.3.4. It was seen that for change in price of carbon dioxide, the demand for ammonia reduced. Cross price elasticity estimations were done to compute the PED of ammonia with respect to changes in carbon dioxide cost.

D.5 Data Used for Price Elasticity of Demand and Supply for Corn (Tables D.10 through D.12)

TABLE D.10

PES for Corn

	Total Corn Supply				
Year	Production (Million Bushels per Year)	$\Delta Q/Q * 100$	Price ($/Bushel)	$\Delta P/P * 100$	PES
1981	8,675	9.65	3.11	−19.61	−0.49
1982	9,511	13.26	2.50	2.00	6.63
1983	10,772	−28.53	2.55	25.88	−1.10
1984	7,699	12.74	3.21	−18.07	−0.71
1985	8,680	21.35	2.63	−15.21	−1.40
1986	10,534	16.46	2.23	−32.74	−0.50
1987	12,267	−2.04	1.5	29.33	−0.07
1988	12,016	−23.52	1.94	30.93	−0.76
1989	9,191	2.98	2.54	−7.09	−0.42
1990	9,464	−1.93	2.36	−3.39	0.57
1991	9,282	−2.87	2.28	3.95	−0.73
1992	9,016	17.40	2.37	−12.66	−1.37
1993	10,584	−19.96	2.07	20.77	−0.96
1994	8,472	28.79	2.50	−9.60	−3.00
1995	10,910	−17.74	2.26	43.36	−0.41
1996	8,974	7.77	3.24	−16.36	−0.48
1997	9,672	4.42	2.71	−10.33	−0.43
1998	10,099	9.77	2.43	−20.16	−0.48
1999	11,085	1.33	1.94	−6.19	−0.21
2000	11,232	3.62	1.82	1.65	2.20
2001	11,639	−1.96	1.85	6.49	−0.30
2002	11,412	−7.31	1.97	17.77	−0.41
2003	10,578	5.77	2.32	4.31	1.34
2004	11,188	14.18	2.42	−14.88	−0.95
2005	12,775	3.60	2.06	−2.91	−1.24
2006	13,235	−5.48	2.00	52.00	−0.11
2007	12,510	14.80	3.04	38.16	0.39
2008	14,362	−4.40	4.20	−3.33	1.32
2009	13,729	7.75	4.06	−12.56	−0.62
2010	14,793	−0.41	3.55	5.63	−0.07
2011	14,733		3.75		
				Average PES	−0.16

Source: USDA, Feed grains database: Yearbook tables, http://www.ers.usda. gov/data/feedgrains/, accessed April 8, 2010, 2010b.

TABLE D.11

PED for Corn in Fuel Use

		Corn Used for Fuel Alcohol Production			
Year	Demand (Million Bushels per Year)	$\Delta Q/Q * 100$	Price ($/Bushel)	$\Delta P/P * 100$	PED
1981	35.00	145.71	3.11	−19.61	−7.43
1982	86.00	62.79	2.50	2.00	31.40
1983	140.00	14.29	2.55	25.88	0.55
1984	160.00	45.00	3.21	−18.07	−2.49
1985	232.00	16.81	2.63	−15.21	−1.11
1986	271.00	7.01	2.23	−32.74	−0.21
1987	289.99	−3.74	1.5	29.33	−0.13
1988	279.15	2.97	1.94	30.93	0.10
1989	287.45	11.83	2.54	−7.09	−1.67
1990	321.45	8.59	2.36	−3.39	−2.53
1991	349.07	14.09	2.28	3.95	3.57
1992	398.26	6.84	2.37	−12.66	−0.54
1993	425.51	7.70	2.07	20.77	0.37
1994	458.26	16.26	2.50	−9.60	−1.69
1995	532.79	−25.73	2.26	43.36	−0.59
1996	395.68	8.35	3.24	−16.36	−0.51
1997	428.72	13.76	2.71	−10.33	−1.33
1998	487.73	6.17	2.43	−20.16	−0.31
1999	517.82	9.27	1.94	−6.19	−1.50
2000	565.85	11.31	1.82	1.65	6.86
2001	629.83	12.29	1.85	6.49	1.89
2002	707.24	40.76	1.97	17.77	2.29
2003	995.50	17.28	2.32	4.31	4.01
2004	1,167.55	13.33	2.42	−14.88	−0.90
2005	1,323.21	21.17	2.06	−2.91	−7.27
2006	1,603.32	32.19	2.00	52.00	0.62
2007	2,119.49	43.87	3.04	38.16	1.15
2008	3,049.21	20.58	4.20	−3.33	−6.18
2009	3,676.88	22.39	4.06	−12.56	−1.78
2010	4,500.00	4.44	3.55	5.63	0.79
2011	4,700.00		3.75		
				Average PED	0.51

Source: USDA, Feed grains database: Yearbook tables, http://www.ers.usda.gov/data/feedgrains/, accessed April 8, 2010, 2010b.

TABLE D.12

PED for Corn in Fuel Use

	Corn Used for Feed and Residual Use				
Year	Demand (Million Bushels per Year)	$\Delta Q/Q * 100$	Price ($/Bushel)	$\Delta P/P * 100$	PED
1981	4232	0.29	3.11	−19.61	−0.01
1982	4245	7.74	2.50	2.00	3.87
1983	4573	−15.24	2.55	25.88	−0.59
1984	3876	6.15	3.21	−18.07	−0.34
1985	4115	−0.01	2.63	−15.21	0.00
1986	4114	13.25	2.23	−32.74	−0.40
1987	4659	2.79	1.50	29.33	0.09
1988	4789	−17.86	1.94	30.93	−0.58
1989	3934	11.40	2.54	−7.09	−1.61
1990	4382	5.17	2.36	−3.39	−1.52
1991	4609	4.10	2.28	3.95	1.04
1992	4798	9.47	2.37	−12.66	−0.75
1993	5252	−10.90	2.07	20.77	−0.52
1994	4680	16.66	2.50	−9.60	−1.74
1995	5460	−14.05	2.26	43.36	−0.32
1996	4692	12.46	3.24	−16.36	−0.76
1997	5277	3.29	2.71	−10.33	−0.32
1998	5450	0.04	2.43	−20.16	−0.00
1999	5452	3.49	1.94	−6.19	−0.56
2000	5643	3.18	1.82	1.65	1.93
2001	5822	0.46	1.85	6.49	0.07
2002	5849	−5.14	1.97	17.77	−0.29
2003	5548	4.20	2.32	4.31	0.97
2004	5781	6.12	2.42	−14.88	−0.41
2005	6135	−0.33	2.06	−2.91	0.11
2006	6115	−9.40	2.00	52.00	−0.18
2007	5540	5.73	3.04	38.16	0.15
2008	5858	−11.14	4.20	−3.33	3.34
2009	5205	6.14	4.06	−12.56	−0.49
2010	5525	−3.17	3.55	5.63	−0.56
2011	5350		3.75		
				Average PED	−0.01

Source: USDA, Feed grains database: Yearbook tables, http://www.ers.usda.gov/data/feedgrains/, accessed April 8, 2010, 2010b.

Appendix E: Chemical Complex Analysis System

E.1 Chemical Complex Analysis System Program Structure

The chemical complex analysis system has been developed at the LSU Minerals Processing Research Institute to determine the best configuration of plants in a chemical complex based on economic, energy, environmental, and sustainable costs. A detailed description of chemical complex analysis system can be obtained from MPRI (2010). The system structure is shown in Figure E.1. It incorporates a flowsheeting component where simulations of the plants in the complex are entered. Each simulation includes the process or block flow diagram with material and energy balances, rate equations, equilibrium relations, and thermodynamic and transport properties for the process units and heat exchanger networks. These equations are entered through a graphical user interface and stored in the database to be shared with the other components of the system.

The objective function is entered as an equation associated with each process with related information for prices and economic, energy, environmental, and sustainable costs that are used in the evaluation of the Total Cost Assessment (TCA) for the complex. The TCA includes the total profit for the complex that is a function of the economic, energy, environmental, and sustainable costs and income from sales of products. Then the information is provided to the mixed-integer nonlinear programming solver to determine the optimum configuration of plants in the complex. Also, sources of pollutant generation are located by the Pollution Index component of the system using the EPA Pollution Index methodology (Cabezas et al., 1997), which is similar to the TRACI system of EPA.

All interactions with the system are through a graphical user interface that is designed and implemented in Visual Basic. As shown in the diagram (Figure E.1), the process flow diagram for the complex is constructed, and equations for the process units and variables for the streams connecting the process units are entered and stored in an access database using interactive data forms as shown on the complex simulation block in Figure E.1. Material and energy balances, rate equations, and equilibrium relations for the plants are entered as equality constraints using the format of the GAMS programming language that is similar to FORTRAN and stored

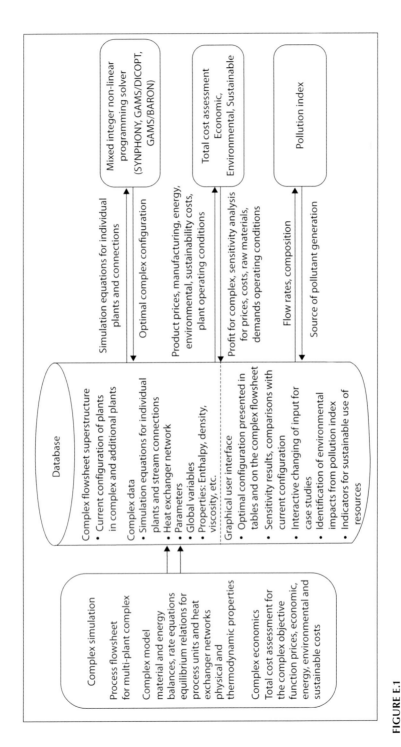

FIGURE E.1
Program structure for the chemical complex analysis system.

in the database. Process unit capacities, availability of raw materials, and demand for product are entered as inequality constraints and stored in the database. The system takes the equations in the database and writes and runs a GAMS program to solve the mixed-integer nonlinear programming problem for the optimum configuration of the complex. Then the important information from the GAMS solution is presented to the user in a convenient format, and the results can be exported to Excel, if desired. Features for developing flowsheets include adding, changing, and deleting the equations that describe units and streams and their properties. Usual Windows features include cut, copy, paste, delete, print, zoom, reload, update, and grid, among others.

The system has the TCA component prepare the assessment model for use with determination of the optimum complex configuration. Economic costs are estimated by standard methods (Garrett, 1989). Environmental costs are estimated from the data provided by Amoco, DuPont, and Novartis in the AIChE/CWRT TCA report. Sustainable costs are estimated from the air pollution data in the AIChE/CWRT TCA report.

A description of the tool is given in the following using actual screenshots from the program.

E.2 Chemical Complex Analysis System Model Development for Optimization

The chemical complex analysis system is a user-friendly tool to develop complex flowsheet and optimize the flowsheet. The flowsheet is developed as shown in Figure E.2. The database stores information from the flowsheet, which includes the design equations, parameters, sets, and tables. The design equations are required to be provided in the GAMS format. This information is then used to write a GAMS program, with the choice of solver provided by the user. There are three choices for the user: to run complex optimization, to run sensitivity analysis, or to perform multicriteria analysis. The development of the flowsheet is discussed next.

The development of the process flowsheet includes defining the sets (similar to GAMS terminology), constants (similar to SCALARS in GAMS), 1-D variables (similar to PARAMETERS in GAMS), and multi-D variables (similar to TABLE in GAMS). These are selected from the Model menu in the flowsheet as shown in Figure E.3.

The species for the biomass processes added for complex extension were created in a set from Chemical Complex Analysis System → Flowsheet Simulation → Model → Sets. The biomass components were added to a new set, "bio" with the description "biomass components in complex" as shown in Figure E.3. The molecular weight of the biomass species was added in

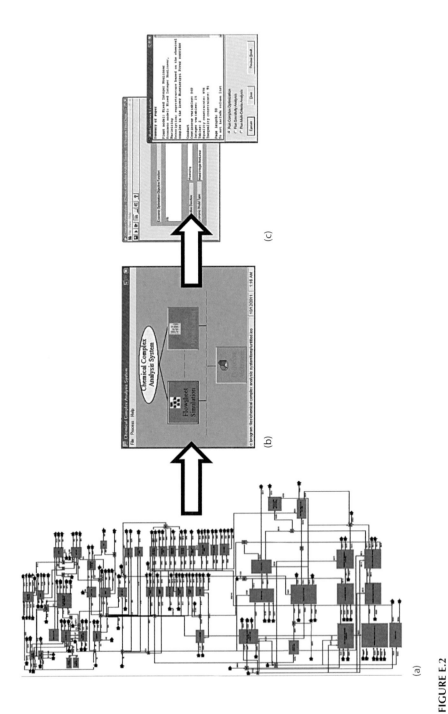

FIGURE E.2

Chemical complex analysis system. (a) process flowsheet, (b) graphical user interface, and (c) complex optimization.

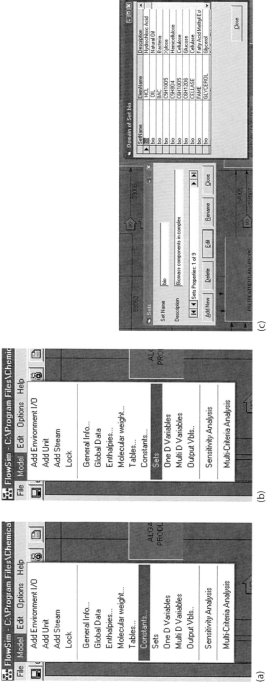

FIGURE E.3

Defining constants, sets, one-D variables, and multi-D variables. (a) define constants, (b) define sets, and (c) define 1-dimensional variables and define multi-dimensional variables.

Chemical Complex Analysis System → Flowsheet Simulation → Model → One-D Variables (Lists). The biomass component molecular weights are added to a new list, "mwbio" defined on the set "bio" with the description "formula weight of biomass components used in the model extension." The parameters for the process were added to a new scalar set, scalar 4 with the description "constant parameters for bioprocesses" in Chemical Complex Analysis System → Flowsheet Simulation → Model → Constants.

The next step was to develop the flowsheet. The development of the pretreatment (corn stover fermentation) is shown in Figure E.4. The block for the pretreatment section was created from Chemical Complex Analysis System → Flowsheet Simulation → Model → Add Unit. The input/output specification was added using Chemical Complex Analysis System → Flowsheet Simulation → Model → Add Environment *I/O*. Then the

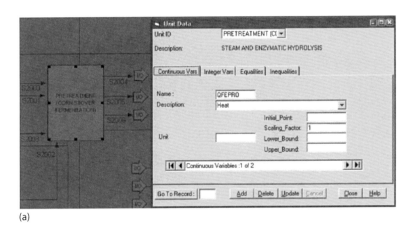

(a)

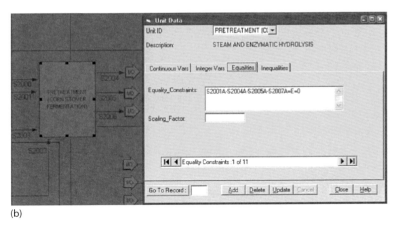

(b)

FIGURE E.4
Model development in chemical complex analysis system. (a) add variables and (b) add constraint equations.

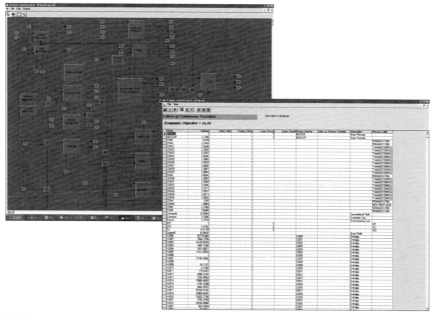

FIGURE E.5
Overall process flowsheet and optimization results.

unit (block) was connected to the *I/O* unit by adding stream using Chemical Complex Analysis System → Flowsheet Simulation → Model → Add Stream. These three (add unit, add *I/O*, and add stream) are also available from the toolbar on the flowsheet. In Figure E.4, the pretreatment (corn stover fermentation) is the block created. The S2000–S2007 are the streams created. S2004 is an output stream, so it is linked to the environment *I/O*.

The next step in the model development was to add variables and equations. The variables for the process may be entered either on the stream or on the units. The variables for the streams (material and energy) were added to the streams on the flowsheet. The variables for external energy were added to the unit, as shown in Figure E.4. The equality constraints for the streams were added in the equalities tab. These were the stream compositions and stream relations. The equality constraints for the process (reactions) were added in the units, as shown in Figure E.4. It may be noted that the upper and lower bounds on variables may be specified from the variable definition tab. Also, the equality constraints are in the form of GAMS programming language. Similarly, inequality constraints may be specified from the inequalities tab. The model was validated using data from HYSYS. The overall flowsheet is shown in Figure E.5. The model was run for complex optimization as shown in Figure E.2, and the results from optimization are shown in Figure E.5.

Appendix F: Detailed Mass and Energy Streams from Simulation Results

F.1 Fermentation of Corn Stover to Ethanol HYSYS Results

Components	Mass Fractions							
		Pretreatment (Fermentation)						
		Inlet Streams				Outlet Streams		
HYSYS Design Stream Name	Water	Biomass (Dry Corn Stover)	HP Steam	Cellulase	Fine Particles	V-101 Steam	V-102 Vapor	E-102 Pretreated Biomass
Model Stream Name	S2000	S2001	S2002	S2003	S2004	S2005	S2006	S2007
H_2O	1.0000	0.0000	1.0000	0.9000	0.5000	0.9974	0.9993	0.5709
Xylose[a]	0.0000	0.0000	0.0000	0.0000	0.0000	0.0000	0.0000	0.0679
Cellulose[a]	0.0000	0.3740	0.0000	0.0000	0.1870	0.0000	0.0000	0.0141
Hemicellulose[a]	0.0000	0.2110	0.0000	0.0000	0.1055	0.0000	0.0000	0.0256
Dextrose	0.0000	0.0000	0.0000	0.0000	0.0000	0.0000	0.0000	0.1525
CO_2	0.0000	0.0000	0.0000	0.0000	0.0000	0.0000	0.0000	0.0000
Ash[a]	0.0000	0.0520	0.0000	0.0000	0.0260	0.0025	0.0006	0.0207
Corn steep liquor	0.0000	0.0000	0.0000	0.0000	0.0000	0.0000	0.0000	0.0000
Other solids[a]	0.0000	0.1830	0.0000	0.0000	0.0915	0.0000	0.0000	0.0740
Lignin[a]	0.0000	0.0000	0.0000	0.0000	0.0900	0.0000	0.0000	0.0728
Nitrogen	0.0000	0.0000	0.0000	0.0000	0.0000	0.0000	0.0000	0.0000
Oxygen	0.0000	0.0000	0.0000	0.0000	0.0000	0.0000	0.0000	0.0000
Ethanol	0.0000	0.0000	0.0000	0.0000	0.0000	0.0000	0.0000	0.0000
Bacteria[a]	0.0000	0.0000	0.0000	0.0000	0.0000	0.0000	0.0000	0.0000
Cellulase[a]	0.0000	0.0000	0.0000	0.1000	0.0000	0.0000	0.0000	0.0014
DAP[a]	0.0000	0.0000	0.0000	0.0000	0.0000	0.0000	0.0000	0.0000

(continued)

Components	Fermentation								Purification (Fermentation)					
Mass Fractions	Inlet Streams					Outlet Streams			Inlet Streams		Outlet Streams			
HYSYS Design Stream Name	E-102 Pretreated Biomass	Air-Seed Production	CSL	DAP	Bacteria	MIX-110 Out	V-105 Vapor	V-105 Liquid	V-105 Vapor	V-105 Liquid	E-106 Ethanol	CO_2 Vent	E-103 Out	MIX-109 Out
Model Stream Name	S2007	S2008	S2009	S2010	S2011	S2012	S2013	S2014	S2013	S2014	S2015	S2016	S2017	S2018
H_2O	0.5709	0.0000	0.0000	0.0000	0.0000	0.0382	0.0251	0.6351	0.0251	0.6351	0.0052	0.0147	0.6210	1.0000
Xylose[a]	0.0679	0.0000	0.0000	0.0000	0.0000	0.0000	0.0000	0.0000	0.0000	0.0000	0.0000	0.0000	0.0000	0.0000
Cellulose[a]	0.0141	0.0000	0.0000	0.0000	0.0000	0.0000	0.0000	0.0157	0.0000	0.0157	0.0000	0.0000	0.0229	0.0000
Hemicellulose[a]	0.0256	0.0000	0.0000	0.0000	0.0000	0.0000	0.0000	0.0285	0.0000	0.0285	0.0000	0.0000	0.0418	0.0000
Dextrose	0.1525	0.0000	0.0000	0.0000	0.0000	0.0000	0.0000	0.0000	0.0000	0.0000	0.0000	0.0000	0.0000	0.0000
CO_2	0.0000	0.0000	0.0000	0.0000	0.0000	0.0000	0.9333	0.0007	0.9333	0.0007	0.0000	0.9841	0.0029	0.0000
Ash[a]	0.0207	0.0000	0.0000	0.0000	0.0000	0.1139	0.0000	0.0231	0.0000	0.0231	0.0000	0.0000	0.0338	0.0000
Corn steep liquor	0.0000	0.0000	1.0000	0.0000	0.0000	0.0000	0.0000	0.0023	0.0000	0.0023	0.0000	0.0000	0.0033	0.0000
Other solids[a]	0.0740	0.0000	0.0000	0.0000	0.0000	0.0000	0.0000	0.0825	0.0000	0.0825	0.0000	0.0000	0.1207	0.0000
Lignin[a]	0.0728	0.0000	0.0000	0.0000	0.0000	0.0000	0.0000	0.0811	0.0000	0.0811	0.0000	0.0000	0.1187	0.0000
Nitrogen	0.0000	0.7425	0.0000	0.0000	0.0000	0.0000	0.0000	0.0000	0.0000	0.0000	0.0000	0.0000	0.0000	0.0000
Oxygen	0.0000	0.2575	0.0000	0.0000	0.0000	0.7134	0.0000	0.0000	0.0000	0.0000	0.0000	0.0000	0.0000	0.0000
Ethanol	0.0000	0.0000	0.0000	0.0000	0.0000	0.1344	0.0415	0.1072	0.0415	0.1072	0.9948	0.0012	0.0000	0.0000
Bacteria[a]	0.0000	0.0000	0.0000	0.0000	1.0000	0.0000	0.0000	0.0220	0.0000	0.0220	0.0000	0.0000	0.0322	0.0000
Cellulase[a]	0.0014	0.0000	0.0000	0.0000	0.0000	0.0000	0.0000	0.0015	0.0000	0.0015	0.0000	0.0000	0.0022	0.0000
DAP[a]	0.0000	0.0000	0.0000	1.0000	0.0000	0.0000	0.0000	0.0004	0.0000	0.0004	0.0000	0.0000	0.0006	0.0000

[a] denotes component created in HYSYS.

	Water	Biomass (Dry Corn Stover)	HP Steam	Cellulase	Fine Particles	V-101 Steam	V-102 Vapor	E-102 Pretreated Biomass
	S2000	S2001	S2002	S2003	S2004	S2005	S2006	S2007
Mass flow rates (kg/h)	8.33E+04	8.33E+04	6.00E+04	2.69E+03	8.33E+03	2.54E+04	0.00E+00	1.96E+05
Heat flow rates (kJ/h)	-1.32E+09	-1.92E+08	-7.85E+08	-4.04E+07	-7.55E+07	-3.35E+08	0.00E+00	-1.89E+09
Mass enthalpy (kJ/kg)	-1.58E+04	-2.30E+03	-1.32E+04	-1.50E+04	-9.06E+03	-1.32E+04	-1.33E+04	-9.69E+03

	E-102 Pretreated Biomass	Air-Seed Production	CSL	DAP	Bacteria	MIX-110 Out	V-105 Vapor
	S2007	S2008	S2009	S2010	S2011	S2012	S2013
Mass flow rates (kg/h)	1.96E+05	5.30E+04	3.97E+02	7.10E+01	1.98E+03	5.52E+04	2.03E+04
Heat flow rates (kJ/h)	-1.89E+09	0.00E+00	-8.50E+05	-1.52E+05	-4.53E+06	-8.37E+07	-1.80E+08
Mass enthalpy (kJ/kg)	-9.69E+03	0	-2.14E+03	-2.14E+03	-2.29E+03	-1.52E+03	-8.89E+03

	V-105 Liquid	V-105 Vapor	E-106 Ethanol	CO₂ Vent	E-103 Out	MIX-109 Out
	S2014	S2013	S2015	S2016	S2017	S2018
Mass flow rates (kg/h)	1.76E+05	2.03E+04	1.98E+05	1.98E+04	1.20E+05	3.72E+04
Heat flow rates (kJ/h)	-1.97E+09	-1.80E+08	-1.20E+08	-1.71E+08	-1.28E+09	-5.74E+08
Mass enthalpy (kJ/kg)	-1.12E+04	-8.89E+03	-6.07E+03	-9.01E+03	-1.07E+04	-1.54E+04

Energy Stream	V-100 Heating	V-102 Heating	P-100 Heating	T-100 Reboiler	T-101 Reboiler	E-105 Heating	V-103 Cooling	V-104 Cooling
Heat flow rates (kJ/h)	2.56E+08	9.83E+07	1.71E+05	1.78E+08	5.56E+07	1.31E+06	1.80E+07	7.14E+07

Energy Stream	V-105 Cooling	T-101 Condenser	E-101 Cooling	E-102 Cooling	E-103 Cooling	E-104 Cooling	E-106 Cooling
Heat flow rates (kJ/h)	2.49E+08	1.77E+08	2.82E+08	1.42E+07	−1.31E+06	5.88E+07	2.12E+07

F.2 Fermentation of Corn to Ethanol
SuperPro Designer® Results

Pretreatment (Corn Fermentation)

	Inlet Streams						Outlet Streams
Corn	Alpha Amylase	Liquid Ammonia	Lime	Caustic	Glucoamylase	Sulfuric Acid	Pretreated Corn
S5001	S5002	S5003	S5004	S5005	S5006	S5007	S5008
0.067	0.000	1.000	1.000	0.050	0.000	0.000	0.031
0.070	0.000	0.000	0.000	0.000	0.000	0.000	0.022
0.034	0.000	0.000	0.000	0.000	0.000	0.000	0.011
0.060	0.000	0.000	0.000	0.000	0.000	0.000	0.019
0.024	0.000	0.000	0.000	0.000	0.000	0.000	0.010
0.595	0.000	0.000	0.000	0.000	0.000	0.000	0.002
0.150	1.000	0.000	0.000	0.950	1.000	0.000	0.696
0.000	0.000	0.000	0.000	0.000	0.000	1.000	0.001
0.000	0.000	0.000	0.000	0.000	0.000	0.000	0.000
0.000	0.000	0.000	0.000	0.000	0.000	0.000	0.003
0.000	0.000	0.000	0.000	0.000	0.000	0.000	0.205
0.000	0.000	0.000	0.000	0.000	0.000	0.000	0.000
0.000	0.000	0.000	0.000	0.000	0.000	0.000	0.000

(continued)

(continued)

	Fermentation (Corn)					Purification (Corn Fermentation)								
	Inlet Streams			Outlet Streams		Inlet Streams				Outlet Streams				
	Pretreated Corn	Yeast	S-190	S-129	S-132	S-129	S-132	Water	Hot Air	Ethanol	PC (Process Condensate)	DDGS	Exhaust (S-178)	CO_2
	S5008	S2009	S5010	S5011	S5012	S5011	S5012	S5013	S5014	S5015	S5016	S5017	S5018	S5019
	0.031	0.000	0.000	0.000	0.043	0.000	0.043	0.000	0.000	0.000	0.000	0.290	0.000	0.000
	0.022	0.000	0.000	0.000	0.025	0.000	0.025	0.000	0.000	0.000	0.000	0.211	0.000	0.000
	0.011	0.000	0.000	0.000	0.012	0.000	0.012	0.000	0.000	0.000	0.000	0.102	0.000	0.000
	0.019	0.000	0.000	0.000	0.021	0.000	0.021	0.000	0.000	0.000	0.000	0.181	0.000	0.000
	0.010	0.000	0.000	0.000	0.015	0.000	0.015	0.000	0.000	0.000	0.000	0.099	0.000	0.000
	0.002	0.000	0.000	0.000	0.002	0.000	0.002	0.000	0.000	0.000	0.000	0.023	0.000	0.000
	0.696	1.000	0.000	0.016	0.770	0.016	0.770	1.000	0.000	0.004	0.994	0.088	0.417	0.007
	0.001	0.000	1.000	0.000	0.001	0.000	0.001	0.000	0.000	0.000	0.000	0.006	0.000	0.000
	0.000	0.000	0.000	0.957	0.002	0.957	0.002	0.000	0.000	0.000	0.000	0.000	0.000	0.993
	0.003	0.000	0.000	0.027	0.111	0.027	0.111	0.000	0.000	0.996	0.005	0.000	0.000	0.000
	0.205	0.000	0.000	0.000	0.000	0.000	0.000	0.000	0.000	0.000	0.000	0.000	0.000	0.000
	0.000	0.000	0.000	0.000	0.000	0.000	0.000	0.000	0.767	0.000	0.000	0.000	0.447	0.000
	0.000	0.000	0.000	0.000	0.000	0.000	0.000	0.000	0.233	0.000	0.000	0.000	0.135	0.000

	Corn	Alpha-Amylase	Liquid Ammonia	Lime	Caustic	Glucoamylase	Sulfuric Acid	Pretreated Corn	Pretreated Corn	Yeast	S-190 (Water)
	S5001	S5002	S5003	S5004	S5005	S5006	S5007	S5008	S5008	S2009	S5010
	4.52E+04	3.15E+01	8.97E+01	5.36E+01	2.26E+03	4.54E+01	8.97E+01	1.44E+05	1.44E+05	1.09E+01	5.42E+01

	S-129	S-132	S-129	S-132	Water	Hot Air	Ethanol	PC (Process Condensate)	DDGS	Exhaust (S-178)	CO_2
	S5011	S5012	S5011	S5012	S5013	S5014	S5015	S5016	S5017	S5018	S5019
	1.41E+04	1.30E+05	1.41E+04	1.30E+05	1.33E+04	2.44E+04	1.44E+04	5.55E+02	1.50E+04	4.16E+04	1.38E+04

F.3 Anaerobic Digestion of Corn Stover to Acetic Acid HYSYS Results

Components		Pretreatment (Anaerobic Digestion)				
Mass Fractions		Inlet Streams				Outlet Streams
HYSYS Design Stream Name	Biomass (Corn Stover)	Pig Manure	Water	Steam	Lime	V-100 Out
Model Stream Name	S4001	S4002	S4003	S4004	S4005	S4006
$Ca(OH)_2$[a]	0.000	0.000	0.000	0.000	1.000	0.038
H_2O	0.000	0.000	1.000	1.000	0.000	0.456
CHI_3	0.000	0.000	0.000	0.000	0.000	0.000
Acetic acid	0.000	0.000	0.000	0.000	0.000	0.000
Xylose[a]	0.000	0.000	0.000	0.000	0.000	0.084
Cellulose[a]	0.374	0.525	0.000	0.000	0.000	0.015
Hemicellulose[a]	0.211	0.000	0.000	0.000	0.000	0.006
Dextrose	0.000	0.000	0.000	0.000	0.000	0.197
Hydrogen	0.000	0.000	0.000	0.000	0.000	0.000
CO_2	0.000	0.000	0.000	0.000	0.000	0.000
Ash[a]	0.052	0.300	0.000	0.000	0.000	0.048
Nutrients[a]	0.000	0.000	0.000	0.000	0.000	0.000
Other solids[a]	0.183	0.175	0.000	0.000	0.000	0.086
Lignin[a]	0.180	0.000	0.000	0.000	0.000	0.069
Ethanol	0.000	0.000	0.000	0.000	0.000	0.000
M-i-B-Ketone	0.000	0.000	0.000	0.000	0.000	0.000
Mixed bacteria[a]	0.000	0.000	0.000	0.000	0.000	0.000

(continued)

(continued)

Components												
	Anaerobic Digestion						Purification (Anaerobic Digestion)					
Mass Fractions	Inlet Streams				Outlet Streams		Inlet Streams			Outlet Streams		
HYSYS Design Stream Name	V-100 Out	Iodoform	Nutrients	Terrestrial Inoculum	CO$_2$ H$_2$ Mix Gas	MIX-103 Out	MIX-103 Out	Solvent	X-104 Steam	X-101 Solids	X-103 Acetic Acid	X-104 Waste Water
Model Stream Name	S4006	S4007	S4008	S4009	S4010	S4011	S4011	S4012	S4013	S4014	S4015	S4016
Ca(OH)$_2$[a]	0.038	0.000	0.000	0.000	0.000	0.043	0.043	0.000	0.000	0.116	0.000	0.000
H$_2$O	0.456	0.000	0.526	0.000	0.000	0.466	0.466	0.000	1.000	0.000	0.000	0.854
CHI$_3$	0.000	0.025	0.000	0.000	0.000	0.000	0.000	0.000	0.000	0.000	0.000	0.000
Acetic acid	0.000	0.000	0.000	0.000	0.000	0.163	0.163	0.000	0.000	0.000	1.000	0.022
Xylose[a]	0.084	0.000	0.000	0.000	0.000	0.022	0.022	0.000	0.000	0.059	0.000	0.000
Cellulose[a]	0.015	0.000	0.000	0.000	0.000	0.017	0.017	0.000	0.000	0.047	0.000	0.000
Hemicellulose[a]	0.006	0.000	0.000	0.000	0.000	0.007	0.007	0.000	0.000	0.020	0.000	0.000
Dextrose	0.197	0.000	0.000	0.000	0.084	0.051	0.051	0.000	0.000	0.137	0.000	0.000
Hydrogen	0.000	0.000	0.000	0.000	0.000	0.000	0.000	0.000	0.000	0.000	0.000	0.000
CO$_2$	0.000	0.000	0.000	0.000	0.916	0.000	0.000	0.000	0.000	0.001	0.000	0.000
Ash[a]	0.048	0.000	0.474	0.000	0.000	0.055	0.055	0.000	0.000	0.147	0.000	0.000
Nutrients[a]	0.000	0.000	0.000	0.000	0.000	0.001	0.001	0.000	0.000	0.001	0.000	0.000
Other solids[a]	0.086	0.000	0.000	0.000	0.000	0.097	0.097	0.000	0.000	0.262	0.000	0.000
Lignin[a]	0.069	0.000	0.000	0.000	0.000	0.077	0.077	0.000	0.000	0.208	0.000	0.000
Ethanol	0.000	0.976	0.000	0.000	0.000	0.000	0.000	0.000	0.000	0.000	0.000	0.000
M-i-B-Ketone	0.000	0.000	0.000	0.000	0.000	0.000	0.000	1.000	0.000	0.000	0.000	0.000
Mixed bacteria[a]	0.000	0.000	0.000	1.000	0.000	0.001	0.001	0.000	0.000	0.003	0.000	0.125

[a] denotes component created in HYSYS.

	Biomass (Corn Stover)	Pig Manure	Water	Steam	Lime	V-100 Out	V-100 Out	Iodoform	Nutrients
	S4001	S4002	S4003	S4004	S4005	S4006	S4006	S4007	S4008
Mass flow rates (kg/h)	8.33E+04	2.08E+04	1.04E+05	2.14E+03	8.33E+03	2.19E+05	2.19E+05	2.29E+00	1.91E+02
Heat flow rates (kJ/h)	-1.92E+08	-4.78E+07	-1.65E+09	-2.84E+07	-6.13E+07	-1.77E+09	-1.77E+09	-1.35E+04	-1.80E+06
Mass enthalpy (kJ/kg)	-2.30E+03	-2.30E+03	-1.58E+04	-1.32E+04	-7.36E+03	-8.08E+03	-8.08E+03	-5.88E+03	-9.40E+03

	Terrestrial Inoculum	CO$_2$ H$_2$ Mix Gas	MIX-103 Out	MIX-103 Out	Solvent	X-104 Steam	X-101 Solids	X-103 Acetic Acid	X-104 Waste Water
	S4009	S4010	S4011	S4011	S4012	S4013	S4014	S4015	S4016
Mass flow rates (kg/h)	1.91E+02	2.52E+04	1.94E+05	1.94E+05	1.34E+04	1.80E+03	7.20E+04	2.92E+04	1.08E+05
Heat flow rates (kJ/h)	-4.37E+05	-2.06E+08	-1.85E+09	-1.85E+09	-4.37E+07	-2.39E+07	-1.83E+08	-2.20E+08	-1.36E+09
Mass enthalpy (kJ/kg)	-2.29E+03	-8.17E+03	-9.54E+03	-9.54E+03	-3.25E+03	-1.32E+04	-2.54E+03	-7.52E+03	-1.26E+04

Energy Stream	V-100 Heating	E-100 Heating	E-101 Heating	V-101 Cooling	E-102 Cooling	X-100 Energy
Heat flow rates (kJ/h)	2.09E+08	2.44E+08	5.79E+06	-2.81E+08	9.98E+07	-5.28E+06

F.4 Transesterification of Natural Oils to Fatty Acid Methyl Esters and Glycerol HYSYS Results

Mass Fractions			Components								
			Inlet Streams						Outlet Streams		
Name	Soybean Oil	Catalyst	Methanol	Water	HCL	NAOH	FAME	Glycerol	X-103 Top	X-104 Top
Model Stream Name	S3001	S3002	S3003	S3004	S3005	S3006	S3020	S3021	S3022	S3023
Glycerol	0.00	0.00	0.00	0.00	0.00	0.00	0.00	0.96	0.00	0.90
HCL	0.00	0.00	0.00	0.00	0.35	0.00	0.00	0.00	0.00	0.00
Methanol	0.00	0.75	1.00	0.00	0.00	0.00	0.00	0.00	0.09	0.00
Trilinolein*	1.00	0.00	0.00	0.00	0.00	0.00	0.01	0.00	0.06	0.10
Sodium methoxide	0.00	0.25	0.00	0.00	0.00	0.00	0.00	0.00	0.00	0.00
H_2O	0.00	0.00	0.00	1.00	0.65	0.00	0.00	0.00	0.00	0.00
Sodium chloride	0.00	0.00	0.00	0.00	0.00	0.00	0.00	0.04	0.00	0.00
M-Linoleate	0.00	0.00	0.00	0.00	0.00	0.00	0.99	0.00	0.00	0.00
NaOH	0.00	0.00	0.00	0.00	0.00	1.00	0.00	0.00	0.00	0.00

430   *Appendix F: Detailed Mass and Energy Streams from Simulation Results*

Name	Soybean Oil	Catalyst	Methanol	Water	HCL	NAOH	FAME	Glycerol	X-103 Top	X-104 Top
Model Stream Name	S3001	S3002	S3003	S3004	S3005	S3006	S3020	S3021	S3022	S3023
Mass flow rates (kg/h)	4.25E03	5.31E+01	4.23E+02	8.55E+01	3.03E+01	1.91E+00	4.26E+03	4.13E+02	1.22E+02	4.87E+01
Heat flow rates (kJ/h)	−7.12E+06	−3.39E+05	−3.16E+06	−1.35E+06	−3.38E+06	−3.99E+03	−8.36E+06	−2.71E+06	−1.66E+06	−3.32E+05
Mass enthalpy (kJ/kg)	−1.68E+03	−6.39E+03	−7.47+03	−1.58E+04	−1.12E+04	−2.09E+03	−1.96E+03	−6.57E+03	−1.36E+04	−6.81E+03

Energy Stream	E-100 Heating	T-101 Reboiler	CRV-100 Cooling	CRV-101 Cooling	CRV-102 Cooling	CRV-103 Cooling	CRV-104 Cooling	T-101 Condenser
Heat flow rates (kJ/h)	2.26E+05	4.60E+06	1.13E+06	8.21E+04	8.77E+04	1.10E+05	2.26E+04	2.19E+06

Energy Stream	P-100 Heating	P-103 Heating	P-100 Heating	P-102 Heating	E-102 Cooling	T-100 Condenser	T-100 Reboiler
Heat flow rates (kJ/h)	2.47E+02	1.66E+02	3.20E+01	5.95E − 01	2.39E+06	6.15E+06	6.59E+06

F.5 Ethanol to Ethylene HYSYS Results

Mass Fraction	Components		
	Inlet Streams	Outlet Streams	
HYSYS Design Stream Name	Ethanol	X-100 Ethylene	MIX-100 Out
Model Stream Name	S2030	S2031	S2032
Ethylene	0.0000	1.0000	0.0015
Ethanol	1.0000	0.0000	0.0251
Water	0.0000	0.0000	0.9734

	Ethanol	X-100 Ethylene	MIX-100 Out
Model Stream Name	S2030	S2031	S2032
Mass flow rates (kg/h)	4.15E+04	2.50E+04	1.65E+04
Heat flow rates (kJ/h)	−2.50E+08	−2.53E+08	4.69E+07
Mass enthalpy (kJ/kg)	−6.04E+03	1.88E+03	−1.54E+04

Energy Stream	CRV-100 Heating	E-100 Heating	E-101 Cooling
Heat flow rates (kJ/h)	5.12E+07	5.13E+07	5.84E+07

F.6 Glycerol to Propylene Glycol HYSYS Results

Mass Fractions	Components			
	Inlet Streams		Outlet Streams	
HYSYS Design Stream Name	Glycerol	Hydrogen	E-104 Propylene Glycol	T-101 Water Vapor
Model Stream Name	S3030	S3031	S3032	S3033
Glycerol	0.76	0.00	0.00	0.00
Hydrogen	0.00	1.00	0.00	0.00
12-C3 diol	0.00	0.00	1.00	0.00
H_2O	0.24	0.00	0.00	1.00

Name	Glycerol	Hydrogen	E-104 Propylene Glycol	T-101 Water Vapor
Model Stream Name	S3030	S3031	S3032	S3033
Mass flow rates (kg/h)	1.48E+04	2.46E+02	9.28E+03	5.74E+03
Mass flow rates (kJ/h)	−1.39E+08	0.00E+00	−6.05E+07	−7.59E+07
Mass enthalpy (kJ/kg)	−9.38E+03	0.00E+00	−6.52E+03	−1.32E+04

Energy Stream	T-101 Condenser	T-101 Reboiler	T-100 Reboiler	T-100 Condenser	E-104 Cooling
Heat flow rates (kJ/h)	2.59E+07	4.11E+07	2.50E+07	2.41E+07	4.60E+06

Energy Stream	E-102 Cooling	E-103 Cooling	CRV-100 Heating	CRV-101 Heating	E-101 Cooling
Heat flow rates (kJ/h)	1.95E+07	2.44E+07	1.63E+07	1.99E+07	1.60E+06

Appendix G: Equipment Mapping and Costs from ICARUS

G.1 Ethanol Production from Corn Stover Fermentation

The equipment from the HYSYS case was mapped in IPE and is given in Table G.1 as default. The equipment in Table G.1, listed as manual, required manual mapping in IPE because appropriate equipment in HYSYS was not available in the design. The perfect separator X-100 was mapped as a solid bowl centrifuge with flow specifications and dimensions automatically calculated in IPE. The perfect separator X-101 in HYSYS was mapped as a molecular sieve adsorber with 13 XMS (IPE code for molecular sieve) sieves. The mixers were manually mapped as default mapping did not identify the cost for the mixers. The reactors V-100, V-102, V-103, V-104, and V-105 were designed as tanks in HYSYS as only tanks allow multiple reactions in HYSYS. The default mappings as tanks for these vessels were deleted in IPE and were mapped as agitated tank reactors. The rest of the equipment with IPE description given as "C" in Table G.1 were not mapped, and the cost was not calculated for them because those were mainly piping components, and the cost of piping was included in installation costs.

G.2 Ethylene Production from Dehydration of Ethanol

The equipments from the HYSYS case were mapped in IPE and are given in Table G.2. The following equipment required manual mapping in IPE as this type of equipment was not available in HYSYS. The reactor CRV-100 was a fluidized bed reactor with the reaction occurring in the vapor phase. Therefore, the reactor type was mapped to a jacketed vertical tower in IPE. The jacketed type was chosen as the reactor was maintained at a temperature of 300°C. The packing material in the CRV-100 was selected as alumina. The quantity of catalyst required was computed based on Takahara et al. (2005), where 25.1 g-cat min/mmol-C_2H_5OH was required to achieve 99.9% yield of ethylene using H-mordenite (HM90) catalyst. Using the same rate of catalyst usage in a fluidized bed reactor and activated alumina catalyst, the amount

TABLE G.1

Equipment Mapping for Ethanol Production from Corn Stover Fermentation

Equipment HYSYS	Equipment ICARUS	IPE Description	Mapping Type	Equipment Cost (Dollars)
E-100	E-100	DHE FLOAT HEAD	Default	383,200
E-101	E-101	DHE FLOAT HEAD	Default	145,400
E-102	E-102	DHE FLOAT HEAD	Default	117,400
E-103	E-103	DHE FLOAT HEAD	Default	137,600
E-104	E-104	DHE FLOAT HEAD	Default	453,200
E-105	E-105	DHE FLOAT HEAD	Default	68,000
E-106	E-106	DHE FLOAT HEAD	Default	129,800
MIX-100	MIX-100	DMX STATIC	Manual	64,200
MIX-101	MIX-101	DMX STATIC	Manual	71,800
MIX-102	MIX-102	DMX STATIC	Manual	29,300
MIX-103	MIX-103	DMX STATIC	Manual	39,000
MIX-104	MIX-104	DMX STATIC	Manual	79,200
MIX-105	MIX-105	DMX STATIC	Manual	58,100
MIX-106	MIX-106	DMX STATIC	Manual	105,100
MIX-107	MIX-107	DMX STATIC	Manual	94,600
MIX-108	MIX-108	DMX STATIC	Manual	71,800
MIX-109	MIX-109	DMX STATIC	Default	0
MIX-110	MIX-110	DMX STATIC	Default	0
P-100	P-100	DCP CENTRIF	Default	57,500
RCY-1	RCY-1	C	Default	0
T-100	T-100-tower	DTW TRAYED	Default	706,800
	T-100-bottoms split	C	Default	0
	T-100-reb	DRB U TUBE	Default	287,000
T-101	T-101-tower	DTW TRAYED	Default	3,119,200
	T-101-cond	DHE FIXED T S	Default	441,100
	T-101-cond acc	DHT HORIZ DRUM	Default	178,300
	T-101-reflux pump	DCP GEN SERV	Default	74,200
	T-101-overhead split	C	Default	0
	T-101-bottoms split	C	Default	0
	T-101-reb	DRB U TUBE	Default	214,700
TEE-100	TEE-100	C	Default	0
TEE-101	TEE-101	C	Default	0
TEE-102	TEE-102	C	Default	0
TEE-103	TEE-103	C	Default	0
TEE-104	TEE-104	C	Default	0
TEE-105	TEE-105	C	Default	0
TEE-106	TEE-106	C	Default	0
V-100	V-100	DAT REACTOR	Manual	296,800
V-101	V-101	DHT HORIZ DRUM	Default	140,200
V-102	V-102	DAT REACTOR	Manual	316,200

TABLE G.1 (continued)

Equipment Mapping for Ethanol Production from Corn Stover Fermentation

Equipment HYSYS	Equipment ICARUS	IPE Description	Mapping Type	Equipment Cost (Dollars)
V-103	V-103	DAT REACTOR	Manual	167,100
V-104	V-104	DAT REACTOR	Manual	204,300
V-105	V-105	DAT REACTOR	Manual	316,200
V-106	V-106	DHT HORIZ DRUM	Default	146,800
X-100	X-100	ECT SOLID BOWL	Manual	2,610,600
X-101	X-101 adsorber	DTW TS ADSORB	Manual	2,221,800
Total equipment cost calculated from ICARUS				13,546,500

TABLE G.2

Equipment Mapping for Ethylene Production from Dehydration of Ethanol

Equipment HYSYS	Equipment ICARUS	IPE Description	Mapping Type	Equipment Cost (Dollars)
E-100	E-100	DHE FLOAT HEAD	Default	130,000
E-101	E-101	DHE FLOAT HEAD	Default	230,200
RCY-1	RCY-1	C	Default	0
T-100	T-100-tower	DTW TRAYED	Default	280,400
X-101	X-101	C	Default	0
X-100	X-100	EE FORCED CIR	Manual	145,000
CRV-100	CRV-100	DVT JACKETED	Manual	300,200
	CRV-100 Packing	EPAKPACKING	Manual	152,500
Total equipment cost calculated from ICARUS				1,238,300

of alumina required was 376,500 kg. The density of alumina is 3.95–4.1 gm/cc (~4000 kg/m^3) (Wikipedia, 2009). The volume of alumina required for the process was 94.125 m^3. This volume was entered in the mapping for packing calculation CRV packing. The equipment X-100 was mapped as a forced circulation drier in IPE.

G.3 Fatty Acid Methyl Ester and Glycerol from Transesterification of Soybean Oil

The equipment from the HYSYS case was mapped in IPE and is given in Table G.3. Some of the equipment required manual mapping in IPE like the centrifuges. They were not available in HYSYS and were designed as perfect

TABLE G.3

Equipment Mapping for Fatty Acid Methyl Ester and Glycerol
from Transesterification of Soybean Oil

Equipment HYSYS	Equipment IPE	IPE Description	Mapping Type	Equipment Cost (Dollars)
E-100	E-100	DHE FLOAT HEAD	Default	248,900
E-101	E-101	DHE FLOAT HEAD	Default	71,100
E-102	E-102	DHE FLOAT HEAD	Default	70,800
E-103	E-103	DHE FLOAT HEAD	Default	92,000
MIX-100	MIX-100	DMX STATIC	Manual	28,400
MIX-101	MIX-101	DMX STATIC	Manual	29,000
MIX-102	MIX-102	DMX STATIC	Manual	29,900
TEE-100	TEE-100	C	Default	—
TEE-101	TEE-101	C	Default	—
P-100	P-100	DCP CENTRIF	Default	21,700
P-101	P-101	DCP CENTRIF	Default	20,500
P-102	P-102	DCP CENTRIF	Default	20,300
P-103	P-103	DCP CENTRIF	Default	20,500
RCY-1	RCY-1	C	Default	—
RCY-2	RCY-2	C	Default	—
T-100	T-100-tower	DTW TRAYED	Default	228,600
	T-100-cond	DHE FIXED T S	Default	80,200
	T-100-cond acc	DHT HORIZ DRUM	Default	76,200
	T-100-reflux pump	DCP GEN SERV	Default	49,200
	T-100-overhead split	C	Default	—
	T-100-bottoms split	C	Default	—
	T-100-reb	DRB U TUBE	Default	67,900
T-101	T-101-tower	DTW TRAYED	Default	223,500
	T-101-cond	DHE FIXED T S	Default	62,800
	T-101-cond acc	DHT HORIZ DRUM	Default	83,600
	T-101-reflux pump	DCP GEN SERV	Default	28,200
	T-101-overhead split	C	Default	—
	T-101-bottoms split	C	Default	—
	T-101-reb	DRB U TUBE	Default	133,800
CRV-100	CRV-100	DAT REACTOR	Default	160,600
CRV-101	CRV-101	DAT REACTOR	Default	160,100
CRV-102	CRV-102	DAT REACTOR	Default	162,300
CRV-103	CRV-103	DAT REACTOR	Default	137,000
CRV-104	CRV-104	DAT REACTOR	Default	138,100
X-100	X-100	ECT SOLID BOWL	Manual	137,800
X-101	X-101	ECT SOLID BOWL	Manual	137,800
X-102	X-102	ECT SOLID BOWL	Manual	137,800
X-103	X-103	ED VAC TRAY	Manual	19,800
X-104	X-104	ECT SOLID BOWL	Manual	139,500
X-106	X-106	C	Default	—
X-107	X-107	C	Default	—
Total equipment cost calculated from ICARUS				3,017,900

separators. This equipment was available for mapping in IPE. The mixers were mapped in IPE, but a cost was not associated with the mixers. So, the mixers were manually mapped as static mixers in IPE.

The perfect separators, X-100, X-101, X-102, and X-104 from HYSYS were mapped as a solid bowl centrifuge in IPE with flow specifications and dimensions automatically calculated in IPE. The perfect separator X-103 in HYSYS was mapped as a vacuum tray dryer with $40\,ft^2$ area of top tray in IPE. The mixers, MIX-100, MIX-101, and MIX-102 from HYSYS were manually mapped as static mixers in IPE. The rest of the equipment with IPE description given as "C" in Table G.3 were piping components, and they were included in installed costs.

TABLE G.4

Equipment Mapping for Propylene Glycol Production from Hydrogenolysis of Glycerol

Equipment HYSYS	Equipment IPE	IPE Description	Mapping Type	Equipment Cost (Dollars)
CRV-100	CRV-100	DAT REACTOR	Default	194,900
CRV-101	CRV-101	DAT REACTOR	Default	195,300
E-100	E-100	DHE FLOAT HEAD	Default	72,700
E-101	E-101	DHE FLOAT HEAD	Default	67,600
E-102	E-102	DHE FLOAT HEAD	Default	81,600
E-103	E-103	DHE FLOAT HEAD	Default	109,900
E-104	E-104	DHE FLOAT HEAD	Default	78,900
MIX-100	MIX-100	C	Default	—
MIX-101	MIX-101	C	Default	—
RCY-1	RCY-1	C	Default	—
T-100	T-100-tower	DTW TRAYED	Default	331,700
	T-100-cond	DHE FIXED T S	Default	88,700
	T-100-cond acc	DHT HORIZ DRUM	Default	88,600
	T-100-reflux pump	DCP GEN SERV	Default	33,500
	T-100-overhead split	C	Default	—
	T-100-bottoms split	C	Default	—
	T-100-reb	DRB U TUBE	Default	853,500
T-101	T-101-tower	DTW TRAYED	Default	495,100
	T-101-cond	DHE FIXED T S	Default	115,900
	T-101-cond acc	DHT HORIZ DRUM	Default	76,000
	T-101-reflux pump	DCP GEN SERV	Default	33,200
	T-101-overhead split	C	Default	—
	T-101-bottoms split	C	Default	—
	T-101-reb	DRB U TUBE	Default	203,600
TEE-100	TEE-100	C	Default	—
X-100	X-100	C	Default	—
Total equipment cost calculated from ICARUS				3,120,700

G.4 Propylene Glycol Production from Hydrogenolysis of Glycerol

The equipment from the HYSYS case was mapped in IPE and is given in Table G.4. The equipment with IPE description given as "C" in Table G.4 were piping components, and they were included in installed costs. Manual mapping was not required in any of the equipment for propylene glycol.

G.5 Acetic Acid Production from Corn Stover Anaerobic Digestion

The equipment from the HYSYS case was mapped in IPE and is given in Table G.5 as default. The equipment in Table G.5, listed as manual, required manual mapping in IPE because appropriate equipment in HYSYS was not available in the design. The perfect separator X-100 was mapped as a vertical

TABLE G.5

Equipment Mapping for Acetic Acid Production from Corn Stover Anaerobic Digestion

Equipment HYSYS	Equipment IPE	IPE Description	Mapping Type	Equipment Cost (Dollars)
E-100	E-100	DHE FLOAT HEAD	Default	231,100
E-101	E-101	DHE FLOAT HEAD	Default	57,800
E-102	E-102	DHE FLOAT HEAD	Default	180,400
MIX-100	MIX-100	C	Default	—
MIX-101	MIX-101	C	Default	—
MIX-102	MIX-102	C	Default	—
MIX-103	MIX-103	C	Default	—
MIX-104	MIX-104	C	Default	—
MIX-105	MIX-105	C	Default	—
RCY-1	RCY-1	C	Default	—
TEE-100	TEE-100	C	Default	—
V-100	V-100	DAT REACTOR	Manual	345,600
V-101	V-101	DAT REACTOR	Manual	345,200
X-100	X-100	C	Default	—
X-101	X-101	ECT SOLID BOWL	Manual	1,275,600
X-102	X-102	DVT CYLINDER	Manual	326,200
X-103	X-103	DVT CYLINDER	Manual	138,800
X-104	X-104	DVT CYLINDER	Manual	264,000
Total equipment cost calculated from ICARUS				3,164,700

cylinder, which simulated the separator for acetic acid and gases. The perfect separator X-101 in HYSYS was mapped as a solid bowl centrifuge for separating the unreacted solids from the acetic acid and water mixture. The separator X-102 was mapped as a vertical cylinder to simulate the liquid–liquid extraction column. The separator X-103 was mapped as a vertical cylinder to simulate the rectification column for acetic acid recovery. The separator X-104 was mapped as a vertical cylinder to simulate stripping column for the solvent recovery.

The reactor V-100 was designed as tank in HYSYS as only tanks allow multiple reactions in HYSYS. The description of the pretreatment reactor in Holtzapple et al. (1999) was comparable to a tank. It was not necessary to map the rest of the equipment with IPE description given as "C" in Table G.5 because those were mainly piping components, and the cost of piping was included in installation costs.

Appendix H: Molecular Weights

H.1 Ethanol from Corn Stover

TABLE H.1

Molecular Weights of Individual Species
in Corn Stover Fermentation

Component	Molecular Weight
H_2O	18.02
Xylose	150.13
Cellulose	162.16
Hemicellulose	132.13
Glucose (dextrose)	180.16
CO_2	44.01
Ash	200.00
Corn steep liquor	200.00
Other solids	180.16
Lignin	342.00
Nitrogen	28.01
Oxygen	32.00
Ethanol	46.07
Bacteria	180.16
Cellulase	18.02
DAP	132.07

Corn Ethanol

TABLE H.2

Molecular Weights of Individual Species
in Corn Ethanol Fermentation

Component	Molecular Weight
NFDS	180.16
Protein (soluble)	180.16
Starch	18.20

Ethylene from Ethanol

TABLE H.3

Molecular Weights of Individual Species
in Ethylene Process

Component	Molecular Weight
H_2O	18.02
Ethanol	46.07
Ethylene	28.05

Transesterification to FAME

TABLE H.4

Molecular Weights of Individual Species
in Transesterification

Component	Molecular Weight
H_2O	18.02
Glycerol	92.09
Oil (trilinolein)	879.40
HCl	36.46
CH_3OH (methanol)	32.04
$NaOCH_3$ (sodium methylate)	54.02
NaCl (sodium chloride)	58.45
FAME (methyl linoleate)	294.5
NaOH (sodium hydroxide)	40.00

Propylene Glycol

TABLE H.5

Molecular Weights of Individual Species in Propylene Glycol Process

Component	Molecular Weight
H_2O	18.02
Glycerol	92.09
Propylene glycol	76.10
Hydrogen	2.02

Acetic Acid from Anaerobic Digestion

TABLE H.6

Molecular Weights of Individual Species in Anaerobic Digestion

Component	Molecular Weight
H_2O	18.02
Acetic acid	60.05
Xylose	150.13
Cellulose	162.16
Hemicellulose	132.13
Glucose (dextrose)	180.16
Hydrogen	2.02
CO_2	44.01

Appendix I: Postscript

Conclusions

A new methodology has been demonstrated for the integration of bioprocesses in an existing industrial complex producing chemicals. The transitioning from nonrenewable resources to renewable resources as feedstock was possible with a triple bottom line profit increasing by 93% was an important conclusion in this research. The increase in carbon dioxide for using renewable resources was utilized for algae oil production and for manufacturing chemicals.

This new methodology was developed for identifying potentially new bioprocesses. Five processes were simulated using Aspen HYSYS® and cost estimations performed in Aspen ICARUS®. Three of these processes converted biomass to chemicals, and two of the processes converted the bioproducts to demonstrate the introduction of biomass feedstock into ethylene and propylene chain. These bioprocesses were integrated into a superstructure that includes plants in the existing chemical production complex in the lower Mississippi River corridor.

Then, the optimal configuration of new and existing plants was determined by optimizing a triple bottom line profit based on economic, environmental, and sustainable costs using the chemical complex analysis system. Sustainable credits were given to processes that consumed carbon dioxide and other greenhouse gases.

The optimal solution gave a triple bottom line profit of $1650 million/year, an increase of 93% from the base case solution of $854 million/year. The bioprocesses increased the pure carbon dioxide sources to 1.07 million MT/year from 0.75 million MT/year for the base case. The pure carbon dioxide vented to the atmosphere was 0.61 million MT/year in the base case and was reduced to zero in the optimal structure. This was achieved by using the carbon dioxide in an algae oil production process and in processes that utilize carbon dioxide as raw material. A total of 0.84 million MT/year of CO_2 was used by the algae oil production process.

Six corn ethanol plants each producing 57,000 MT/year of ethanol were required to meet the demand for ethylene. A total of 200,000 MT/year of ethylene was produced from the complex; 480,000 MT of fatty acid methyl esters were produced from the complex (15 plants each producing 33,000 MT/year), which can be used for manufacturing polymers. The total glycerol produced from the complex was 59,000 MT/year, which was used to produce 37,000 MT of propylene glycol.

The existing plants for ethylbenzene and styrene were excluded in the optimal structure. The high cost of benzene, a raw material for the process, was the main reason for the exclusion of the plants. All of the acetic acid plants were excluded from the optimal structure. The selling price of acetic acid was the main reason for the exclusion of the plants from the superstructure.

The total energy required by the optimal complex was 6405 TJ/year. The power plant in the complex supplied 2340 TJ/year; the rest was purchased utility from steam for corn ethanol process. The total utility costs for the complex increased to $46 million/year from $12 million/year in the base case.

The sustainable costs to the society decreased to $10 million/year from $18 million/year in the base case. This was a 44% decrease in the costs to the society, achieved by consumption of pure carbon dioxide in the complex for algae and chemical production. The raw material costs decreased to $470 million/year from $685 million/year (31%) in the base case due to the exclusion of the costly ethylbenzene process. This reduced the environmental costs by 31% for decreased raw material use.

Multicriteria optimization of the complex gave Pareto optimal solutions. A profit of $1660 million/year and sustainable costs of $10 million/year were obtained when the decision was to have maximum weight on profit. When the decision was to have maximum sustainability credit, the profit reduced to $1193 million/year and sustainable credits of $26 million/year were obtained.

Monte Carlo simulations of the complex for price sensitivity using 1000 iterations gave an average triple bottom line of 1898 million/year. The standard deviation for the triple bottom line was $311 million/year. The corn stover ethanol process was selected in 23% of the 1000 iterations, and the corn ethanol process was selected in 77% of the iterations. The ethylbenzene process was selected in almost half of the total iterations (47% times). The new acetic acid process from the consumption of CO_2 was selected in 36% of the iterations. The complex was able to curb the pure carbon dioxide emissions to zero level in almost all of the 1000 iterations (995 times out of 1000).

Five case studies were presented that demonstrated the use of chemical complex optimization for sustainability analysis. The first case demonstrated how changes to the model of the superstructure can be made for scenario analysis. The second, third, and fifth cases demonstrated the optimization of the triple bottom line by changing parameters for sustainable credits/costs, utility costs (production costs), and raw material costs, respectively. The fourth case demonstrated multicriteria optimization of the complex based on parameter changes for technology and policy changes in the future.

Thus, this methodology evaluated integrating new plants that use renewable feedstock into the existing infrastructure of plants in a chemical production complex. This methodology can be used by any concerned decision maker. With this system, engineers can convert company's goals and capital into viable projects that meet economic, environmental, and sustainable requirements. The method can be used by government organizations to evaluate the emissions and life cycle of greenhouse gases in any industrial complex. The methodology can be used by engineering groups to design and evaluate energy efficient and environmentally acceptable plants and have new products from greenhouse gases.

Based on these results, the methodology could be applied to other chemical complexes in the world for reduced emissions and energy savings. The model for the superstructure can be obtained from www.mpri.lsu.edu

Future Directions

The methodology for chemical complex optimization is general and applicable for all chemical complexes. The bioprocess plant model formulation is specific for different types of chemicals that the processes produce. Chapters 2 and 3 provided detailed review of chemicals that may be produced from biomass and have industrial significance. All those chemicals could not be included as the focus of this research was to introduce renewable resources to the most important chain of chemicals (ethylene and propylene). Further work may be undertaken to include other chemicals in the complex. Binary variables may be associated with processes to select the chemicals that compete with each other for raw materials.

In this research, the emphasis was on integrating technologies for bioprocesses and demonstration of the methodology to determine sustainability on a quantitative basis. It was assumed that the raw materials were available on a continuous basis. Further research may be conducted to determine and include constraints related to supply-chain optimization of the raw materials. These may include constraints related to crop cycles and transportation costs among others.

Price elasticities can be included in the model. The availability of data on price elasticity for bioethanol and ethylene calculations is very limited, as was shown in Appendix D. With available data, price elasticities can be used as leading indicators to estimate future prices of chemicals in the complex and have optimization over time periods.

The bioprocess plant model formulation requires extensive knowledge of the process for developing the plant model. The thermodynamic information of biomass species was determined using sophisticated simulation tools, but the accuracy could not be validated due to lack of experimental data. Research is ongoing to determine thermodynamic properties of biomass species, and these could be included in future if any discrepancy is found in the model.

The algae oil production process and the gasification process from biomass were black box models constructed with limited information.

The corn-to-ethanol process was a SuperPro Designer® model, without information of thermodynamics for individual streams. These processes may be modeled in detail in Aspen HYSYS to compare all the processes on the same platform.

References

ACES (2010) H.R.2454: American Clean Energy and Security Act of 2009, http://www.opencongress.org/bill/111-h2454/show (accessed May 8, 2010).

Aden A (2008) *Biochemical Production of Ethanol from Corn Stover: 2007 State of Technology Model*, National Renewable Energy Laboratory, Golden, CO, NREL/TP-510-43205.

Aden A, Ruth M, Ibsen K, Jechura J, Neeves K, Sheehan J, Wallace B (2002) *Lignocellulosic Biomass to Ethanol Process Design and Economics Utilizing Co-Current Dilute Acid Prehydrolysis and Enzymatic Hydrolysis for Corn Stover*, National Renewable Energy Laboratory, Golden, CO, NREL/TP-510-32438.

Agboola AE (2005) Development and model formulation of scalable carbon nanotube processes: HiPCO and CoMoCAT process models, MS thesis, Louisiana State University, Baton Rouge, LA.

AGMRC (2010) Soybean oil and biodiesel usage projections & balance sheet, http://www.extension.iastate.edu/agdm/crops/outlook/biodieselbalancesheet.pdf (accessed August 4, 2010).

Aiello-Mazzarri C, Agbogbo FK, Holtzapple MT (2006) Conversion of municipal solid waste to carboxylic acids using a mixed culture of mesophilic microorganisms, *Bioresource Technology*, 97(1): 47–56.

Anderson J (2010) Communicating the cost of product and process development, *Chemical Engineering Progress*, 106(2): 46–51.

Arakawa H, Aresta M, Armor JN, Barteau MA, Beckman EJ, Bell AT, Bercaw JE et al. (2001) Catalysis research of relevance to carbon management: Progress, challenges, and opportunities, *Chemical Reviews*, 101, 953–996.

Arnaud C (2008) Algae pump out hydrocarbon biofuels, *Chemical and Engineering News*, 86(35): 45.

Arnold RA (2008) *Economics*, Cengage Learning, Stamford, CT, ISBN 0-324-59542-5.

Austin GT (1984) *Shreve's Chemical Process Industries*, 5th edn., McGraw-Hill Book Company, New York, ISBN 0070571473.

Axelsson L (2004) Lactic acid bacteria: Classification and physiology. In: Salmien S, Wright AV, Ouwehand A (eds.) *Lactic Acid Bacteria: Microbiological and Functional Aspects*, Marcel Dekker, Inc., New York.

Aydogan S, Kusefoglu S, Akman U, Hortacsu O (2006) Double-bond depletion of soybean oil triglycerides with $KMnO_4/H_2$ in dense carbon dioxide, *Korean Journal of Chemical Engineering*, 23(5): 704–713.

Banerjee A (2010) The advent of corn-based ethanol: A re-examination of the competition for grains, RIS Policy Briefs, No. 46, March 2010, http://www.ris.org.in/images/RIS_images/pdf/pb46.pdf

Banholzer WF, Watson KJ, Jones ME (2008) How might biofuels impact the chemical industry? *Chemical Engineering Progress*, 104(3): S7–S14.

Bickel P, Friedrich R (2005) *ExternE Externalities of Energy Methodology 2005 Update*, Office for Official Publications of the European Communities, Luxembourg, ISBN 92-79-00423-9.

Blanco N, Kelly A (2009), The Biosequestration Challenge, World Resource Institute, http://www.wri.org/stories/2009/05/biosequestration-challenge (accessed May 10, 2012).

Bohringer C, Jochem PEP (2007) Measuring the immeasurable—A survey of sustainability indices, *Ecological Economics*, 63(1): 1–8.

Bonivardi AL, Chiavassa DL, Baltanas MA (1998) Promoting effect of calcium addition to Pd/SiO_2 catalysts in CO_2 hydrogenation to methanol, *Advances in Chemical Conversions for Mitigating Carbon Dioxide, Studies in Surface Science and Catalysis,* 114: 533–536.

Bourne Jr. JK (2007) Green dreams. *National Geographic Magazine*, http://ngm.nationalgeographic.com/print/2007/10/biofuels/biofuels-text (accessed May 8, 2010).

Brown RC (2003) *Biorenewable Resources: Engineering New Products from Agriculture*, Iowa State Press, Iowa City, IA, ISBN 0813822637.

Burridge E (2006) Formic acid, *ICIS Chemical Business*, 1(29).

Byers P (2006) Louisiana Department of Economic Development, Private Communication.

C&E News (2007a) Dow to make polyethylene from sugar in Brazil, *Chemical and Engineering News*, 85(30): 17.

C&E News (2007b) ConocoPhillips funds biofuel research, *Chemical and Engineering News*, 85(18): 24.

C&E News (2009a) Industry considers carbon options, *Chemical and Engineering News*, 87(13): 16–19.

C&E News (2009b) Chemical output slipped in most regions, *Chemical and Engineering News*, 87(27): 51.

Cabezas H, Bare JC, Mallick SK (1997) Pollution prevention with chemical process simulators: The general waste reduction algorithm, *Computers and Chemical Engineering*, 21(Suppl.) S305–310.

Cameron DC, Koutsky JA (1994) Conversion of glycerol from soy diesel production to 1,3-propanediol. Final Report prepared for National Biodiesel Development Board, Department of Chemical Engineering, UW-Madison, Madison, WI.

CCFX (2007) Chicago climate futures exchange, http://www.ccfe.com/ (accessed October 8, 2007).

CCX (2007) Chicago climate exchange, http://www.chicagoclimatex.com/ (accessed October 8, 2007).

CEP (2007) $100-Million plant is first to produce propanediol from corn sugar, *Chemical Engineering Progress*, 103(1): 10.

CFTC (2012) A Guide to the Language of the Futures Industry, U.S. Commodity Futures Trading Commission, http://www.cftc.gov/consumerprotection/educationcenter/cftcglossary/glossary_f (accessed May 10, 2012).

Ciferno JP, Marano JJ (2002) Benchmarking biomass gasification technologies for fuels, chemicals and hydrogen production, U.S. Department of Energy National Energy Technology Laboratory, Pittsburgh, PA, http://www.netl.doe.gov/technologies/coalpower/gasification/pubs/pdf/BMassGasFinal.pdf (accessed August 1, 2010).

Constable D, Hunter J, Koch D, Murphy J, Mathews J, Heim L, Tyls L et al. (July 22, 1999) *Total Cost Assessment Methodology; Internal Managerial Decision Making Tool*, American Institute of Chemical Engineers, AIChE/CWRT, New York, ISBN 0-8169-0807-9.

Constantinou L, Gani R (1994) New group contribution method for estimating properties of pure compounds, *AIChE J.*, 40(10), 1697–1710.

CYPENV (2010) Sequestration, http://www.cypenv.org/Files/sequest.htm#Chemical%20sequestration (accessed August 30, 2010).

Dasari MA, Kiatsimkul PP, Sutterlin WR, Suppes GJ (2005) Low-pressure hydrogenolysis of glycerol to propylene glycol, *Applied Catalysis: A General*, 281(1–2): 225–231.

DDBST (2009) Online property estimation, http://www.ddbst.com/new/OnlineEstimation.htm (accessed November 17, 2009).

De Dietrich (2010) The recovery of acetic acid by liquid-liquid extraction, http://www.ddpsinc.com/Downloads/ProcessProfiles/pp50-2.pdf (accessed January 10, 2010).

DOE (2007) DOE selects six cellulosic ethanol plants for up to $385 million in federal funding, http://www.energy.gov/print/4827.htm (accessed October 2, 2007).

DOE (2010a) National algal biofuels technology roadmap, energy efficiency and renewable energy (U.S. DOE), Draft document: https://e-center.doe.gov/iips/faopor.nsf/UNID/79E3ABCACC9AC14A852575CA00799D99/$file/AlgalBiofuels_Roadmap_7.pdf (accessed May 8, 2010).

DOE (2010b) Biomass multi-year program plan March 2010, Energy efficiency and renewable energy (U.S. DOE), http://www1.eere.energy.gov/biomass/pdfs/mypp.pdf (accessed May 8, 2010).

DOE (2010c) Biomass energy databook, United States Department of Energy, http://cta.ornl.gov/bedb/biofuels.shtml (accessed May 8, 2010).

Dutta A, Phillips SD (2009) *Thermochemical Ethanol via Direct Gasification and Mixed Alcohol Synthesis of Lignocellulosic Biomass*, National Renewable Energy Laboratory, Golden, CO, NREL/TP-510-45913.

D'Aquino R (2007) Cellulosic ethanol—Tomorrow's sustainable energy source, *Chemical Engineering Progress*, 103(3): 8–10.

EIA (2010a) Annual energy outlook 2010, Energy Information Administration, Report No. DOE/EIA-0383(2010).

EIA (2010b) Weekly United States Spot Price FOB Weighted by Estimated Import Volume (Dollars per Barrel), Energy Information Administration, http://tonto.eia.doe.gov/dnav/pet/hist/LeafHandler.ashx?n=PET&s=WTOTUSA&f=W (accessed May 8, 2010).

EIA (2010c) Total carbon dioxide emissions from the consumption of energy (million metric tons), Energy Information Administration, http://tonto.eia.doe.gov/cfapps/ipdbproject/IEDIndex3.cfm?tid=90&pid=44&aid=8 (accessed May 8, 2010).

EIA (2010d) United States natural gas industrial price (dollars per thousand cubic feet), http://tonto.eia.doe.gov/dnav/ng/hist/n3035us3m.htm (accessed April 10, 2010).

Elkington J (1998) *Cannibals with Forks—The Triple Bottom Line of 21st Century Business*, New Society Publishers, British Columbia, Canada, ISBN 0-86571-392-8.

Energetics (2000) Energy and Environmental Profile of the U.S. Chemical Industry, Energy Efficiency and Renewable Energy (US DOE), http://www1.eere.energy.gov/industry/chemicals/pdfs/profile_chap1.pdf (accessed May 8, 2010).

EPA (2010) Mandatory reporting of greenhouse gases rule, United States Environmental Protection Agency, http://www.epa.gov/climatechange/emissions/ghgrulemaking.html (accessed May 8, 2010).

Ethanol Producers Magazine (2010) Plants list, *Ethanol Producers Magazine*, http://www.ethanolproducer.com/plant-list.jsp (accessed May 8, 2010).

Ethanol Producers Magazine (Spring 2009) Fuel ethanol plant map.

Faden M (2010), Taxes-Excise Tax-Carbon Dioxide Emissions, http://www.montgomerycountymd.gov/content/council/pdf/agenda/col/2010/100427/20100427_3-1.pdf (accessed May 10, 2012).

Freedman B, Pryde EH, Mounts TL (1984) Variables affecting the yields of fatty esters from transesterified vegetable oils, *JAOCS*, 61(10): 1638–1643.

Fukuda H, Kondo A, Noda H (2001) Biodiesel production by the transesterification of oils, *Journal of Bioscience and Bioengineering*, 92(5): 405–416.

GAMS (2010) GAMS solvers, http://www.gams.com/solvers/index.htm (accessed August 1, 2010).

Garrett DE (1989) *Chemical Engineering Economics*, Van Nostrand Reinhold, New York.

Glazer AW, Nikaido H (1995) *Microbial Biotechnology: Fundamentals of Applied Microbiology*, W.H. Freeman & Company, San Francisco, CA, ISBN 0-71672608-4.

Granda C (2007) The MixAlco process: Mixed alcohols and other chemicals from biomass, *Seizing Opportunity in an Expanding Energy Marketplace, Alternative Energy Conference*, LSU Center for Energy Studies, http://www.enrg.lsu.edu/Conferences/altenergy2007/granda.pdf (accessed August 1, 2010).

Gray P, Hart W, Painton L, Phillips C, Trahan M, Wagner J (1997) A survey of global optimization methods, Sandia National Laboratories, Livermore, CA, http://www.cs.sandia.gov/opt/survey/(accessed August 1, 2010).

Grossmann I, Caballero JA, Yeomans H (1999) Mathematical programming approaches to the synthesis of chemical process systems, *Korean Journal of Chemical Engineering*, 16(4): 407–426.

Guettler MV, Jain MK, Soni BK (1996) Process for making succinic acid, microorganisms for use in the process and methods of obtaining the microorganisms. U.S. Patent No. 5,504,004.

Haas MJ, McAloon AJ, Yee WC, Foglia TA (2006) A process model to estimate biodiesel production costs, *Bioresource Technology*, 97(4): 671–678.

Hayes DJ, Fitzpatrick S, Hayes MHB, Ross JRH (2006) The biofine process—Production of levulinic acid, furfural and formic acid from lignocellulosic feedstock. In: Kamm B, Gruber PR, Kamm M (eds.) *Biorefineries—Industrial Processes and Products*, Vol. 1, Wiley-VCH Verlag GmbH & Co. KGaA, Weinheim, Germany, ISBN 3-527-31027-4.

Heinzle E, Biwer AP, Cooney CL (2007) *Development of Sustainable Bioprocesses: Modeling and Assessment*, John Wiley & Sons, England, U.K, ISBN 978-0-470-01559-9.

Hertwig TA (2004) Mosaic (formerly IMC Phosphate), Uncle Sam, LA, Private Communication.

Hertwig TA (2006) Mosaic (formerly IMC Phosphate), Uncle Sam, LA, Private Communication.

Hileman B (February 19, 2007) A dubious way out of CO_2 emissions, *Chemical and Engineering News*, 31.

Hitchings MA (2007) *Algae: The Next Generation of Biofuels. Fuel Fourth Quarter 2007*, Hart Energy Publishing, Houston, TX.

Hofstrand D (2010) Natural gas and coal measurements and conversions, http://www.extension.iastate.edu/agdm/wholefarm/pdf/c6-89.pdf (accessed August 8, 2010).

Holmgren J, Gosling C, Couch K, Kalnes T, Marker T, McCall M, Marinangeli R (2007) Refining biofeedstock innovations, *Petroleum Technology Quarterly*, 12(4): 119–124.

Holtzapple MT, Davison RR, Ross MK, Aldrett-Lee S, Nagwani M, Lee CM, Lee C et al. (1999) Biomass conversion to mixed alcohol fuels using the Mixalco process, *Applied Biochemistry and Biotechnology*, 79(1–3): 609–631.

Humbird D, Aden A (2009) *Biochemical Production of Ethanol from Corn Stover: 2008 State of Technology Model*, National Renewable Energy Laboratory, Golden, CO, NREL/TP-510-46214.

Hunt J (August 23–30, 2004) Phosphoric acid prices increase as market tightens, *Chemical Market Reporter*, 16, http://www.icis.com/Articles/2004/08/20/606991/phosphoric-acid-prices-increase-as-market-tightens.html (accessed May 10, 2012).

ICIS (2006b) Mosaic's Q2 earnings fall short of forecast, *ICIS Chemical Business*, 1(3): 16, January 23, 2007.

ICIS (2009a) Acetic acid prices and pricing information, http://www.icis.com/v2/chemicals/9074786/acetic-acid/pricing.html (accessed February 11, 2010).

ICIS (2009b) Ethylene, *ICIS Chemical Business*, 276(15): 40.

ICIS (April 6, 2009e) Key indicators, *ICIS Chemical Business*, 275(14).

ICIS (April 7, 2008d) Key indicators, *ICIS Chemical Business*, 273(14).

ICIS (August 24, 2008a) MIBK, *ICIS Chemical Business*, 276(6): 36.

ICIS (December 18, 2006a) Markets watch, *ICIS Chemical Business*, 1(47).

ICIS (February 23, 2009g) Styrene, *ICIS Chemical Business*, 275(8).

ICIS (January 21, 2008b) Key indicators, *ICIS Chemical Business*, 273(3).

ICIS (January 4, 2010a) Key indicators, *ICIS Chemical Business*, 277(1).

ICIS (January 5, 2009c) Key indicators, *ICIS Chemical Business*, 275(1).

ICIS (July 14, 2008e) Key indicators, *ICIS Chemical Business*, 274(2).

ICIS (July 20, 2009f) Key indicators, *ICIS Chemical Business*, 276(2).

ICIS (June 11, 2007) Market news and views, *ICIS Chemical Business*, 2(70): 8.

ICIS (March 1, 2010b) Key indicators, *ICIS Chemical Business*, 277(8).

ICIS (March 15, 2010c) Ethylbenzene, *ICIS Chemical Business*, 277(10).

ICIS (October 5, 2009d) Key indicators, *ICIS Chemical Business*, 276(12).

ICIS (October 6, 2008c); Key indicators, *ICIS Chemical Business*, 274(13).

Indala S (2004) Development and integration of new processes consuming carbon dioxide in multi-plant chemical production complexes, MS thesis, Louisiana State University, Baton Rouge, LA.

Intelligen (2009) Modeling the process and costs of fuel ethanol production by the corn dry-grind process, http://www.intelligen.com/downloads/CornToEthanol.zip (accessed October 21, 2009).

IPCC (2007) Climate change 2007: Synthesis report, http://www.ipcc.ch/pdf/assessment-report/ar4/syr/ar4_syr.pdf (accessed May 8, 2010).

Ito T, Nakashimada Y, Senba K, Matsui T, Nishio N (2005) Hydrogen and ethanol production from glycerol-containing wastes discharged after biodiesel manufacturing process, *Journal of Bioscience and Bioengineering*, 100(3): 260–265.

Johnson DL (2006) The corn wet milling and corn dry milling industry—A base for biorefinery technology developments. In: Kamm B, Gruber PR, Kamm M (eds.) *Biorefineries—Industrial Processes and Products*, Vol. 1, Wiley-VCH Verlag GmbH & Co. KGaA, Weinheim, Germany, ISBN 3-527-31027-4.

Jun Ki-Won, Mi-Hee Jung, Rama Rao KS, Myoung-Jae Choi, Kyu-Wan Lee (1998) Effective conversion of CO_2 to methanol and dimethyl ether over hybrid catalysts, *Advances in Chemical Conversions for Mitigating Carbon Dioxide, Studies in Surface Science and Catalysis*, 114: 447–450.

Kaar WE, Holtzapple MT (2000) Using lime pretreatment to facilitate the enzymic hydrolysis of corn stover, *Biomass and Bioenergy*, 18(3): 189–199.

Kamm B, Kamm M, Gruber PR, Kromus S (2006) Biorefinery systems—An overview. In: Kamm B, Gruber PR, Kamm M (eds.) *Biorefineries—Industrial Processes and Products*, Vol. 1, Wiley-VCH Verlag GmbH & Co. KGaA, Weinheim, Germany, ISBN 3-527-31027-4.

Karinen RS, Krause AOI (2006) New biocomponents from glycerol, *Applied Catalysis A: General*, 306: 128–133.

Katzen R, Schell DJ (2006) Lignocellulosic feedstock biorefinery: History and plant development for biomass hydrolysis. In: Kamm B, Gruber PR, Kamm M (eds.) *Biorefineries—Industrial Processes and Products*, Vol. 1, Wiley-VCH Verlag GmbH & Co. KGaA, Weinheim, Germany, ISBN 3-527-31027-4.

Katzer J (2008) The future of coal-based power generation, *CEP Magazine*, 104(3): S15–S22.

Kebanli ES, Pike RW, Culley DD, Frye JB (1981) Fuel gas from dairy farm waste, *Agricultural Energy Vol. II Biomass Energy Crop Production*, ASAE Publication 4-81 (three volumes), American Society of Agricultural Engineers, St. Joseph, MI.

Kho J (2009) Big oil bets on biofuels, Renewable Energy World http://www.renewableenergyworld.com/rea/news/article/2009/07/bio-oil-bets-on-biofuels (accessed May 8, 2010).

Klass DL (1998) *Biomass for Renewable Energy, Fuels and Chemicals*, Academic Press, San Diego, CA, ISBN 0124109500.

Kment R (2009) Commodities report, *Ethanol Producers Magazine*, p. 21, May 2009, p. 21 June 2009, p. 23, July 2009, September, 2009, p. 19, October, 2009, p. 19, November, 2009, p. 19, 2010.

Knopf FC (April 1, 2007) Lousiana State University, Private Communication.

Koch D (2002) Dow chemical pilot of total "business" cost assessment methodology: A tool to translate EH&S "…right things to do" into economic terms (dollars), *Environmental Progress*, 21(1): 20–28.

Koutinas AA, Du C, Wang RH, Webb C (2008) Production of chemicals from biomass. In: Clark JH, Deswarte FEI (eds.) *Introduction to Chemicals from Biomass*, Wiley, Great Britain, ISBN 978-0-470-05805-3.

Krotscheck C, Narodoslawsky M (June 1996) The sustainable process index a new dimension in ecological evaluation, *Ecological Engineering*, 6(4): 241–258.

Kuehler GP (October 10, 2003) Exxon Mobil Baton Rouge Refinery, Private Communication.

Kwiatkowski JR, McAloon AJ, Taylor F, Johnston DB (2006) Modeling the process and costs of fuel ethanol production by the corn dry-grind process, *Industrial Crops and Products*, 23(3): 288–296.

LaCapra LT (March 1, 2007) Take my emissions, please, *The Wall Street Journal*.

Laurin L (2007) Keeping the competitive edge, *Chemical Engineering Progress*, 103(6): 44–46.

LCA (2007) Louisiana chemical association, www.lca.org (accessed October, 2007).

Lehni M (October 2000) *Eco-efficiency-Creating More Value with Less Impact*, World Business Council for Sustainable Development, WBCSD, North Yorkshire, U.K., ISBN 2-940240-17-5

Lindo Systems, Inc. (2007) LINDOGlobal, http://www.gams.com/dd/docs/solvers/lindoglobal.pdf (accessed October 15, 2007).

Liu D, Liu H, Sun Y, Lin R, Hao J (2010) Method for producing 1,3-propanediol using crude glycerol, a by-product from biodiesel production, http://www.freepatentsonline.com/20100028965.pdf, Pub. No. 2010/0028965 A1.

Louisiana Department of Economic Development (1998) Louisiana chemicals and petroleum products list, Office of Policy and Research, Louisiana Department of Economic Development, Baton Rouge, LA.

Luchansky MS, Monks J (2009) Supply and demand elasticities in the U.S. ethanol fuel market, *Energy Economics*, 31: 403–410.

Lucia LA, Argyropolous DS, Adamopoulos L, Gaspar AR (2007) Chemicals, materials and energy from biomass: A review. In: Argyropoulos DS (ed.) *Materials, Chemicals and Energy from Forest Biomass*, American Chemical Society, Washington, DC, ISBN 978-0-8412-3981-4.

Ma F, Hanna MA (1999) *Biodiesel production: A review*, *Bioresource Technology*, 70(1): 1–15.

McGowan TF (2009) *Biomass and Alternate Fuel Systems*, John Wiley & Sons Inc., Hoboken, NJ, ISBN 978-0-470-41028-8.

Meher LC, Vidya Sagar D, Naik SN (2006) Technical aspects of biodiesel production by transesterification—A review, *Renewable and Sustainable Energy Review*, 10(3): 248–268.

Methanex (2010) Methanol price, http://www.methanex.com/products/methanolprice.html (accessed August 8, 2010).

Metz B, Davidson O, de Coninck H, Meyer MLL (2005) *IPCC Special Report on Carbon Dioxide Capture and Storage*, Cambridge University Press, New York, ISBN 978-0-521-68551-1.

Monteiro, JGMS, Silvaa PAC, Araújoa OQF, Medeiros JLD (2010) Pareto optimization of an industrial ecosystem: Sustainability maximization, *Brazilian Journal of Chemical Engineering*, 27(3): 429–440.

Moreira AR (1983) Acetone-butanol fermentation. In: Wise DL (ed.) *Organic Chemicals from Biomass*, The Benjamin Cummind Publishing Company, Inc., Menlo Park, CA, ISBN 0-8053-9040-5.

MPRI (2010) Chemical complex analysis system, http://www.mpri.lsu.edu/chemcomplex.html (accessed September 5, 2010).

Mu Y, Teng H, Zhang D, Wang W, Xiu Z (2006) Microbial production of 1,3-propanediol by *Klebsiella pneumoniae* using crude glycerol from biodiesel preparations, *Biotechnology Letters*, 28(21): 1755–1759.

Myint LL, El-Halwagi MM (2009) Process analysis and optimization of biodiesel production, *Clean Technology and Environmental Policy*, 11: 263–267.

NAS (2009) *Liquid Transportation Fuels from Coal and Biomass: Technological Status, Costs, and Environmental Impacts*, National Academy of Sciences, Washington, DC, ISBN: 978-0-309-13712-6.

NBB (2008a) Commercial biodiesel production plants, National biodiesel board, website, http://www.biodiesel.org/pdf_files/fuelfactsheets/Producers%20Map%20-%20existing.pdf (accessed November 2008).

NBB (2008b) Specification for biodiesel blends B6-B20, National biodiesel board, website, http://www.biodiesel.org/resources/fuelqualityguide/files/B20_Specification_Nov2008.pdf (accessed March 2009).

NEOS (2010) NEOS server for optimization, http://www-neos.mcs.anl.gov/ (accessed August 1, 2010).

Nerlov J, Chorkendorff I (1999), Methanol synthesis from CO_2, CO, and H_2 over Cu(100) and Ni/Cu(100), *Journal of Catalysis*, 181: 271–279.

Neumaier A, Shcherbina O, Huyer W, Vinkó T (2007) A comparison of complete global optimization solvers, http://www.mat.univie.ac.at/~neum/ms/comparison.pdf (accessed October 15, 2007).

Niederl A, Narodoslawsky M (December 2004) *Life Cycle Assessment of Biodiesel from Tallow and Used Vegetable Oil*, Institute for Resource Efficient and Sustainable Systems Process Evaluation, Graz, Austria.

NIST (2009) Trilinolein, http://webbook.nist.gov/cgi/cbook.cgi?ID=C537406&Units=SI (accessed November 17, 2009).

Oilgae (2010) CO_2 capture using algae—A comprehensive guide from oilgae, http://www.oilgae.com/ref/report/co2_capture/co2_capture.html (accessed July 27, 2010).

Ondrey G (2007a) Coproduction of cellulose acetate promises to improve economics of ethanol production, *Chemical Engineering*, 114(6): 12.

Ondrey G (2007b) Propylene glycol, *Chemical Engineering*, 114(6): 10.

Ondrey G (2007c) A vapor-phase glycerin-to-PG process slated for its commercial debut, *Chemical Engineering*, 114(8): 12.

Ondrey G (2007d) A sustainable route to succinic acid, *Chemical Engineering*, 114(4): 18.

Ondrey G (June 1, 2009) A one step process for extracting oil from algae, *Chemical Engineering*, http://www.che.com/chementator/4832.html (accessed August 10, 2010).

Optimal Methods, Inc. and OptTek System, Inc. (2007) OQNLP and MSNLP, http://www.gams.com/dd/docs/solvers/msnlp.pdf (accessed October 15, 2007).

Paster M, Pellegrino JL, Carole TM (2003) Industrial bioproducts: Today and tomorrow, Department of Energy Report prepared by Energetics, Inc, http://www.energetics.com/resourcecenter/products/studies/Documents/bioproducts-pportunities.pdf (accessed May 8, 2010).

Pater JE (February 2006) *A Framework for Evaluating the Total Value Proposition of Clean Energy Technologies*, National Renewable Energy Laboratory, Golden, CO, NREL/TP-620-38597.

Pellegrino JL (2000) Energy and environmental profile of the U.S. chemical industry, Office of Industrial Technologies, Department of Energy prepared by Energetics, Inc., Washington, DC, May 2000.

Perlack RD, Turhollow AF (2002) Assessment of options for the collection, handling, and transport of corn stover, http://www.osti.gov/bridge, ORNL/TM-2002/44. (accessed August 1, 2010).

Perlack RD, Wright LL, Turhollow AF, Graham RL (2005) Biomass as feedstock for a bioenergy and bioproducts industry: The technical feasibility of a billion-ton annual supply, USDA document prepared by Oak Ridge National Laboratory, Oak Ridge, TN, ORNL/TM-2005/66.

Perry RH, Green DW (1997) *Perry's Chemical Engineers' Handbook*, 7th edn., McGraw-Hill, New York, ISBN 978-0070498419.

Peterson RW (2000) *Giants on the River*, Homesite Company, Baton Rouge, LA.

Peters SM, Timmerhaus KD (1991) *Plant Design and Economics for Chemical Engineers*, 4th edn., McGraw-Hill Inc., Singapore, ISBN 00071008713.

Petrides D (2002) Bioprocess design. In: Harrison RG, Todd PW, Rudge SR, Petrides D (eds.) *Bioseparations Science and Engineering*, Oxford University Press, Oxford, U.K., ISBN 0-19-512340-9.

Petrides D (October 11, 2008) Corn Stover to EtOH-SuperPro Designer® Model, Private Communication.

Phillips CB, Datta R (1997) Production of ethylene from hydrous ethanol on H-ZSM-5 under mild conditions, *Industrial and Engineering Chemistry Research*, 36(11): 4466–4475.

Phillips S, Aden A, Jechura J, Dayton D, Eggeman T (2007) *Thermochemical Ethanol via Indirect Gasification and Mixed Alcohol Synthesis of Lignocellulosic Biomass*, National Renewable Energy Laboratory, Golden, CO, NREL/TP-510-41168.

Pienkos PT, Daezins A (2009) The promise and challenges of microalgal-derived biofuels, *Biofuels, Bioproducts and Biorefining*, 3(4): 431–440.

Pike RW (1986) *Optimization for Engineering Systems*, Van Nostrand Reinhold, New York, ISBN 9780442275815.

Pike RW, Knopf FC (April 2010) Private Communication.

Pokoo-Aikins G, Nadim A, El-Halwagi MM, Mahalec V (2010) Design and analysis of biodiesel production from algae grown through carbon sequestration, *Clean Technologies and Environmental Policy*, 12(3): 239–254.

Pulz O (2007) Evaluation of Greenfuel's 3D matrix algae growth engineering scale unit, http://moritz.botany.ut.ee/~olli/b/Performance_Summary_Report.pdf, APS Red Hawk Power Plant, AZ.

Ritter S (2006) Biorefineries get ready to deliver the goods, *Chemical and Engineering News*, 84(34): 47.

Rossell CEV, Mantelatto PE, Agnelli JAM, Nascimento J (2006) Sugar-based biorefinery—Technology for integrated production of poly(3-hydroxybutyrate), sugar, and ethanol. In: Kamm B, Gruber PR, Kamm M (eds.) *Biorefineries—Industrial Processes and Products*, Vol. 1, Wiley-VCH Verlag GmbH & Co. KGaA, Weinheim, Germany, ISBN 3-527-31027-4.

Sahinidis N, Tawarmalani M (2007) BARON, http://www.gams.com/dd/docs/solvers/baron.pdf (accessed October 15, 2007).

SAIC (May 2006) *Life Cycle Assessment: Principles and Practice*, Scientific Applications International Corporation, National Risk Management Research Laboratory, Cincinnati, OH, EPA/600/R-06/060.

Sass R, Meier W (2010) Contact for Information Systems and Databases, Dechema, Private Communication at *AIChE Spring Meeting*, San Antonio, TX, March 23, 2010.

SciFinder Scholar (2009) *SciFinder Research Tool*, American Chemical Society, Washington, DC.

Seay J (April 2009) Evonik Industries, Private Communication.

Sengupta D (2010) Integrating bioprocesses into industrial complexes for sustainable development, PhD dissertation, Louisiana State University, Baton Rouge, LA, December 2010.

Sengupta D, Pike RW, Hertwig TA, Lou HH (2009) Integration of industrial scale processes using biomass feedstock in the petrochemical complex in the Lower Mississippi River Corridor, Paper No. 20c, *American Institute of Chemical Engineers Annual Meeting*, November 8–13, 2009, Nashville, TN.

Sheehan J, Dunahay T, Benemann J, Roessler P (1998) *A Look Back at the U.S. Department of Energy's Aquatic Species Program—Biodiesel from Algae*, National Renewable Energy Laboratory, Golden CO, NREL/TP-580-24190.

Shima M, Takahashi T (2006) Method for producing acrylic acid. U.S. Patent No. 7,612,230.

Short PL (2007) Small French firm's bold dream, *Chemical and Engineering News*, 85(35): 26–27.

Sikdar SK (August 2003) Sustainable development and sustainability metrics, *AIChE Journal*, 49(8): 1928–1932.

Singh A, Lou HH, Pike RW, Agboola A, Li X, Hopper JR, Yaws CL (2008) Environmental impact assessment for potential continuous processes for the production of carbon nanotubes, *American Journal of Environmental Sciences*, 4(5): 522–534.

Singh A, Lou HH, Yaws CL, Hopper JR, Pike RW (August 2007) Environmental impact assessment of different design schemes of an industrial ecosystem, *Resources, Conservation and Recycling*, 51(2): 294–313.

Smith RA (2005) Analysis of a petrochemical and chemical industrial zone for the improvement of sustainability, MS thesis, Lamar University, Beaumont, TX.

Snell KD, Peoples OP (2009) PHA bioplastic: A value-added coproduct for biomass biorefineries, *Biofuels, Bioproducts and Biorefining*, 3(4): 456–467.

Snyder, SW (2007) Overview of biobased feedstocks. *Twelfth New Industrial Chemistry and Engineering Conference on Biobased Feedstocks*, June 11–13, 2007, Council for Chemical Research, Argonne National Laboratory, Chicago, IL.

Spath PL, Dayton DC (2003) *Preliminary Screening—Technical and Economic Feasibility of Synthesis Gas to Fuels and Chemicals with the Emphasis on the Potential for Biomass-Derived Syngas*, National Renewable Energy Laboratory, Golden, CO, NREL/TP-510-34929, http://www.nrel.gov/docs/fy04osti/34929.pdf (accessed August 1, 2010).

Takahara I, Saito M, Inaba M, Murata K (2005) Dehydration of ethanol into ethylene over solid acid catalysts, *Catalysis Letters*, 105(3–4): 249–252.

Tanzil D, Ma G, Beloff BR (2003) Sustainability metrics, innovating for sustainability, *11th International Conference of Greening of Industry Network*, October 12–15, 2003, San Francisco, CA.

Teter SA, Xu F, Nedwin GE, Cherry JR (2006) Enzymes for biorefineries. In: Kamm B, Gruber PR, Kamm M (eds.) *Biorefineries—Industrial Processes and Products*, Vol. 1, Wiley-VCH Verlag GmbH & Co. KGaA, Weinheim, Germany, ISBN 3-527-31027-4.

Thanakoses P, Alla Mostafa NA, Holtzapple MT (2003a) Conversion of sugarcane bagasse to carboxylic acids using a mixed culture of mesophilic microorganisms, *Applied Biochemistry and Biotechnology*, 107(1–3): 523–546.

Thanakoses P, Black AS, Holtzapple MT (2003b) Fermentation of corn stover to carboxylic acids, *Biotechnology and Bioengineering*, 83(2): 191–200.

Thrasher L (2006) *Mosaic Announces Non-Binding Letter of Intent for Ammonia Offtake*, Press Release, Mosaic, Plymouth, MN.

Tolan JS (2006) Iogen's demonstration process for producing ethanol from cellulosic biomass. In: Kamm B, Gruber PR, Kamm M (eds.) *Biorefineries—Industrial Processes and Products*, Vol. 1, Wiley-VCH Verlag GmbH & Co. KGaA, Weinheim, Germany, ISBN 3-527-31027-4.

Tsao U, Zasloff HB (1979) Production of ethylene from ethanol, U.S. Patent No. 4,134,926.

Tullo AH (2007) Eastman pushes gasification, *Chemical and Engineering News*, 85(32): 10.

Tullo AH (2007a) Firms advance chemicals from renewable resources, *Chemical and Engineering News*, 85(19): 14.

Tullo AH (2007b) Soy rebounds, *Chemical and Engineering News*, 85(34): 36–39.

Tullo AH (2008) Growing plastics, *Chemical and Engineering News*, 86(39): 21–25.

United Nations (1987) Report of the World Commission on Environment and Development, *96th Plenary Meeting*, December 11, 1987, http://www.un.org/documents/ga/res/42/ares42-187.htm (accessed August 1, 2010).

USDA (2010a) Oil crops yearbook, http://usda.mannlib.cornell.edu/MannUsda/viewDocumentInfo.do?documentID=1290 (accessed August 4, 2010).

USDA (2010b) Feed grains database: Yearbook tables, http://www.ers.usda.gov/data/feedgrains/ (accessed August 4, 2010).

Ushikoshi, Kenji, Kouzou Mori, Taiki Watanabe, Masami Takeuchi, Masahiro Saito (1998) A 50 Kg/day class test plant for methanol synthesis from CO_2 and H_2, *Advances in Chemical Conversions for Mitigating Carbon Dioxide, Studies in Surface Science and Catalysts*, 114: 357–362.

Vaca-Garcia C (2008) Biomaterials. In: Clark JH, Deswarte FEI (eds.) *Introduction to Chemicals from Biomass*, Wiley, West Sussex, U.K., ISBN 978-0-470-05805-3.

Varisli D, Dogu T, Dogu G (2007) Ethylene and diethyl-ether production by dehydration reaction of ethanol over different heteropolyacid catalysts, *Chemical Engineering Science*, 62(18–20): 5349–5352.

Voith M (2009) Dow plans algae biofuels pilot, *Chemical and Engineering News*, 87(27): 10.

Wall-Markowski CA, Kicherer A, Saling P (December 2004) Using eco-efficiency analysis to assess renewable-resource–based technologies, *Environmental Progress*, 23(4): 329–333.

Wang JJ, Jing YY, Zhang CF, Zhao JH (2009) Review on multi-criteria decision analysis aid in sustainable energy decision-making, *Renewable and Sustainable Energy Reviews*, 13(9): 2263–2278.

Wells GM (1999) *Handbook of Petrochemicals and Processes*, 2nd edn., Ashgate Publishing Company, Brookfield, VT, ISBN 9780566080463.

Werpy T, Peterson G, Aden A, Bozell J, Holladay J, White J, Manheim A (2004) Top value added chemicals from biomass: Volume 1 results of screening for potential candidates from sugars and synthesis gas, Energy Efficiency and Renewable Energy (U.S. DOE) http://www1.eere.energy.gov/biomass/pdfs/35523.pdf (accessed May 8, 2010).

WGPL (2010) http://www.worldwidegraphite.com/property.htm#data (accessed August 10, 2010).

Wikipedia (2009) Aluminium oxide, http://en.wikipedia.org/wiki/Aluminium_oxide (accessed August 1, 2010).

Wikipedia (2010) Elasticity (economics), http://en.wikipedia.org/wiki/Elasticity_%28economics%29 (accessed June 21, 2010).

Wilke T, Pruze U, Vorlop KD (2006) Biocatalytic and catalytic routes for the production of bulk and fine chemicals from renewable resources. In: Kamm B, Gruber PR, Kamm M (eds.) *Biorefineries—Industrial Processes and Products*, Vol. 1, Wiley-VCH Verlag GmbH & Co. KGaA, Weinheim, Germany, ISBN 3-527-31027-4.

Wilson SR (February 16, 2006) Testimony of Steven R. Wilson, Chairman and CEO, CF Industries before the Senate Committee on Energy and Natural Resources.

Wool RP, Sun XS (2005) *Bio-Based Polymers and Composites*, Elsevier Academic Press, Amsterdam, the Netherlands, ISBN 0-12-763952-7.

Wyman CE, Dale BE, Elander RT, Holtzapple M, Ladisch MR, Lee YY (2005) Coordinated development of leading biomass pretreatment technologies, *Bioresource Technology*, 96 (18): 1959–1966.

Wyman CE et al. (2005) Coordinated development of leading biomass pretreatment technologies, *Bioresource Technology*, 96(18): 1959–1966.

Xu A (2004) Chemical production complex optimization, pollution reduction and sustainable development, PhD dissertation, Louisiana State University, Baton Rouge, LA.

Zaworski F (2010) ICIS pricing chemical price reports, http://www.icispricing.com/il_shared/il_splash/chemicals.asp (accessed August 8, 2010).

Zelder O (November 21–22, 2006) Fermentation—A versatile technology utilizing renewable resources. In: *Raw Material Change: Coal, Oil, Gas, Biomass—Where Does the Future Lie?* Ludwigshafen, Germany, http://www.basf.com/group/corporate/en/function/conversions:/publish/content/innovations/events-presentations/raw-material-change/images/BASF_Expose_Dr_Zelder.pdf (accessed August 1, 2010).

Zhang ZY, Jin B, Kelly JM (2007) Production of lactic acid from renewable materials by *Rhizopus* fungi, *Biochemical Engineering Journal*, 35(3): 251–263.

Index